D1637150

Hybrid and Incompatible Finite Element Methods

CRC Series: Modern Mechanics and Mathematics

Series Editors: David Gao and Ray W. Ogden

PUBLISHED TITLES

Beyond Perturbation: Introduction to The Homotopy Analysis Method
by Shijun Liao

Mechanics of Elastic Composites
by Nicolaie Dan Cristescu, Eduard-Marius Craciun, and Eugen Soós

Continuum Mechanics and Plasticity
by Han-Chin Wu

Hybrid and Incompatible Finite Element Methods
by Theodore H.H. Pian and Chang-Chun Wu

FORTHCOMING TITLE

Microstructural Randomness in Mechanics of Materials
by Martin Ostroja Starzewski

Hybrid and Incompatible Finite Element Methods

Theodore H. H. Pian
Chang-Chun Wu

Chapman & Hall/CRC
Taylor & Francis Group
Boca Raton London New York

Published in 2006 by
Chapman & Hall/CRC
Taylor & Francis Group
6000 Broken Sound Parkway NW, Suite 300
Boca Raton, FL 33487-2742

Printed in the United States of America on acid-free paper
10 9 8 7 6 5 4 3 2 1

International Standard Book Number-10: 1-58488-276-X (Hardcover)
International Standard Book Number-13: 978-1-58488-276-3 (Hardcover)
Library of Congress Card Number 2004063406

Library of Congress Cataloging-in-Publication Data

Pian, Theodore H. H., 1919-
Hybrid and incompatible finite element methods / Theodore H.H. Pian, Chang-Chun Wu.
p. cm. -- (Modern mechanics and mathematics ; 4)
Includes bibliographical references and index.
ISBN 1-58488-276-X (alk. paper)
1. Finite element method. I. Wu, Chang-Chun. II. Title. III. CRC series--modern mechanics and mathematics ; 4.

TA347.F5P53 2005
620'.001'51825--dc22 2004063406

Visit the Taylor & Francis Web site at
http://www.taylorandfrancis.com

and the CRC Press Web site at
http://www.crcpress.com

Preface

The purpose of this book is to introduce the advancement of the theory and applications of incompatible and multivariable finite element methods. The majority of the content of this book is the result of the authors' research.

The theory and application of finite element methods are not limited to ordinary compatible displacement methods. The incompatible element, hybrid element, and mixed element, developed since the early 1960s have been central topics for engineers and mathematicians. However, one important issue in this area is to determine the reliability for the solution of the incompatible multivarible elements (that is, the uniqueness, convergence, and adoptivity to the computing environment). For example, the convergence problem of incompatible elements was resolved in the early 1970s, but until now there have not been any available equations for determining functions for incompatible elements. Another example is that many engineers and mathematicians consider the uniqueness of multivariable (B-B condition) too abstract and not practical. Thus, it is a very important task to reduce the numerical stability theory and abstract mathematics to a tool in mechanics.

Chapter 1 is an introduction of the variational formulation of finite element methods in solid mechanics. Chapters 2 through 6 discuss fundamental theories, systematically demonstrating the theoretical foundations of incompatible elements and their application to different typical problems in the theory of elasticity and plasticity. These chapters also introduce new ideas in the development of hybrid finite elements, and studies the numerical stability of the hybrid and mixed elements, establishing the theory of zero-energy deformation mode. Chapter 7 addresses the application to fracture problems, where a bound analysis for fracture parameters is presented. An implementation of a finite element analysis program is provided in Chapter 8.

Chapter 9 presents the application to the computational materials such as composite plates, piezoelectric materials, and bimaterial interface problems. The homoginization approach is also presented. In Chapter 10, a finite element program is completely presented, in which a series of advanced imcompatible elements and the hybrid ones are included.

Acknowledgments

We appreciate Dr. Ying-Qing Huang providing the finite element computer program that carries out the analyses for the theory presented in Chapters 2 through 6. We would like to thank Dr. Lei Li who composed the manuscripts. We sincerely value Liu Jinghua's work on this book, especially his assistance with the editing. We also thank Dr. Qi-Zi Xiao and Dr. Miao-Lin Feng for providing new meaningful materials.

The support provided by the National Science Foundation of China is also gratefully acknowledged.

Authors

Theodore H. H. Pian, Ph.D., is Professor Emeritus in Aeronautics and Astronautics at the Massachusetts Institute of Technology (MIT). Dr. Pian obtained his undergraduate degree from Tsing Hua University in 1940 and later his Master's and Ph.D. degrees from the Massachusetts Institute of Technology in Aeronautical Engineering. After graduating from MIT, Dr. Pian worked for several aircraft manufacturing companies in both China and the United States.

In 1948, Dr. Pian began teaching at MIT, retiring as a full professor in 1989. During his long career, Dr. Pian traveled widely as a visiting professor in the United States, Japan, China, and Europe. His most important research contribution was developed in the 1950s when he developed the Hybrid Element Method for the derivation of element stiffness matrices. Dr. Pian is a Fellow of the American Society of Aeronautics and Astronautics and a winner of the Von Karman Memorial prize in 1974. He holds honorary doctoral degrees from Beijing University and Shanghai University in China.

Chang-Chun Wu, Ph.D., received a Bachelor of Science degree in civil engineering and Master's of Science in solid mechanics at Hefei Polytechnic University. He was appointed lecturer at the University of Science and Technology of China (USTC) in 1982, where he received a Ph.D. in physics in 1987. His supervisors were Professor T. H. H. Pian at MIT in the United States and Professor M. G. Huang at USTC.

Dr. Wu was honored by the Alexander von Homboldt Foundation (AvH). In 1982, he was appointed associate professor at USTC and in 1993 he was appointed as full professor. Dr. Wu spent many years as a visiting professor in many international institutes, including the Institute of Structural Engineering with Professor Bufler and the Institute of Civil Engineering with Professor E. Ramm. On numerous occasions he was invited to visit the University of Hong Kong to work with Professor Y. K. Cheung. In 1996 and 2002, he was invited as a JSPS-Research Fellow to visit the University of Tokyo by Professor G. Yagawa to work in the field of computational fractures and computational materials. He once again worked with Professor E. Ramm and his institute at the University of Stuttgart on incompatible numerical models. Dr. Wu has authored 150 papers in the field of computational engineering science and an academic book on finite elements.

In 2001, he moved to Shanghai Jiao Tong University (SJTU), where he has rebuilt his Computational Engineering Sciences Laboratory. His current research interests include the field of high-performance numerical models and numerical approaches; computational materials and computational fractures, which include failure analysis and engineering safety estimation.

Dr. Wu has been the recipient of a number of academic honors and awards. He is a former member of the Council on Chinese Society of Theoretical and Applied Mechanics, and a current member and editor of *Computational Mechanics*. He is also a founding member of the Institute of Soc. Comput. Eng. and Sci., a member of GAMM and EURMECH, and a senior member of IMS, NSU in Singapore. He received the 1992 T. H. H. Pian Medal from ICES, the 1988 and 1993 Science Awards from the Chinese Academy of Sciences, and in 1997 he received the National Nature Science Award in China.

Contents

1

Variational Formulation of Finite Element Methods in Solid Mechanics

1.1 Introduction

In a finite element method for solid mechanics a domain is represented by a finite number of connecting elements. The displacements are represented by a finite number of nodes that are located along the interelement boundaries and possibly also within the elements. The property of each element is represented by the element stiffness matrix **k** that defines the relation between the nodal forces and nodal displacements. The global stiffness matrix **K** can then be established by assembling the element stiffness matrices. A static problem, for example, is to reduce the determination of the displacements **q** of the unrestrained nodes, by a matrix equation of the form

$$\mathbf{K}\,\mathbf{q} = \mathbf{Q} \tag{1.1}$$

where **Q** represents the applied loads at the element nodes. Thus, the key step in the formulation of the finite element method in solid mechanics is the determination of the element stiffness matrices **k**.

Finite element methods are based on approximate modeling of individual elements. The accuracy of finite element methods depends on the number of the elements used and the method used for modeling the element properties. Mathematical modeling of finite elements in solid mechanics can be based on variational principles in elasticity.

During the early days of the development of finite element methods for solid mechanics, the assumed displacement method was used [1]. The formulation of an element stiffness matrix is based on the principle of minimum potential energy. It is a primal principle that contains only displacements as the field variables. Hence, it is a direct procedure to apply this principle for the formulation of an element stiffness matrix. Element stiffness matrices can also be formulated by alternative conventional variational principles [2, 3] and modified principles for relaxed continuity conditions along interelement boundaries [4]. Such principles are multivariate. In using these principles for finite element formulations, additional variables are involved, but at the end only nodal displacements are left in the final matrix equations. Such elements are called hybrid finite elements.

This chapter presents the various variational principles in solid mechanics. In this discussion vector notations will be used for the field variables. The equations for the problem of three-dimensional (3-D) elasticity will be presented first.

1.2 Equations for 3-D Elasticity

The three sets of field variables in 3-D elasticity are the displacements **u**, strains ε, and stresses σ. They are expressed in vector form as follows:

$$\mathbf{u} = \{u, v, w\} \tag{1.2}$$

$$\varepsilon = \{\varepsilon_x, \varepsilon_y, \varepsilon_z, \gamma_{yz}, \gamma_{xz}, \gamma_{xy}\} \tag{1.3}$$

$$\sigma = \{\sigma_x, \sigma_y, \sigma_z, \tau_{yz}, \tau_{xz}, \tau_{xy}\} \tag{1.4}$$

The components of body force $\boldsymbol{p}$(per unit volume) and the boundary traction $\boldsymbol{T}$(per unit area) are

$$\boldsymbol{p} = \{p_x, p_y, p_z\} \tag{1.5}$$

$$\mathbf{T} = \{T_x, T_y, T_z\} \tag{1.6}$$

The equilibrium equations for the boundary tractions are

$$\begin{bmatrix} \nu_x & 0 & 0 & 0 & \nu_z & \nu_y \\ 0 & \nu_y & 0 & \nu_z & 0 & \nu_x \\ 0 & 0 & \nu_z & \nu_y & \nu_x & 0 \end{bmatrix} \begin{Bmatrix} \sigma_x \\ \vdots \\ \tau_{xy} \end{Bmatrix} = \begin{bmatrix} T_x \\ T_y \\ T_z \end{bmatrix} \tag{1.7}$$

or

$$\nu\sigma = \boldsymbol{T} \tag{1.8}$$

where ν_x, ν_y, and ν_z are the directional cosines.

The stress–strain relation may be written as

$$\varepsilon = \boldsymbol{S}\sigma \tag{1.9}$$

$$\text{or } \sigma = \boldsymbol{C}\varepsilon \tag{1.10}$$

The strain-displacement relation is given by

$$\begin{Bmatrix} \varepsilon_x \\ \varepsilon_y \\ \varepsilon_z \\ \gamma_{yz} \\ \gamma_{zx} \\ \gamma_{xy} \end{Bmatrix} = \begin{bmatrix} \partial/\partial x & 0 & 0 \\ 0 & \partial/\partial y & 0 \\ 0 & 0 & \partial/\partial z \\ 0 & \partial/\partial z & \partial/\partial y \\ \partial/\partial z & 0 & \partial/\partial x \\ \partial/\partial y & \partial/\partial x & 0 \end{bmatrix} \begin{Bmatrix} u \\ v \\ w \end{Bmatrix} \tag{1.11}$$

or

$$\varepsilon = \boldsymbol{D}\boldsymbol{u} \tag{1.12}$$

The equilibrium equations are

$$\begin{bmatrix} \partial/\partial x & 0 & 0 & 0 & \partial/\partial z & \partial/\partial y \\ 0 & \partial/\partial y & 0 & \partial/\partial z & 0 & \partial/\partial x \\ 0 & 0 & \partial/\partial z & \partial/\partial y & \partial/\partial x & 0 \end{bmatrix} \begin{Bmatrix} \sigma_x \\ \vdots \\ \tau_{xy} \end{Bmatrix} + \begin{Bmatrix} F_x \\ F_y \\ F_z \end{Bmatrix} = \boldsymbol{0}$$

or (1.13)

$$\boldsymbol{D}^T\sigma + \boldsymbol{p} = 0 \tag{1.14}$$

In the formulation of variational principles two energy densities are used. They are

1. Strain energy density in terms of strains, that is,

$$A(\varepsilon_x \ldots\ldots\ldots \tau_{xy}) = A(\varepsilon) = (1/2)\, \varepsilon^T \boldsymbol{C}\varepsilon \tag{1.15}$$

2. Complementary energy density in terms of stresses, that is,

$$B(\sigma_x \ldots\ldots\ldots\ldots \tau_{xy}) = B(\sigma) = (1/2)\sigma^T \mathbf{S}\sigma \tag{1.16}$$

1.3 Conventional Variational Principles in Solid Mechanics

The conventional variational principles listed here are the principle of minimum potential energy, the principle of minimum complementary energy, the Hellinger–Reissner variational principle, and the Hu–Washizu variational principle.

The principle of minimum potential energy is a one-field principle and is stated as

$$\Pi_p(\boldsymbol{u}) = \int_V \left[\frac{1}{2} \varepsilon^T \boldsymbol{C} \varepsilon - \bar{p}^T \boldsymbol{u} \right] dV - \int_{S_\sigma} \bar{\boldsymbol{T}}^T \boldsymbol{u} dS = \text{Minimum} \tag{1.17}$$

with

$$\varepsilon = \boldsymbol{D}\boldsymbol{u} \text{ in V} \tag{1.18}$$

and

$$\boldsymbol{u} = \bar{\boldsymbol{u}} \text{ on } S_u \tag{1.19}$$

where $\boldsymbol{V}$ = the volume of the continuum, S_σ = the boundary over which the tractions are prescribed, and $\bar{\boldsymbol{u}}$ = prescribed displacements over boundary S_u.

The Hu–Washizu principle [5, 6] is a three-field principle and is stated as

$$\begin{aligned} \Pi_{HW}(\varepsilon, \sigma, \boldsymbol{u}) = & \int_V \left[\frac{1}{2} \varepsilon^T \boldsymbol{C} \varepsilon - \sigma^T \varepsilon + \sigma^T (\boldsymbol{D}\boldsymbol{u}) - \bar{\boldsymbol{p}}^T \boldsymbol{u} \right] dV \\ & - \int_{S_\sigma} \bar{\boldsymbol{T}}^T \boldsymbol{u} dS - \int_{S_u} \boldsymbol{T}^T (\boldsymbol{u} - \bar{\boldsymbol{u}}) dS = \text{stationary} \end{aligned} \tag{1.20}$$

where S_σ and S_u are, respectively, boundaries over which tractions and displacements are prescribed.

In Equation 1.20, when the strains ε are expressed in turns of stresses σ through the stress–strain relation, a two-field variational principle results. It is the Hellinger–Reissner principle [7, 8] given by

$$\begin{aligned} \Pi_{HR}(\sigma, \boldsymbol{u}) = & \int_V \left[-\frac{1}{2} \sigma^T \boldsymbol{S} \sigma + \sigma^T (\boldsymbol{D}\boldsymbol{u}) - \bar{\boldsymbol{p}}^T \boldsymbol{u} \right] dV - \int_{S_\sigma} \bar{\boldsymbol{T}}^T \boldsymbol{u} dS \\ & - \int_{S_u} \boldsymbol{T}^T (\boldsymbol{u} - \bar{\boldsymbol{u}}) dS = \text{stationary} \end{aligned} \tag{1.21}$$

When equilibrium conditions for the stresses σ and the prescribed tractions along the element boundary are satisfied, Equation 1.21 is reduced to the principle of minimum complementary energy with stresses σ as the only field variable:

$$\Pi_C(\sigma) = \int_V \frac{1}{2}\sigma^T S\sigma dV - \int_{S_u} \boldsymbol{T}^T \bar{\boldsymbol{u}} ds = \text{minimum} \tag{1.22}$$

1.4 Modified Variational Principles for Relaxed Continuity or Equilibrium Conditions along Interelement Boundaries

In applying the various variational principles for a solid continuum that is divided into finite elements, certain discontinuity conditions along the interelement boundaries are allowed to a degree that the variational functionals exist. However, an even broader extension of the variational methods lies in the possibility of further relaxation of the continuity conditions along the interelement boundaries by the introduction of interelement constraint conditions and the corresponding Lagrange multipliers. The resulting variational principles thus contain the Lagrange multipliers as additional variables along the element boundaries [4].

For example, for the principle of minimum complementary energy the constraining conditions for the stresses within each element V_n are given by Equations 1.8 and 1.14. Also along an element boundary S_σ, where the boundary tractions are prescribed,

$$\boldsymbol{T} - \bar{\boldsymbol{T}} = 0 \tag{1.23}$$

and on the interelement boundary S_{ab} between two elements a and b

$$\boldsymbol{T}^a + \boldsymbol{T}^b = \boldsymbol{0} \tag{1.24}$$

By introducing the traction reciprocity condition as equations of constraint with corresponding Lagrange multipliers, one obtains the following modified variational functional:

$$\Pi_{mc} = \sum_n \left[\int_{V_n} \frac{1}{2}\sigma^T S\sigma dV - \int_{S_{un}} \boldsymbol{T}^T \bar{\boldsymbol{u}} dS + \int_{S_{\sigma n}} (\boldsymbol{T} - \bar{\boldsymbol{T}})^T \lambda dS \right] + \sum_{ab} \int_{S_{ab}} (\boldsymbol{T}^a - \boldsymbol{T}^b)\lambda_{ab} dS \tag{1.25}$$

where V_n is the volume of the n-th element and S_{u_n} and S_{σ_n} are, respectively, the element boundaries over which the displacements and tractions are prescribed.

It is recognized that the Lagrange multipliers are negative values of the boundary displacements $\tilde{\boldsymbol{u}}$. Thus, the modified variational principle can be written as

$$\Pi_{mc}(\sigma,\tilde{\boldsymbol{u}}) = \sum_n \left[\int_{V_n} \frac{1}{2}\sigma^T S\sigma dV - \int_{\partial V_n} \boldsymbol{T}^T \tilde{\boldsymbol{u}} dS + \int_{S_{\sigma n}} \bar{\boldsymbol{T}}^T \tilde{\boldsymbol{u}} dS \right] = \text{stationary} \tag{1.26}$$

This modified variational principle is a two-field variational principle in which the element stresses σ and element boundary displacements $\tilde{\boldsymbol{u}}$ are independent. This is the variational principle used in the original formulation of the hybrid-stress finite element [9]. Since the stress parameters for one element are independent from those of the other elements, they can be eliminated from the element level. With the element boundary displacements interpolated in terms of nodal displacements, the resulting equations are in terms of only nodal displacements. The resulting method is based on element stiffness matrices.

Another example is a modified potential energy principle for which the continuity of displacement between neighboring elements a and b are relaxed [10]. The scheme is to treat boundary displacements $\tilde{\boldsymbol{u}}$ and interior displacements $\boldsymbol{u}$ as independent and then to introduce boundary tractions $\boldsymbol{T}$ as Lagrange multipliers to maintain the compatibility at the element boundary in an integral sense. This becomes a three-field principle. However, since both $\boldsymbol{T}$ and $\boldsymbol{u}$ in one element are independent of $\boldsymbol{T}$ and $\boldsymbol{u}$ in the neighboring elements, the corresponding parameters for $\boldsymbol{T}$ and $\boldsymbol{u}$ can be eliminated in the element level. The resulting method is also based on element stiffness matrices.

1.5 Assumed Displacement Finite Elements

When a solid continuum is discretized into a finite number of elements, the principle of minimum potential energy can be restated as

$$\Pi_p = \sum_n \left[\int_{V_n} \left\{ \frac{1}{2}(Du)^T C(Du) - u^T \bar{F} \right\} dV - \int_{S_{\sigma n}} u^T \bar{T} dS \right] = \text{minimum} \tag{1.27}$$

In the finite element formulation the element displacements $\boldsymbol{u}$ are interpolated in terms of nodal displacements that may be at both boundary nodes and internal nodes. For example, irregularly shaped quadrilateral plane elements formulated by using isoparametric coordinates [11] can be classified as Lagrange elements and serendipity elements. The shape functions for serendipity elements are in terms of only nodal points along the boundary of the element, but Lagrange elements (especially higher-order ones) usually contain nodal points located in the interior of the element. Elements can also be formulated by adding to the original element displacements $\boldsymbol{u}_q$, which are in terms of nodal displacements $\boldsymbol{q}$, and higher-order displacements $\boldsymbol{u}_\lambda$, which are not expressed in terms of nodal displacements of the boundary nodes. For example, the displacements u and v for the four-node quadrilateral element, Q4, are based on bilinear interpolation functions. They are incomplete in quadratic terms. Improvement of the performance of a four-node element can be made by adding terms such that the displacements are complete in quadratic terms. Pian [12] suggested the addition of a bubble function that vanishes along all boundaries. Wilson et al. [13] suggested the addition of incompatible displacements that vanish at all corner nodes. In these cases the element displacements $\boldsymbol{u}$ can be expressed as

$$\boldsymbol{Du} = \begin{bmatrix} \boldsymbol{B}_q^* & \boldsymbol{B}_\lambda \end{bmatrix} \begin{Bmatrix} \boldsymbol{q} \\ \lambda \end{Bmatrix} \tag{1.28}$$

Substituting into Equation 1.27 one obtains

$$\Pi_p = \sum_n \left[\frac{1}{2} \boldsymbol{q}^T \boldsymbol{k}_{qq} \boldsymbol{q} + \lambda^T \boldsymbol{k}_{\lambda q} \boldsymbol{q} + \frac{1}{2} \lambda^T \boldsymbol{k}_{\lambda\lambda} \lambda - \boldsymbol{q}^T \bar{\boldsymbol{Q}}_q - \lambda^T \bar{\boldsymbol{Q}}_\lambda \right] \tag{1.29}$$

where

$$k_{qq} = \int_{V_n} \boldsymbol{B}_q^T \boldsymbol{C} \boldsymbol{B}_q dV;$$

$$k_{\lambda q} = \int_{V_n} \boldsymbol{B}_\lambda^T \boldsymbol{C} \boldsymbol{B}_q dV$$

$$k_{\lambda\lambda} = \int_{V_n} \boldsymbol{B}_\lambda^T \boldsymbol{C} \boldsymbol{B}_\lambda dV$$

$$\bar{\boldsymbol{Q}}_q = \int_{V_n} \boldsymbol{N}_q^T \boldsymbol{F} dV + \int_{S_{\sigma n}} \mathrm{N}_q^T \bar{\boldsymbol{T}} dS$$

$$\bar{\boldsymbol{Q}}_\lambda = \int_{V_n} \boldsymbol{N}_\lambda^T \bar{\boldsymbol{F}} dV + \int_{S_{\sigma n}} \mathrm{N}_\lambda^T \bar{\boldsymbol{T}} dS \tag{1.30}$$

Since the displacements λ in one element are independent of the displacement of other elements, one can set $\partial\Pi_p/\partial\lambda$ to zero within each element and obtain λ in terms of $\boldsymbol{q}$. The total potential energy Π_p then is

$$\Pi_p = \sum_n \left[\frac{1}{2} \boldsymbol{q}^T \boldsymbol{k} \boldsymbol{q} - \boldsymbol{q}^T \boldsymbol{Q}_n \right] \tag{1.31}$$

Here, the element stiffness matrix is given by

$$\boldsymbol{k} = \boldsymbol{k}_{qq} - \boldsymbol{k}_{\lambda q}^T \boldsymbol{k}_{\lambda\lambda}^{-1} \boldsymbol{k}_{\lambda q} \tag{1.32}$$

and the equivalent nodal force vector is

$$\bar{\boldsymbol{Q}}_n = \bar{\boldsymbol{Q}}_q - \boldsymbol{k}_{\lambda q}^T \boldsymbol{k}_{\lambda\lambda}^{-1} \bar{\boldsymbol{Q}}_\lambda \tag{1.33}$$

Obviously, $\boldsymbol{k}_{qq}$ and $\bar{Q}_q$ correspond to elements that contain only boundary nodes. After assembling $\boldsymbol{k}$ and $\boldsymbol{Q}_n$, respectively, into global stiffness matrix $\boldsymbol{K}$ and nodal load $\boldsymbol{Q}$, Π_p becomes

$$\Pi_p = \frac{1}{2} \boldsymbol{q}^T \boldsymbol{K} \boldsymbol{q} - \boldsymbol{q}^T \bar{\boldsymbol{Q}} \tag{1.34}$$

and the operation $\partial\Pi_p / \partial q = 0$ yields the equations for finite element analysis:

$$\boldsymbol{K}\boldsymbol{q} = \bar{\boldsymbol{Q}} \tag{1.35}$$

Assumed displacement finite elements are derived by the potential energy variational principle, which has only displacements as field variables. Thus, its derivation is most straightforward. For problems that require only $\mathbf{C}^0$ continuity along interelement boundaries, such as plane stress and 3-D solid problems, one can easily obtain compatible shape functions by using isoparametric coordinates. In that case, the invariance property of the resulting elements can always be maintained.

The conventional assumed displacement method, however, has several shortcomings:

1. From problems that require a $\mathbf{C}^1$ continuity condition, such as thin plates and shells under Kirchhoff conditions, it is very difficult to construct shape functions that maintain compatibility of the normal derivatives along the interelement boundaries [14].
2. Lower-order elements such as four-node plane elements and eight-node solid elements for which displacement distributions are linear

along the edges will experience shear locking phenomena when they are used to model bending problems.

3. Locking phenomena may also appear when plane strain and solid elements are used to model nearly incompressible materials.

A remedy for item 1 is to include transverse shear strains in the formulation. However, such a scheme will lead to shear locking difficulties.

The shear locking phenomena of assumed displacement elements can be explained by the simple fact that under bending, the upper and lower edges of the quadrilateral element should be deformed as a parabola, while the element boundary displacements are linearly distributed. Wilson et al. [13] suggested the addition of higher-order incompatible displacements as a remedy.

Consider a quadrilateral plane element in isoparametric coordinates (ξ,η) as shown in Figure 1.1. The displacements $\boldsymbol{u}_q$ and $\boldsymbol{u}_\lambda$ of Equation 1.28 are, respectively,

$$u_q = \sum_1^4 (1+\xi_i\xi)(1+\eta_i\eta)\begin{Bmatrix} u_i \\ v_i \end{Bmatrix} \tag{1.36}$$

which is compatible along the interelement boundary, and

$$\boldsymbol{u}_\lambda = \begin{Bmatrix} \lambda_1(1-\xi^2)+\lambda_2(1-\eta^2) \\ \lambda_3(1-\xi^2)+\lambda_4(1-\eta^2) \end{Bmatrix} \tag{1.37}$$

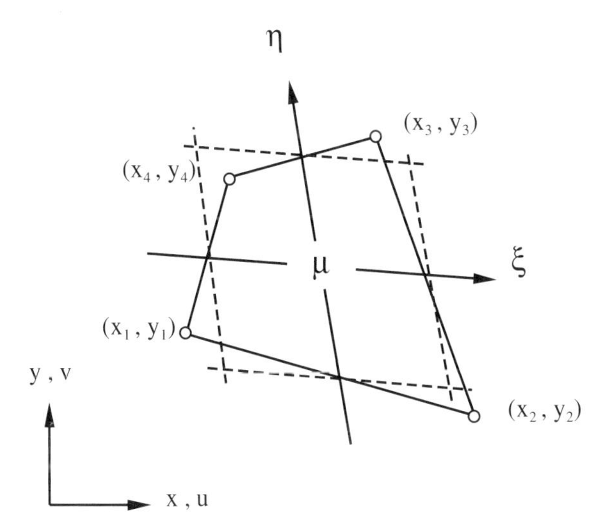

FIGURE 1.1
Plane quadrilateral isoparametric element.

which provide quadratic displacement distribution and are incompatible along the interelement boundary.

Wilson et al. have found that the resulting elements behave excellently if they are parallelograms. But if an element is distorted, it cannot maintain the solution convergence. Chapter 2 will discuss the foundation in the construction of incompatible elements.

1.6 Assumed Stress Hybrid Finite Elements

1.6.1 Formulation Based on the Principle of Minimum Complementary Energy

For an element with prescribed displacements $\bar{u}$ along the entire boundary, the principle of minimum complementary energy (Equation 1.26) is reduced to

$$\Pi_C^e = \int_{V^e} \frac{1}{2}\sigma^T S\sigma dV - \int_{S_u} T^T \bar{u} dS = \text{minimum} \tag{1.38}$$

where the displacements u of the entire boundary S_u of the element are prescribed.

In the finite element formulation, we begin by

(a) expressing the stress distribution as

$$\sigma = P\beta \tag{1.39}$$

(b) interpolating the boundary displacements $\bar{u}$ along the edges of the element in terms of nodal displacements q by

$$\bar{u} = Lq \tag{1.40}$$

(c) expressing the corresponding boundary tractions T in terms of stress parameters β:

$$T = R\mathrm{b} \tag{1.41}$$

Then, the functional Π_C^e is

$$\Pi_C^e = \frac{1}{2}\beta^T H\beta - \beta G q \tag{1.42}$$

For Π_C^e to be stationary,

$$\beta = \mathbf{H}^{-1}\mathbf{G}\mathbf{q} \tag{1.43}$$

where

$$H = \int_{V_n} P^T S P dV \tag{1.44}$$

and

$$G = \int_{A_s} R^T L dA \tag{1.45}$$

Thus,

$$\Pi_C^e = -\frac{1}{2} q^T G^T H^{-1} G q \tag{1.46}$$

and the element stiffness matrix k is given by

$$k = \mathbf{G}^T \mathbf{H}^{-1} \mathbf{G} \tag{1.47}$$

Initially, this form of hybrid element formulation was recognized by its advantage of constructing Kirchhoff plate elements on account of its avoidance of the difficult task of constructing element shape functions that should meet the C^1 continuity requirement. The early applications of this element were mainly for Kirchhoff plate problems [15, 16].

Another early application of the hybrid stress finite element method is the use of special elements at the tip of a crack for plane elasticity problems [17]. When the assumed stress terms in these elements include those of the expected singularity, the stress intensity factors for a crack can then be determined directly from the finite element solution.

Other advantages of hybrid stress finite elements are the avoidance locking for nearly incompressible materials [18]. It has also been shown that hybrid stress elements can be constructed to avoid shear locking difficulties both in beam analyses using rectangular plane stress elements and in thin plate analyses using Reissner–Mindlin theory [19].

1.6.2 Elements with *a priori* Satisfaction of Equilibrium and Compatibility Conditions

1.6.2.1 Hybrid Trefftz Elements

In analyzing solid mechanics problems by the principle of minimum complementary energy, if the assumed stresses satisfy the homogeneous equilibrium condition and the corresponding strains satisfy the constitutive relation, then in Equation 1.38, $S\sigma$ can be replaced by $\boldsymbol{Du}$, and

$$\int_{V^e} \sigma^T (\boldsymbol{D}\boldsymbol{u}) dV = \int_{V^e} (\boldsymbol{D}^T \sigma)^T \boldsymbol{u}\, dV + \int_{\partial V^e} \boldsymbol{T}^T \boldsymbol{u}\, dS \tag{1.48}$$

Thus, the complementary energy functional can be written as

$$\Pi_C^e = \int_{V_n} -\frac{1}{2} (\boldsymbol{D}^T \sigma)^T \boldsymbol{u}\, dV + \int_{\partial V_n} \left(\frac{1}{2} \boldsymbol{T}^T \boldsymbol{u} - \boldsymbol{T}^T \bar{\boldsymbol{u}} \right) dS \tag{1.49}$$

Since the stresses σ satisfy the equilibrium equation

$$\boldsymbol{D}^T \sigma = \boldsymbol{0} \tag{1.50}$$

the variational principle can be restated as

$$\Pi_{mC}^e = \oint_{\partial V_n} \left(\frac{1}{2} \boldsymbol{T}^T \boldsymbol{u} - \boldsymbol{T}^T \bar{\boldsymbol{u}} \right) dS = \text{stationary} \tag{1.51}$$

Here, the boundary traction vector $\boldsymbol{T}$ and the boundary displacement vector $\boldsymbol{u}$ are all related to the stress field of the element, while the boundary displacements $\bar{\boldsymbol{u}}$ are independently prescribed.

In the finite element formulation, let

$$\sigma = \boldsymbol{P}\beta \tag{1.52}$$

$$\boldsymbol{u} = \boldsymbol{U}\beta \tag{1.53}$$

$$\bar{\boldsymbol{u}} = \boldsymbol{L}\boldsymbol{q} \tag{1.54}$$

and

$$\boldsymbol{T} = \nu\sigma = \boldsymbol{R}\beta \tag{1.55}$$

Substituting Equation 1.52 through Equation 1.55 into Equation 1.51, we obtain

$$\Pi_{mC}^e = \frac{1}{2} \beta^T \boldsymbol{H} \beta - \beta^T \boldsymbol{G}\boldsymbol{q} \tag{1.56}$$

where

$$H = \frac{1}{2} \int_{\partial V_n} (\boldsymbol{R}^T \boldsymbol{U} + \boldsymbol{U}^T \boldsymbol{R}) dS \tag{1.57}$$

and

$$G = \int_{\partial V_n} R^T L dS \tag{1.58}$$

The matrix H is expressed this way in order to maintain its symmetry.

Equations 1.57 and 1.58 can be used to determine the parameter β and the element stiffness matrix k.

$$\beta = H^{-1} G q \tag{1.59}$$

$$k = G^T H^{-1} G \tag{1.60}$$

The initial application of this method is to construct special super-elements with embedded cracks to be used jointly with conventional elements for the determination of stress intensity factors of plane cracks [20]. The assumed stress terms and the corresponding displacement terms satisfy not only all the governing equations but also the traction-free conditions along the surface of the crack. They also include those with the proper singularity behavior. Thus, the stress intensity factors can be evaluated directly. This method has also been used to construct a plane element that contains elliptical voids or inclusions [21, 22, 23].

Jirousek has developed a finite element method that is based on an appropriate parametric displacement field that satisfies the governing differential equations of the problem. He named it the hybrid-Trefftz element [24, 25] and intensively studies plate bending and plane stress problems. The basic approach of this method is equivalent to the method used in Ref. 20.

1.6.3 Formulation Based on the Hellinger–Reissner Variational Principle

There are limitations in the construction of hybrid stress finite elements by the complementary energy principle. In dynamics and initial stress problems, it is not possible to construct assumed stresses to satisfy the equilibrium conditions. Also, for the formulation of elements of distorted geometry, it is necessary to express the assumed stresses in terms of isoparametric coordinates. In that case, the stress equilibrium conditions cannot be satisfied readily. An alternative in these cases is to use the Hellinger–Reissner variational principle, for which the stress equilibrating condition is no longer required.

The Hellinger–Reissner variational for an element is

$$\Pi_{HR}^{e} = \int_{V^e} \left[-\frac{1}{2} \sigma^T S \sigma + \sigma^T (Du) \right] dV - \int_{\partial V} T^T (u - \bar{u}) ds = \text{stationary} \tag{1.61}$$

where σ = stresses, S = elastic compliance, T = boundary tractions that are related to σ, u = displacements, $\bar{u}$ = prescribed boundary displacements, V = the volume of the element, and ∂V = the entire boundary of the element. The strain displacement relation is expressed as

$$\varepsilon = \mathbf{D}u \tag{1.62}$$

In the finite element formulation, if the interpolation of displacements is compatible, that is, $u = \bar{u}$, then with

$$u = Nq \tag{1.63}$$

and assumed stresses given by Equation 1.39, Π_R can be expressed as

$$\Pi_{HR} = -(1/2)\beta^T H\beta + + \beta^T \mathrm{G}q \tag{1.64}$$

where

$$H = \int_{V_n} P^T SP dV; \quad G = \int_{V_n} P^T (DN) dV \tag{1.65}$$

are, respectively, the flexibility and leverage matrices. From variation with respect to β within the element level one can obtain

$$\beta = H^{-1} Gq \tag{1.66}$$

and the element stiffness matrix is given by

$$\mathrm{k} = G^T H^{-1} G \tag{1.67}$$

Two further remarks can be made about the use of the Hellinger–Reissner variational functional for the construction of hybrid elements.

- Kirchhoff plate problems can be analyzed by including the last term in Equation 1.61 in the element formulation. Because the boundary displacements $\bar{u}$ can be independently assumed, it is no longer required to construct the internal displacements u to satisfy the C^1 continuity condition.
- The use of the Hellinger–Reissner variational functional permits the choice of the expansions of assumed displacements and stresses to be balanced and complete, which is a guideline for the construction of elements of optimum performance [26].

1.7 Hybrid Strain Finite Elements

Hybrid strain finite elements [27] can also be constructed based on a modified Hellinger–Reissner variational principle with strains ε and displacements $\boldsymbol{u}$ as field variables:

$$\Pi_{HR}^{e}(\varepsilon,\boldsymbol{u}) = \int_{V^e}\left[-\frac{1}{2}\varepsilon^T \boldsymbol{C}\,\varepsilon + (\boldsymbol{C}\,\varepsilon)^T(\boldsymbol{D}\boldsymbol{u})\right]dV = \text{stationary} \tag{1.68}$$

Here again, the displacements $\boldsymbol{u}$ are interpolated in terms of nodal displacements, the strains ε are expressed in terms of internal parameters that can be eliminated in the element level, and the element stiffness matrix can be obtained. This method has been used by Lee and associates to construct plate and shell elements [27, 28].

1.8 Hybrid Finite Elements by the Hu–Washizu Principle

The Hu–Washizu principle for an element is

$$\begin{aligned}\Pi_{HW}^{e}(\varepsilon,\sigma,\boldsymbol{u}) = \int_{V^e}\left[\frac{1}{2}\varepsilon^T C\varepsilon - \sigma^T\varepsilon + \sigma^T(\boldsymbol{D}\boldsymbol{U}) - \bar{\boldsymbol{p}}^T\boldsymbol{u}\right]dV \\ - \int_{S_\sigma^e}\bar{\boldsymbol{T}}^T\boldsymbol{u}dS = \text{stationary}\end{aligned} \tag{1.69}$$

In the finite element formulation the stresses, strains, and displacements are expressed as

$$\varepsilon = \psi\,\alpha \tag{1.70}$$

$$\sigma = \varphi\,\beta \tag{1.71}$$

$$\boldsymbol{u} = \boldsymbol{N}\,\boldsymbol{q} \tag{1.72}$$

Equation 1.69 becomes

$$\Pi_{HW}^{e}(\alpha,\beta,\boldsymbol{q}) = \frac{1}{2}\alpha^T\boldsymbol{A}\alpha + \alpha^T\boldsymbol{F}\beta + \beta^T\boldsymbol{G}\boldsymbol{q} - \boldsymbol{q}^T\bar{\boldsymbol{Q}} \tag{1.73}$$

in which

$$A = \int_{V^e} \psi^T C \psi dV \tag{1.74}$$

$$F = \int_{V^e} \psi^T \varphi dV \tag{1.75}$$

$$G = \int_{V^e} \phi^T (DN_q) dV \tag{1.76}$$

$$\bar{Q} = \int_{V^e} N_q^T \bar{p} dV + \int_{S_\sigma^e} N_q^T \bar{T} dS \tag{1.77}$$

From the stationary condition of the functional $\Pi_{HW}^e(\alpha,\beta,q)$, one can obtain the following system of equations:

$$\begin{bmatrix} A & F & 0 \\ F & 0 & G \\ 0 & G & 0 \end{bmatrix} \begin{Bmatrix} \alpha \\ \beta \\ q \end{Bmatrix} = \begin{Bmatrix} 0 \\ 0 \\ \bar{Q} \end{Bmatrix} \tag{1.78}$$

Because A is invertible, the stress parameters α can be statically condensed, and the following equations can be obtained:

$$\left.\begin{array}{c} \begin{bmatrix} -F^T A^{-1} F & G \\ G^T & 0 \end{bmatrix} \begin{Bmatrix} \beta \\ q \end{Bmatrix} = \begin{Bmatrix} 0 \\ \bar{Q} \end{Bmatrix} \\ \alpha = -A^{-1} F \beta \end{array}\right\} \tag{1.79}$$

When $F^T A^{-1} F$ is invertible, the stress parameters β can also be statically condensed, and the following set of equations can be obtained with the nodal displacements q as the only variables:

$$\left.\begin{array}{c} [G^T (F^T A^{-1} F)^{-1} G] q = \bar{Q} \\ \beta = (F^T A^{-1} F)^{-1} G q \\ \alpha = -A^{-1} F \beta \end{array}\right\} \tag{1.80}$$

The element stiffness matrix is given by

$$k = G^T (F^T A^{-1} F)^{-1} G \tag{1.81}$$

For the 3-field hybrid element, there exists the following condition, which should be met for the absence of zero energy modes [35, 36].

$$n_a + n_q \geq n_\beta \geq n_q \tag{1.82}$$

where

$$n_a = \dim(\alpha),\ n_\beta = \dim(\beta),\ n_q = \dim(q) - n_o$$

(n_o is the element rigid-body DOF).

Hu [29] has suggested some rational conditions for the elements by $\Pi_{HW}(\varepsilon, \sigma, \mu)$. Pian and Sumihara [30] applied this approach for the construction of shell elements. Tang et al. [31] applied this approach to the construction of an element stiffness matrix for plane stress problems.

1.9 Hybrid Displacement Finite Elements

The hybrid displacement finite element method, which was suggested by Tong [32], is formulated based on the modified potential energy principle presented in Section 1.4:

$$\Pi_{mP} = \int_{V_n} \frac{1}{2}(\boldsymbol{Du})^T \boldsymbol{C}(\boldsymbol{Du})dV - \int_{\partial V} \boldsymbol{T}^T(\boldsymbol{u} - \bar{\boldsymbol{u}})dS = \text{stationary} \tag{1.82}$$

The original motivation for this method was for the formulation of Kirchhoff plate problems for which the element displacements $\boldsymbol{u}$ and boundary displacements $\bar{u}$ can be independently assumed. The boundary tractions $\boldsymbol{T}$ are another variable. In the finite element formulation, $\bar{\boldsymbol{u}}$ is interpolated in terms of nodal displacements, while $\boldsymbol{T}$ and $\boldsymbol{u}$ are expressed in terms of internal parameters that can be eliminated in the element level, and the element stiffness matrix can be obtained. This method has been used for shell analysis [33] and for the construction of special elements with prescribed singular behavior to be used to study crack tip stress intensity factors [34].

References

1. Melosh, R.J., Basis for the derivation of matrices for the direct stiffness method, *AIAA J.*, 1, 1631, 1963.
2. Washizu, K., *Variational Methods in Elasticity and Plasticity*, 3rd ed., Pergamon Press, Oxford, 1982.

3. Hu, H.C., *Variational Principles of Theory of Elasticity with Applications*, Science Press, Beijing, 1984.
4. Pian, T.H.H. and Tong, P., Basis of finite element methods for solid continua, *Int. J. Numer. Methods Eng.*, 1, 3, 1969.
5. Hu, H.C., On some variational principles in the theory of elasticity and the theory of plasticity, *Scintia Sin.*, 4, 33, 1955.
6. Washizu, K., *On the Variational Principles of Elasticity*, Aeroelastic and Structures Research Laboratory, Massachusetts Institute of Technology, Technical Report 25–18, 1955.
7. Hellinger, E., Der allgermeine Ansatz der Mechanik der Kontinun, *Encycl. Math. Wiss.*, Vol. 4, Part 4, p. 602, 1914.
8. Reissner, E., On a variational theorem in elasticity, *J. Math. Phys.*, 29, 90, 1950.
9. Pian, T.H.H., Derivation of element stiffness matrices by assumed stress distributions, *AIAA J.*, 2, 1333, 1964.
10. Tong, P., New displacement hybrid finite element methods for solid continua, *Int. J. Numer. Methods Eng.*, 2, 73, 1970.
11. Ergatoudis, I., Irons, B.M., and Zienkiewicz, O.C., Curved isoparametric, 'quadrilateral' elements for finite element analysis, *Int. J. Solids Struct.*, 4, 31, 1968.
12. Pian, T.H.H., Derivation of element stiffness matrices, *AIAA J.*, 2, 576, 1964.
13. Wilson, E.L., Taylor, R.L., Doherty, W.P., and Ghaboussi, J., Incompatible displacement models, in *Numerical and Computer Methods in Structural Mechanics*, Fenves, S.J., et al., Eds., Academic Press, New York, 1973, pp. 43–57.
14. Irons, B.M. and Draper, K.J., Inadequacy of nodal connections in a stiffness solution for plate bending, *AIAA J.*, 3, 961, 1965.
15. Pian, T.H.H., Element stiffness matrices for boundary compatibility and for prescribed boundary stresses, *Proc. Third Conf. Matrix Methods Struct. Mech.*, AFFDL TR-66-80, Wright Patterson Air Force Base, pp. 457–477, 1966.
16. Severn, R.T. and Taylor, P.R., The finite element method for flexure of slabs when stress distributions are assumed, *Proc. Inst. Civil Eng.*, 34, 153, 1966.
17. Pian, T.H.H., Tong, P., and Luk, C.H., Elastic crack analysis by a finite element hybrid method, in *Proc. Third Conf. Matrix Methods in Struct. Mech.*, AFFDL-TR-41-160, pp. 661–682, 1973.
18. Pian, T.H.H. and Lee, S.W., Notes on finite elements for nearly incompressible materials, *AIAA J.*, 14, 824, 1976.
19. Spilker, R.T. and Murir, N.I., The hybrid-stress model for thin plates, *Int. J. Numer. Methods Eng.*, 15, 1239, 1980.
20. Tong, P., Pian, T.H.H., and Lasry, S., A hybrid-element approach to crack problems in plane-elasticity, *Int. J. Numer. Methods Eng.*, 7, 297, 1973.
21. Piltner, R., Special finite elements with holes and internal cracks, *Int. J. Numer. Methods Eng.*, 21, 1471, 1985.
22. Zhang, J. and Katsube, N., A finite element method for heterogeneous materials with randomly dispersed elastic inclusions, *Finite Elem. Analysis Design*, 19, 45, 1995.
23. Zhang, J. and Katsube, N., A finite element method for heterogeneous materials with randomly dispersed rigid inclusions, *Int. J. Numer. Methods Eng.*, 38, 1635, 1995.
24. Jirousek, J., Hybrid-Trefftz plate bending elements with p-method capabilities, *Int. J. Numer. Methods Eng.*, 24, 1367, 1987.
25. Jirousek, J. and Venkatesh, A., Hybrid-Trefftz plane elasticity elements with p-method capabilities, *Int. J. Numer. Methods Eng.*, 35, 1443, 1992.

26. Pian, T.H.H., Finite elements based on consistently assumed stresses and displacements, *Finite Elem. Analysis Design*, 1, 131, 1985.
27. Lee, S.W. and Pian, T.H.H., Improvement of plate and shell elements by mixed formulations, *AIAA J.*, 16, 29, 1978.
28. Lee, S.W., Wong, S.C., and Rhiu, J.J., Study of a nine-node mixed formulation finite element for thin plates and shells, *Int. J. Numer. Methods Eng.*, 21, 1325, 1985.
29. Hu, H.C., Necessary and sufficient conditions for correct use of generalized variational principles of elasticity in appropriate solutions, *Sci. China (A)*, 33, 196–205, 1989.
30. Pian, T.H.H. and Sumihara, K., Hybrid semiLoof elements for plates and shells based upon a modified Hu-Washizu principle, *Comput. Struct.*, 19, 165, 1984.
31. Tang, L.M., Chen, W.J., and Liu, Y.X., Formulation of quasi-conforming element and Hu-Washizu principle, *Comput. Struct.*, 19, 247, 1984.
32. Tong, P., New displacement hybrid finite element model for solid continua, *Int. J. Numer. Methods Eng.*, 2, 78, 1970.
33. Atluri, S.N. and Pian, T.H.H., Theoretical formulation of finite element methods in linear-elastic analysis of general shells, *J. Struct. Mech.*, 1, 1, 1972.
34. Atluri, S.N., Kobayashi, A., and Nakagaki, M., Fracture mechanics application of an assumed displacement hybrid finite element procedure, *AIAA J.*, 13, 734, 1975.
35. Pain, T.H.H. and Chen, D.P., On the suppression of zero energy deformation modes, *Int. J. Num. Meth. Eng.*, 19, 1741–1752, 1983.
36. Wu, C.C., Dual zero energy modes in mixed/hybrid elements-definition, analysis and control, *Comp. Meth. Appl. Mech. Eng.*, 81, 39–56, 1990.

2

Foundation of Incompatible Analysis

2.1 Introduction

This chapter discusses the problems of stability (existence and uniqueness) and convergence of finite element solutions. For these problems there are strict mathematics theories [1–3] and different numerical tests [4, 5] for finite elements that are formulated based on incompatible discretization. However, it is still desirable to provide the principles of mechanics that are both very clear in concept and reliable for the design of incompatible finite elements.

The following questions concerning principles of mechanics are to be answered: What are the basic restrictions for the interelement discontinuity? What is the basic requirement for the uniqueness of incompatible numerical solutions? What is the energy basis of the constant stress patch test (PTC)? What is its meaning in mechanics? What is the form of the PTC for inclined coordinates? How are the incompatible functions created? When the function of the element does not satisfy the PTC, how can an incompatible element be developed?

This chapter is self-contained. It is not only a theoretical basis of various incompatible elements in elasticity, it also contains preparatory material for the study of hybrid finite element methods.

2.2 Energy Inequality and Elliptic Conditions

We define the inner product of vectors $\boldsymbol{a}$ and $\boldsymbol{b}$ in region V as

$$\langle \boldsymbol{a},\boldsymbol{b}\rangle = \int_V \boldsymbol{a}^T\boldsymbol{b}dV$$

and define the vector norm as

$$\|\boldsymbol{a}\| = \langle \boldsymbol{a},\boldsymbol{a}\rangle^{1/2}$$

and recognize that positive definiteness is the fundamental property of a norm. Let A be an assembly of $\boldsymbol{a}$, then $\forall a \in \mathrm{A}$, and often, $\|\boldsymbol{a}\| \geq 0$, and $\|\boldsymbol{a}\| = 0 \Rightarrow \boldsymbol{a} = \mathbf{0}$. However, in general, the norm $\|\boldsymbol{F}\|$ of a linear function $\boldsymbol{F}(\boldsymbol{a})$ may not be the norm of $\boldsymbol{a}$. This is to say,

$$\forall \boldsymbol{a} \in \mathbf{A}, \text{it is not necessary that } \|\boldsymbol{F}\| \geq 0 \text{ and } \|\boldsymbol{F}\| = 0 \Rightarrow \boldsymbol{a} = \boldsymbol{0}.$$

However, if $\|\boldsymbol{F}\|$ is always positive–definite for $\boldsymbol{a}$; that is,

$$\forall \boldsymbol{a} \in \mathrm{A}, \|\boldsymbol{F}\| \geq 0 \text{ and } \|\boldsymbol{F}\| = 0 \Rightarrow \boldsymbol{a} = \boldsymbol{0}. \tag{2.1}$$

Then $\|\boldsymbol{F}\|$ is the norm of $\boldsymbol{a}$. By noting that $\|\boldsymbol{F}\|$ is not negative, it is easy to prove that Equation 2.1 is equivalent to the following argument: With a positive constant c, let

$$\forall \boldsymbol{a} \in \mathbf{A}, \|\boldsymbol{F}\| \geq c\|\boldsymbol{a}\| \tag{2.2}$$

It is clear that the condition for the norm $\|\boldsymbol{F}\|$ of the vector $\boldsymbol{F}(\boldsymbol{a})$ to be the norm of $\boldsymbol{a}$ is

$$\|\boldsymbol{F}\| = \mathbf{0} \Rightarrow \boldsymbol{a} = \mathbf{0} \tag{2.3}$$

For the problem of three-dimensional (3-D) elasticity, the displacement vector $\boldsymbol{u}$, the stress–strain relation and the strain-displacement relation are given, respectively, by Equation 1.2, Equation 1.10, and Equation 1.12.

If in the finite element discretization the element volume is V^e, then the volume of the entire system is $V = \bigcup_e V^e$ and the set of the displacements in V^e and V are designated respectively as $\mathrm{U}(V^e)$ and $\mathrm{U}(V)$.

From the theory of elasticity, the rigid body displacements are the nontrivial solutions of the zero strain condition $\varepsilon = \boldsymbol{D}\boldsymbol{u} = \mathbf{0}$. For a rigid body displacement system, one must have

$$\varepsilon = \boldsymbol{D}\boldsymbol{u} = \mathbf{0} \Rightarrow \boldsymbol{u} = \mathbf{0} \tag{2.4}$$

According to the Ritz method (including the classical Ritz method and the compatible finite element methods), which is based on the principle of minimum potential energy, the condition for maintaining uniform convergence of the potential energy functional and displacements to exact solutions is that the secondary of the functional is elliptic [6, 7]; that is, the system strain energy should be of

$$U(\boldsymbol{u}) = \frac{1}{2}\langle \sigma(\boldsymbol{u}), \varepsilon(\boldsymbol{u}) \rangle \geq \gamma \|\boldsymbol{u}\|^2 \quad (\gamma \text{ is a positive constant}) \tag{2.5}$$

However, according to Lax-Milgram lemma [2], if the elliptic condition (Equation 2.5) holds, the corresponding generalized variational solution exists uniquely [8]. Therefore, the key in studying the stability of variational approximation is to prove the ellipticity of the functional. Chien and Mikhlin [6, 9] have extensively studied the ellipticity of the strain energy of different boundary value problems in elasticity. The following discusses this problem based on principles of mechanics.

According to thermal dynamics, for an elastic body under constant temperature and insulated condition to be stable, the strain energy function $A(\boldsymbol{u})$ must be stationary [11, 12]; that is:

$$A(\boldsymbol{u}) = \frac{1}{2}\varepsilon^T \mathbf{C}\varepsilon \geq 0 \quad \text{(equality sign applies only when } \varepsilon = \mathbf{0}) \tag{2.5a}$$

Therefore, the elastic matrix $\boldsymbol{C}$ must be symmetric and positive–definite, and can be split into

$$\mathbf{C} = \mathbf{V} \wedge \mathbf{V}^T \tag{2.5b}$$

in which $\wedge = \text{diag}\,(\lambda)$ is a diagonal matrix constructed by a set of characteristic value, while $\boldsymbol{V}$ is an orthogonal matrix formed by the eigenvector corresponding to λ, by $\boldsymbol{V}\boldsymbol{V}^T = \boldsymbol{I}$. Thus,

$$A(\boldsymbol{u}) = \frac{1}{2}(\varepsilon^T \mathbf{V}) \wedge (\mathbf{V}^T \varepsilon) \tag{2.5c}$$

For $\boldsymbol{C}$ to be positive–definite it shows the smallest characteristic value $\lambda_{\min} > 0$. By replacing $\wedge$ in Equation 2.5c with the diagonal matrix $\lambda_{\min}\boldsymbol{I}$, we have

$$A(\boldsymbol{u}) \geq \frac{1}{2}\lambda_{\min}\, \varepsilon^T \varepsilon \tag{2.5d}$$

Integrating Equation 2.5d, we obtain the following unequilibrium condition for energy:

$$\int_V A(\boldsymbol{u})\, dV \geq c\|\varepsilon\|^2 \tag{2.6}$$

the constant in which is $c = \frac{1}{2}\lambda_{\min}$. The corresponding strain norm is

$$\|\varepsilon\| = \|\boldsymbol{D}\boldsymbol{u}\| = \left(\int_V \varepsilon^T \varepsilon\, dV\right)^{1/2} \tag{2.7}$$

For an elastic body without rigid body motion and with geometric boundary conditions satisfied, Equation 2.4 is applicable, and $\|\varepsilon\| = 0 \Rightarrow \boldsymbol{u} = 0$. Thus, $\|\varepsilon\|$ is also the norm of $\boldsymbol{u}$. Based on Equation 2.2, it can be concluded that there exists a positive constant α, such that

$$\forall \boldsymbol{u} \in \mathrm{U}(V), \ \ \|\varepsilon\| \geq \alpha \|\boldsymbol{u}\| \tag{2.8}$$

Based on this, the inequality (Equation 2.6) can be further expressed as a positive constant $\beta = c\alpha^2$, such that

$$\forall \boldsymbol{u} \in \mathrm{U}(V), \int_V A(\boldsymbol{u})\,dV \geq \beta \|\boldsymbol{u}\|^2 \tag{2.9}$$

is the elliptic condition for strain energy. Mikhlin [9] suggested that strain energy that satisfies this requirement is positive bounded below. In mathematics, the development from Equation 2.6 to Equation 2.9 is the result of the combined application of Korn inequality and Poincare inequality.

The ellipticity of strain energy leads to its positive definiteness. If Equation 2.9 is established, the following is also established:

$$\forall \boldsymbol{u} \in \mathrm{U}(V), \int_V A(\boldsymbol{u})\,dV \geq 0 \quad \text{(equal sign only applies to } \boldsymbol{u} = 0) \tag{2.10}$$

However, it is not possible to develop the energy inequality (Equation 2.9). The ellipticity of strain energy is stronger than its positive definiteness. The uniqueness of a variational solution in elasticity can be deducted from the positive definiteness of the strain energy [11]. But the ellipticity of strain energy must be used as the foundation in considering the problem of existence and convergence [1, 9].

The following is a discussion of the stability of finite elements. That is, the problem of the existence of unique solutions. First, for a compatible element that is based on a test solution of assumed displacements, $\boldsymbol{u} \in C^0(V)$, the displacements and strains can be squarely integrated within the entire domain, that is, $\boldsymbol{u}, \varepsilon \in L_2(V)$. Therefore, in the process of discretization the ellipticity of the system's strain energy can always be maintained. Accordingly, for an entire system with finite elements under constraint along its geometric boundaries, the stiffness matrix is positive–definite. Therefore, the stability of the solution of compatible elements can always be maintained.

However, when the test solution of incompatible displacements $\boldsymbol{u}, \varepsilon \in L_2(V)$. is adopted, the strain energy of the discretized system can be expressed as

$$\int_V A(\boldsymbol{u})\,dV = U(\boldsymbol{u}) = U_*(\boldsymbol{u}) + \Delta(\boldsymbol{u}) \tag{2.11}$$

where

$$U_*(\boldsymbol{u}) = \sum_e \int_{V^e} A(\boldsymbol{u})\,dV \tag{2.12}$$

is the sum of the strain energy of all elements, and $\Delta(\boldsymbol{u})$ is the sum of singular integrals along the interelement boundaries resulting from the discontinuity of $\boldsymbol{u}$ along neighboring element boundaries. The appearance of $\Delta(\boldsymbol{u})$ will break the boundedness and ellipticity of the system functional $U(\boldsymbol{u})$. Historically, in handling incompatible elements [1], people always use $U_*(u)$ to approximately replace the original functional $U(u)$ in finite element analysis. The resulting difficulty is that the approximate functional $U_*(u)$ is unnecessarily elliptical, so that the stability of the solution of incompatible elements becomes questionable. Furthermore, even if the solution to incompatible elements is unique, it may not converge to the exact solution. When adapting $U_*(u)$, certain additional conditions must be considered in order to maintain the stability and convergence of incompatible elements.

2.3 Weak Connection Condition of Incompatible Elements

For an incompatible discretizing model, the test solution along interelements always involves a form of discontinuity. But in order to maintain the numerical stability of the solution by incompatible elements there must be a restriction of the functional discontinuity along interelement boundaries. In Reference 13 this idea is referred to as the "minimum requirement for displacement continuity" or "minimum requirement for partial compatible element."

Definition 2.1

The minimum requirement for necessary connections on the element interface, such that the relative rigid body motion between neighboring elements can be prevented, is termed the *weak connection condition* [14] of incompatible elements.

In this section we consider only nodal point incompatible elements that do not contain internal parameters. That is, the incompatible displacement function is defined only by the nodal displacements along the element boundary. A theorem for this type of incompatible element is as follows.

Theorem 2.1

In the node-type incompatible element system, which requires prescribed boundary displacements to prevent the rigid body motion, the existence and uniqueness of a discrete solution can be ensured by the weak connection condition. For a nodal point incompatible element discretizing system that has no rigid body motion, the uniqueness of the finite element solution is guaranteed by a weak connecting condition of incompatible elements.

Proof

It is only necessary to prove the ellipticity of the total strain energy of the incompatible element system. From the definition of the weak connection condition for incompatible elements, we can see that with such condition satisfied, the rigid body displacements of the individual elements of the incompatible system and that of the entire system are identical. If the system has no rigid body movement, there will be no rigid body movement for each individual element. But for element e, the strain norm $\|\varepsilon\|$ is also the norm of u; hence, for element e, the elliptic condition (Equation 2.9) is also satisfied. This means that with a positive constant c that is not related to the shape of the element, one can have

$$\forall \boldsymbol{u} \in \mathrm{U}(V^e), \int_{V^e} A(\boldsymbol{u})dV \geq c_e \|\boldsymbol{u}\|_e^2 \quad e = 1,2,\ldots,n, \tag{2.13}$$

in which n is the number of elements in the system. By summing Equation 2.13 for all elements, one obtains

$$U_*(\boldsymbol{u}) = \sum_e \int_{V^e} A(\boldsymbol{u})dV \geq \alpha \sum_e \|\boldsymbol{u}\|_e^2 \tag{2.14}$$

in which $\alpha = \min\ (c_e,\ e = 1,2, \ldots\ n)$ is a positive constant that is not related to the pattern of discretization.

The right-hand side of Equation 2.14 is always positive–definite with respect to $\boldsymbol{u}$:

$$\left(\sum_e \|u\|_e^2\right)^{1/2} = 0 \Rightarrow (\boldsymbol{u} = \boldsymbol{0} \text{ in } V^e,\ e = 1,2,\ldots,n) \Leftrightarrow \boldsymbol{u} = 0 \text{ within } V$$

Therefore, for the system V, the norm of $\boldsymbol{u}$ is given by $(\sum_e \|\boldsymbol{u}\|_e^2)^{1/2}$, and Equation 2.13 can be expressed further by

$$\forall \boldsymbol{u} \in U(V), U_*(\boldsymbol{u}) = \sum_e \int_{V^e} A(\boldsymbol{u})dV \geq \beta \|u\|^2 \tag{2.15}$$

in which β is a positive constant that is not related to the pattern of discretization. The inequality (Equation 2.15) indicates that the strain energy for a system of incompatible elements with weak connections is elliptic. Hence, a unique solution exists for incompatible elements.

Some fundamental concepts for the weak connecting condition of incompatible elements is presented in the following figures. Consider, first, the equilibrium of a simple thin membrane. Along the boundary S_n, the weak connection is the nodal displacement w (Figure 2.1). Similarly, for the problem of a 2-D heat-transfer problem, the weak connection along the boundary of the element S_n is the nodal temperature ϕ_i.

For plane elasticity problems, the weak connection condition of finite elements is maintained by the three nodal parameters u, v_i, and the inplane rotation θ (Figure 2.2a). An alternative is to use the weak connection

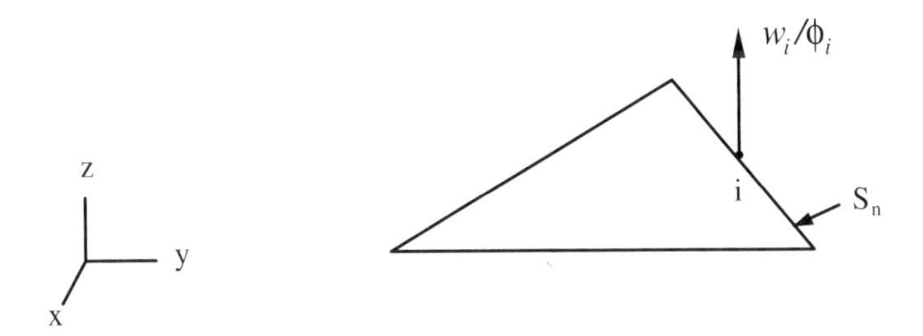

FIGURE 2.1
Weak connection pattern for membrane element and temperature element.

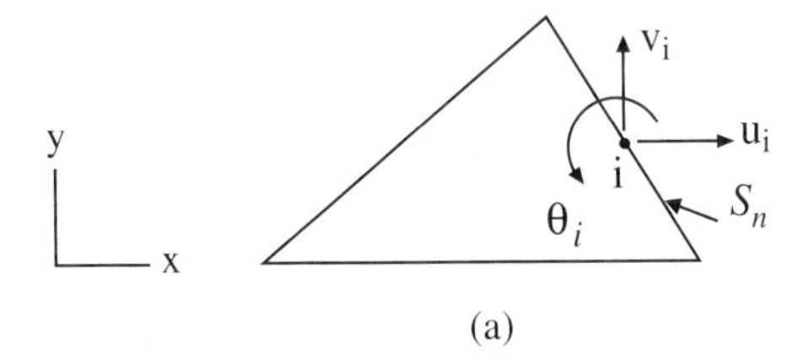

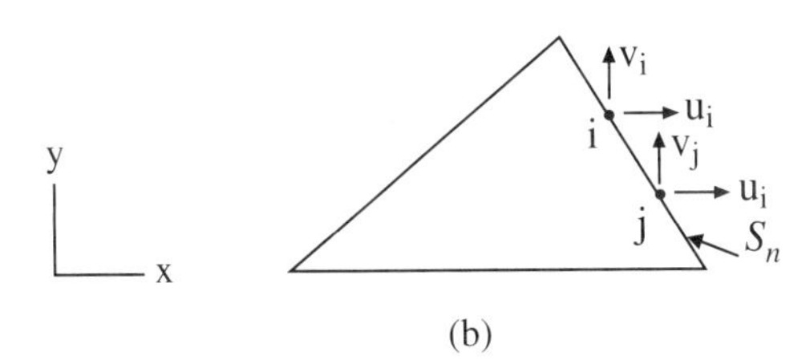

FIGURE 2.2
Weak connection patterns for plane stress and strain elements.

condition shown in Figure 2.2b for which the inplane rotations at the nodes can be avoided.

Figure 2.3a shows the weak connection patterns for 3-D elements. At the center of each surface S_n a node with six nodal parameters u_i, v_i, w_i, θ_{xi}, θ_{yi}, and θ_{zi} is required.

An alternative is to use the weak connection pattern shown in Figure 2.3b with the three nodes (not lined up on one straight line).

For thin plate elements in bending, there are two weak connection patterns shown in Figure 2.4. Similarly, for thick plates for which shear effect is included, there are two weak connection patterns shown in Figure 2.4. To obtain the weak connection pattern for thick plates it is only necessary to replace the nodal parameters $(\partial w/\partial x)_i$, $(\partial w/\partial y)_j$, and $(\partial w/\partial n)_k$ by the nodal rotations, θ_{xi}, θ_{yi}, and θ_{nk}, and along the midsurface of the plate.

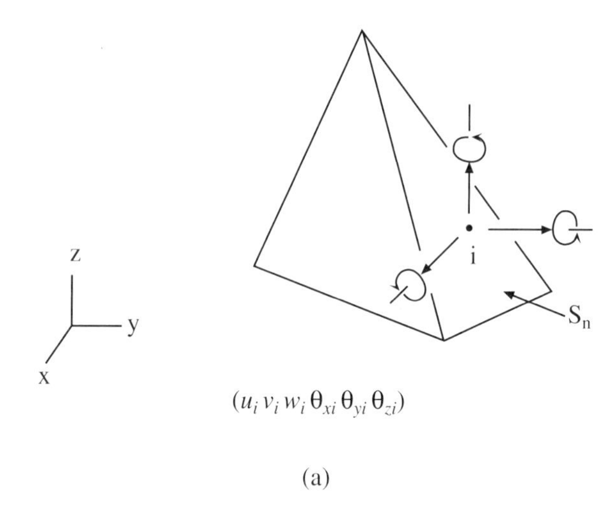

(a)

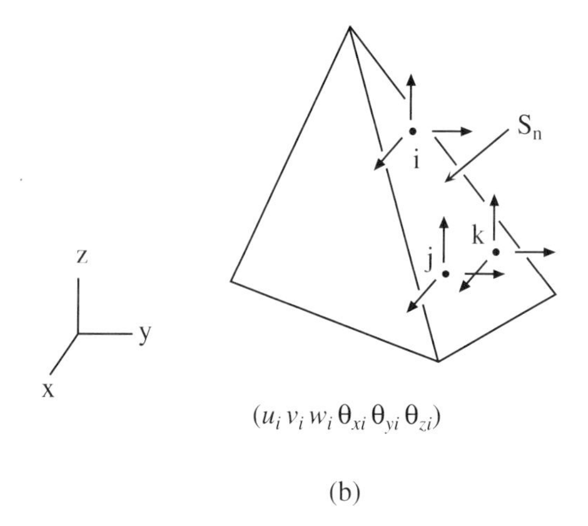

(b)

FIGURE 2.3
Weak connection pattern for 3-D elements.

Here, the nodal rotations must be independent of the lateral displacement w of the midsurface. Finally, the weak connection conditions for shell elements can be obtained by combining the weak connection parameters of the plane stress element and the weak connection parameters of the plate bending element.

For 1-D problems, the weak connection of rod elements is the axial displacement at the end of the rod, while the weak connection for beams are the lateral displacement w_i and the rotation (dw/dx) or θ_{xi}.

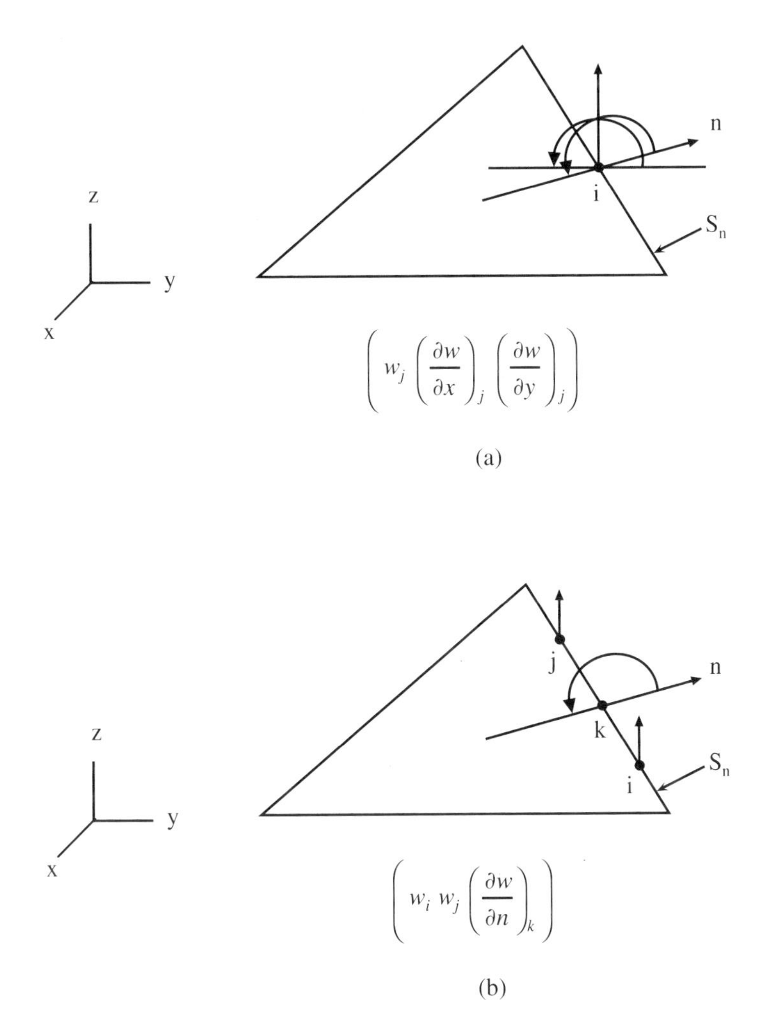

FIGURE 2.4
Weak connection patterns of plate bending element.

2.4 Numerical Stability of Incompatible Elements

The λ-type incompatible elements of the trial function consists of the following two parts [1–3]:

$$u = u_q + u_\lambda \tag{2.16}$$

where $\mathbf{u}_q = \mathrm{N}_q \boldsymbol{q}$ is the ordinary and complete interpolation defined by the nodal displacement $\boldsymbol{q}$. It may be continuous or discontinuous along the element surface S_n, but it should at least satisfy the weak connection conditionals mentioned above. The internal displacement $\boldsymbol{u}_\lambda = \mathbf{N}_\lambda \lambda$ is defined by element internal parameters λ. It is, in general, incompatible with respect to neighboring elements and does not contain rigid body displacements. The corresponding strains for the displacements given by Equation 2.16 are

$$\varepsilon = \varepsilon_q + \varepsilon_\lambda = \boldsymbol{D}\boldsymbol{N}_q\boldsymbol{q} + \boldsymbol{D}\boldsymbol{N}_\lambda\lambda = B_q\boldsymbol{q} + B_\lambda\lambda$$

For this we have the following theorem.

Theorem 2.2 [14]

The solution of the incompatible element system is only unique when the following independent condition for strain basis can be satisfied:

$$\boldsymbol{B}_\lambda = [\boldsymbol{B}_{\lambda 1}\boldsymbol{B}_{\lambda 2}\ldots] \text{ and } \boldsymbol{B}_q = [\boldsymbol{B}_{q1}\boldsymbol{B}_{q2}\ldots] \text{ are linearly independent.} \quad (2.17)$$

To apply the elliptic condition (Equation 2.9) to the λ-type incompatible elements, we must show whether under the trial solution (Equation 2.16), the norm $\|\varepsilon\|$ of the element strain remains to be the norm of $\boldsymbol{u}$. If so, we have

$$\forall \boldsymbol{u} \in U(V^e), \quad \|\varepsilon\|_e = 0 \Rightarrow \boldsymbol{u}_q = 0 \quad \text{and} \quad \boldsymbol{u}_\lambda = 0 \quad (2.18)$$

By noting that under Equation 2.16, ε_q and ε_λ are linearly independent,

$$\|\varepsilon\|_e = 0 \Rightarrow \begin{cases} \varepsilon_q = \boldsymbol{D}\boldsymbol{u}_q = \boldsymbol{0} \quad \forall\ \boldsymbol{u}_q\ U(V^e) & (2.18a) \\ \varepsilon_\lambda = \boldsymbol{D}\boldsymbol{u}_\lambda = \boldsymbol{0} \quad \forall\ \boldsymbol{u}_\lambda\ U(V^e) & (2.18b) \end{cases}$$

Since $\boldsymbol{u}_q$ satisfies the weak connection condition and the system has no rigid body displacement, and the individual element also has no rigid body displacement, Equation 2.18a leads to $\boldsymbol{u}_q = \boldsymbol{0}$. Furthermore, because $\boldsymbol{u}_\lambda$ contains no rigid body displacement, Equation 2.18b leads to $\boldsymbol{u}_\lambda = 0$ and Equation 2.17 applies. Conversely, if the two strain-based functions are linearly dependent, such as the relation $\boldsymbol{B}_{\lambda i} + \alpha\boldsymbol{B}_{qj} = 0$, while α is a constant, then from the condition $\|\varepsilon\|_e = 0$, one can get a relation $\alpha\lambda_i = q_j$. Since, in general, λ_i and q_j are not equal to zero, we will have

$$\boldsymbol{u}_\lambda = \boldsymbol{N}_{\lambda_i}\lambda_i \neq 0, \quad \boldsymbol{u}_q = \boldsymbol{N}_{qj}\boldsymbol{q}_j \neq \boldsymbol{0}$$

and Equation 2.17 will no longer apply.

Under Equation 2.16, the strain norm $\|\varepsilon\|_e$ of incompatible elements with internal parameters is the norm of $\boldsymbol{u}$; hence, the ellipticity condition (Equation 2.9) is applicable to individual elements. Thus, we can again obtain the inequalities in Equation 2.12, Equation 2.13, and Equation 2.14. In general, under conditions of independent strain basis (Equation 2.16), the incompatible element system with internal parameters retains the ellipticity of strain energy and, correspondingly, the uniqueness of the finite element solution.

We now present an example to illustrate Theorem 2.1 and Theorem 2.2. Figure 2.5a is a bar element of length l. Under the local coordinate shown in the figure, we use a three-point, second-order interpolation function:

$$w = \begin{bmatrix} \frac{1}{2}\xi(\xi-1) & 1-\xi^2 & \frac{1}{2}\xi(\xi+1) \end{bmatrix} \begin{Bmatrix} w_1 \\ w_2 \\ w_3 \end{Bmatrix} \tag{2.18c}$$

where w_1 and w_3 are nodal displacements along the axial direction of the element, and w_2 is the displacement at the midpoint of the element. This trial function for the element displacements satisfies the weak connection condition. Now, consider the straight bar of Figure 2.5b, with uniform stretching stiffness EA. The constraining conditions at the ends of the bar are $w(0) = w(2l) = 0$. For this two-element system, we can obtain a stationary stiffness matrix as follows:

$$\mathbf{K} = \frac{2EA}{3l} \begin{bmatrix} 8 & -4 & 0 \\ -4 & 7 & -4 \\ 0 & -4 & 8 \end{bmatrix} \tag{2.18d}$$

Therefore, there is a unique solution for this system.

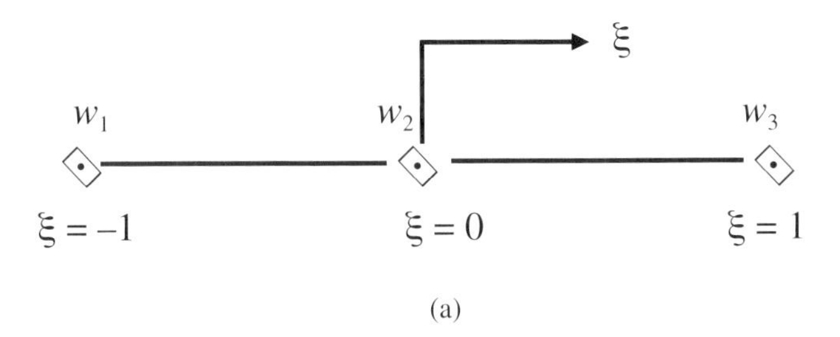

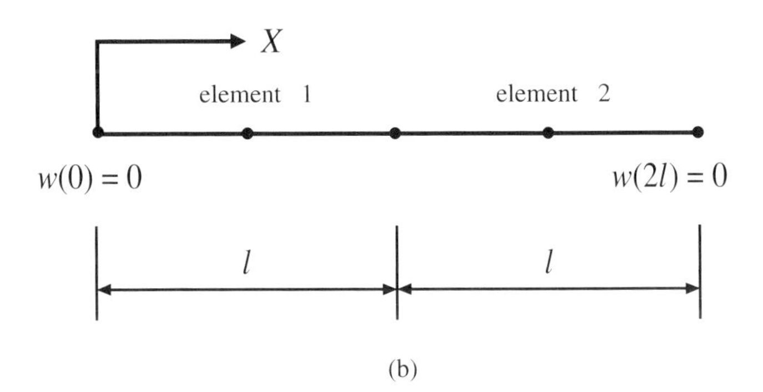

FIGURE 2.5
Illustrations of Theorem 2.1 and Theorem 2.2. (a) Local coordinate for 1-D elements. (b) Assembled system combined by two elements.

If Equation 2.18c is used as the trial function for the lateral displacement of a beam element, then, because of the lack of a rotational connecting element, the weak connecting condition is no longer satisfied. If the bending stiffness of the beam is EI, and the boundary constraints are $w(0) = w(2l) = 0$, the resulting stiffness matrix of the beam structure shown in Figure 2.5b is

$$K = \frac{16EI}{l^3}\begin{bmatrix} 1 & -2 & 1 \\ -2 & 4 & -2 \\ 1 & -2 & 1 \end{bmatrix} \tag{2.18e}$$

Here and hence the resulting solutions are not stable.

We can construct the incompatible bar element by adding an internal displacement $w_\lambda = \xi\lambda$ to the trial solution in Equation 2.18c. In the formulation of the element stiffness matrix we need to eliminate the internal parameter λ by static condensation. Now, by using this element stiffness matrix, we can determine the structural stiffness matrix of the two-element system in Figure 2.5b; again, the result is

$$K = \frac{2EA}{3l}\begin{bmatrix} 8 & -4 & 0 \\ -4 & 4 & -4 \\ 0 & -4 & 8 \end{bmatrix} \tag{2.18f}$$

Here, $|K| = 0$. We can see that the structural stiffness matrix is no longer positive–definite when the internal displacement $w_\lambda = \xi\lambda$ is introduced. The incompatible basic function ξ and the original interpolation functions $\xi(\xi-1)$ and $\xi(\xi+1)$ are linearly dependent. The corresponding strain bases are also linearly dependent. If we take $w_\lambda = \xi^3\lambda$ instead, we can show that, because of its satisfaction of independency of the strain basis, the corresponding stiffness matrix of the bar system must be positive–definite.

Theorem 2.1 and Theorem 2.2 provide two basic conditions for the unique solution of incompatible elements: the weak connection condition and the independent condition for strain basis. These conditions are important considerations for the design and research of incompatible elements. Further discussion of the stability of incompatible elements can be found in Reference 14. For some structures that are supported continuously (such as structures on an elastic foundation), the above two conditions are not important for the stability of finite element solutions.

Finally, when reduced integration is used in the construction of incompatible elements with internal parameters, a zero energy deformation mode associated with the internal parameter λ may appear, as shown in Reference 14. In this case it is not possible to static-condense the λ parameters. Also, such modes $\mathring{u}_\lambda$ cannot be controlled through either assembly of elements or geometric constraining along the element boundary. The only remedy is to eliminate the expression of $\mathring{u}_\lambda$ from u_2 during the element formulation.

In general, $\overset{o}{u}_\lambda$ corresponds to higher-order element stresses. Its elimination will not change the constant stress state and will not affect the convergence of the solution.

2.5 Consistency and Patch Test Condition (PTC)

For compatible finite elements that are based on the potential energy principle, the potential energy functional of a discretized system can be expressed as the sum of the potential of individual elements, that is,

$$\Pi_p = \sum_e \Pi_p^e = \sum_e \left[\int_{V^e} \left(\frac{1}{2} \varepsilon^T C \varepsilon - \bar{f}^T u \right) dV - \int_{S_\sigma^e} \bar{T}^T u dS \right] \tag{2.19}$$

where the element strains $\varepsilon = \boldsymbol{D}\boldsymbol{u}$, $\bar{f}$ and $\bar{T}$ are, respectively, the known loading distribution in V^e and along the element boundary S_σ^e.

In the incompatible element method, Equation 2.19 can still be used for discretization. Here, to maintain solution stability, incompatible elements must satisfy a weak connection condition and independence of strain basis. But the existence of a solution by incompatible elements does not guarantee that it will converge to an exact solution. To guarantee solution convergence, certain restrictions of the incompatible test functions for the element must be made. This is described in the following.

Let the test function for element displacements be

$$u = u^0 + u^\Delta \tag{2.20}$$

where $\boldsymbol{u}^0$ is the compatible part of the test function and $\boldsymbol{u}^\Delta$ is the incompatible part. In the finite element formulation, the geometric boundary condition of the system is maintained by the interpolation function $\boldsymbol{u}^0$, and the known boundary force along S_σ^e. can be reduced to an equivalent nodal load based on known surface force $\bar{\boldsymbol{T}}$ and the shape function based on displacement $\boldsymbol{u}^0$. This implies that the surface loading term is replaced by $\int_{S_\sigma^e} \bar{T}^T u^0 dS$. Then, by substituting Equation 2.20 into Equation 2.19 and by applying the stationary condition of the potential energy of the incompatible system, we can obtain the following variational equation:

$$\begin{aligned}\delta\Pi_p(u^0, u^\Delta) = &\sum_e \int_{V^e} -(D^T\sigma + \bar{f})^T (\delta u^0 + \delta u^\Delta) dV \\ &+ \sum_n \int_{S_{ab}} (T^{(a)} + T^{(b)})^T \delta u^0 dS + \sum_e \int_{S_\sigma^e} (T - \bar{T})^T \delta u^0 dS \\ &+ \sum_e \oint_{\partial V^e} T^T \delta u^\Delta dS = 0\end{aligned} \tag{2.21}$$

where $\sigma = C\varepsilon$ is element stress, $T = n\sigma$ is element surface load, and $S_{ab} = V^e_{(a)} \cap V^e_{(b)}$ is the common surface of a pair of elements (a) and (b). Because of the arbitrariness of virtual displacements δu^0 and δu^Δ, we can obtain from the variational Equation 2.21 the stress equilibrium equations within the elements, the boundary force equilibrium equations along the interelement boundary S_{ab}, and the mechanical boundary condition along S^e_σ. We can also obtain the following energy consistency condition for an incompatible element system [15]:

$$\sum_e \oint_{\partial V^e} T^T \delta u^\Delta dS = 0 \tag{2.22}$$

The meaning of this equation is that the system virtual work done by the element surface tractions $T = n\sigma$ to the incompatible displacements must vanish. The above consistency condition is the theoretical basis of incompatible element methods.

In considering the convergence of the incompatible element analysis, the condition (Equation 2.22) may be softened further. Considering the limiting situation in refining finite element meshes, that is, the mesh parameter $h \to 0$, the element stresses σ will tend to reach a constant σ_c, and the corresponding limiting expression for the above consistency condition will be

$$\sum_e \oint_{\partial V^e} \sigma_c^T n^T \delta u^\Delta dS \to 0 \quad \text{for} \quad h \to 0 \tag{2.23}$$

This is the convergence criterion of incompatible elements. The application of this criterion in this form requires a complicated mathematical analysis. However, we can take its closed homogeneous form of Equation 2.23 and replace δu^Δ by u^Δ. The resulting equation is [15]

$$\sum_e \oint_{\partial V^e} \sigma_c^T n^T u^\Delta dS = 0 \tag{2.24}$$

This simplified convergence criterion is an explicit mathematical expression of Iron's constant stress patch condition [1, 16]. For incompatible elements, the trial functions of which satisfy the completeness requirement and stability condition, the PTC (Equation 2.24) is a sufficient condition for the numerical solutions to converge to exact values.

There are some incompatible elements that do not pass the PTC in Equation 2.24, but they can still yield an exact converging solution [17–19]. The reason is that under a specific mesh arrangement the condition (Equation 2.23) is satisfied. This situation is the so-called weak patch test, which is referred to in References 5 and 20.

Since the condition (Equation 2.24) is dependent on the mesh division and the type of element patch, it is still inconvenient to use it. Therefore, a strong form can be taken by requiring that for every element the following condition should be satisfied:

$$\oint_{\partial V^e \partial} \sigma_c^T \boldsymbol{n}^T \delta \boldsymbol{u}^\Delta dS = 0 \tag{2.25a}$$

and its equivalence:

$$\int_{V^e} \sigma_c^T \varepsilon^\Delta dV = 0 \tag{2.25b}$$

in which $\varepsilon^\Delta = \boldsymbol{D}\boldsymbol{u}^\Delta$ is the incompatible strains. Reference 21 has called Equation 2.25b the energy orthogonal condition. By removing the constant stress factor, the practical form for the PTC of an element becomes, simply

$$\oint_{\partial V^e} \boldsymbol{n}^T \boldsymbol{u}^\Delta dS = \int_{V^e} \varepsilon^\Delta dV = \boldsymbol{0} \tag{2.26}$$

There are even stronger requirements than Equation 2.26, for example, by requiring that along each segment of the element boundary $S^e \subset \partial V^e$, the following condition is satisfied:

$$\int_{S^e} \boldsymbol{u}^\Delta dS = \boldsymbol{0} \tag{2.27}$$

For this issue, Sander and Beckers [17] have provided a more detailed discussion. Shi [19] has suggested another testing method for the incompatible element. Other discussions about the patch test can be found in References 5, 23, and 24.

2.6 Generation of Incompatible Functions: General Formulation

The key step for the formulation of an incompatible element is to find the incompatible shape functions that satisfy the PTC. Based on Equation 2.25 or Equation 2.26, several incompatible functions that can satisfy the PTC have been presented in References 5, 18, and 25–28. General equations used for the formulation of such functions are given in Reference 15. We illustrate this process by considering a 3-D elasticity problem. Let $\boldsymbol{n} = \{l \ m \ n\}$ be the direction cosines of the outward normal on element sides and u^Δ be any component of $\boldsymbol{u}^\Delta$. The PTC in Equation 2.26 can equivalently be expressed as

$$\oint \begin{Bmatrix} l \\ m \\ n \end{Bmatrix} u^{\Delta} dS = \mathbf{0} \quad \text{or the equivalence} \quad \int_{V^e} \begin{Bmatrix} \frac{\partial}{\partial x} \\ \frac{\partial}{\partial y} \\ \frac{\partial}{\partial z} \end{Bmatrix} u^{\Delta} dV = \mathbf{0} \tag{2.28}$$

However, if element nodal displacements and internal parameters are denoted as $\boldsymbol{q}$ and λ, respectively, an incompatible displacement can always be written as

$$u^{\Delta} = u_q^{\Delta} + u_{\lambda} = N_q^{\Delta}\boldsymbol{q} + N_{\lambda}\lambda \tag{2.29}$$

In order to introduce the restriction (Equation 2.28), a modifying term should be added to Equation 2.29, and we have

$$u^{\Delta} = [\mathbf{N}_q^{\Delta} \quad \mathbf{N}_{\lambda}] \begin{Bmatrix} \boldsymbol{q} \\ \lambda \end{Bmatrix} + \boldsymbol{N}^* \lambda^* \tag{2.30}$$

where λ^* includes three virtual parameters. The selection of $\boldsymbol{N}^*$ is an easy matter. It is required that the following matrix be nonsingular:

$$\boldsymbol{P}_* = \oint \begin{Bmatrix} l \\ m \\ n \end{Bmatrix} \mathbf{N}^* dS \tag{2.31}$$

Similarly, we define matrix $\boldsymbol{P}$ as

$$\boldsymbol{P} = \oint \begin{Bmatrix} l \\ m \\ n \end{Bmatrix} [\mathbf{N}_q^{\Delta} \quad \mathbf{N}_{\lambda}]\, dS \tag{2.32}$$

Then Equation 2.28 can be expressed as

$$\boldsymbol{P} \begin{Bmatrix} \boldsymbol{q} \\ \lambda \end{Bmatrix} + P_* \lambda^* = \mathbf{0}$$

From this, the virtual parameter λ^* can be expressed as

$$\lambda^* = -P_*^{-1} \boldsymbol{P} \begin{Bmatrix} \boldsymbol{q} \\ \lambda \end{Bmatrix}$$

Substituting into Equation 2.30, we obtain the desirable general formulation of incompatible shape function (WHP formulation [15]):

$$u^{\Delta} = ([N_q^{\Delta} \quad N_{\lambda}] - N^{*}\boldsymbol{P}_{*}^{-1}\boldsymbol{P})\begin{Bmatrix} q \\ \lambda \end{Bmatrix} \tag{2.33}$$

If the element nodal interpolating function does not contain any incompatible components, that is, $u_q^{\Delta} = 0$, Equation 2.33 will be simplified to the following form that contains only the internal parameter λ:

$$u^{\Delta} = u_{\lambda} = (N_{\lambda} - N^{*}\boldsymbol{P}_{*}^{-1}\boldsymbol{P})\lambda = N_{\lambda}^{*}\lambda \tag{2.34}$$

In applying Equation 2.33 and Equation 2.34 to 2-D incompatible elements, it is only necessary to change $\boldsymbol{P}_{*}$ and $\boldsymbol{P}$ to

$$\boldsymbol{P}_{*} = \oint \begin{Bmatrix} l \\ m \end{Bmatrix} \boldsymbol{N}^{*} dS \tag{2.35}$$

$$\boldsymbol{P} = \oint \begin{Bmatrix} \boldsymbol{l} \\ \boldsymbol{m} \end{Bmatrix} [N_q^{\Delta} \quad N_{\lambda}]\, dS \tag{2.36}$$

$$\text{or} \quad \boldsymbol{P} = \oint \begin{Bmatrix} l \\ m \end{Bmatrix} \boldsymbol{N}_{\lambda} dS \tag{2.37}$$

As an example, we take a four-node plane element shown in Figure 1.1. The displacements are in Cartesian coordinates (x,y), but they are interpolated in terms of natural or isoparametric coordinates (ξ,η). In the formulation of compatible elements, the interpolation functions for element coordinates and element displacement are, respectively,

$$\begin{Bmatrix} x \\ y \end{Bmatrix} = \sum_{i=1}^{4} N_i(\xi,\eta)\begin{Bmatrix} x_i \\ y_i \end{Bmatrix} \tag{2.38}$$

$$\begin{Bmatrix} u \\ v \end{Bmatrix} = \sum_{i=1}^{4} N_i(\xi,\eta)\begin{Bmatrix} u_i \\ v_i \end{Bmatrix} \tag{2.39}$$

in which $N_i(\xi,\eta) = \frac{1}{4}(1+\xi_i\xi)(1+\eta_i\eta)$ is the so-called double linear interpolation function.

In the next chapter we shall apply the general equations for the formulation of incompatible shape functions to construct incompatible elements of several other problems in elasticity. References 29–34 can also be referred to for further study of this problem.

2.7 Relaxation of the PTC: The Revised-Stiffness Approach

The general equations for the creation of incompatible functions presented in the previous section are an introduction of certain modifications to the incompatible test solutions made in order to satisfy the PTC. This section discusses the replacement of the convergence requirement PTC through a modification of the energy equation of the element. As a result, the selection and modification of incompatible shape functions can be avoided.

This alternative way to create the incompatible model is called the Revised-stiffness (R-s) method [34], for which only a simple revision to the currently created stiffness matrix is needed, while the initially assumed displacement need not be changed.

For a given incompatible element with assumed displacement $\boldsymbol{u} = \boldsymbol{u}_q + \boldsymbol{u}_\lambda$, where $\boldsymbol{u}_q$ is the compatible part and $\boldsymbol{u}_\lambda$ is the incompatible one, the element potential energy is expressed as

$$\Pi_p^e = \int_{V^e} \frac{1}{2} \varepsilon^T \boldsymbol{C} \varepsilon \, dV \tag{2.40}$$

in which the element strain is $\varepsilon = \boldsymbol{D}(\boldsymbol{u}_q + \boldsymbol{u}_\lambda)$, and $\boldsymbol{C}$ is the elasticity matrix. The stationary condition of the functional (Equation 2.40), $\delta\Pi_p^e(\boldsymbol{u}_q, \boldsymbol{u}_\lambda) = 0$, leads to an additional energy consistency constraint, that is, the PTC (Equation 2.25a) with $\boldsymbol{u}^\Delta$, for 2-D/3-D problems. For the purpose of releasing the energy constraint for $\boldsymbol{u}_\lambda$, a modified functional is obtained by deleting the nonzero energy integral Equation 2.25 from the functional (Equation 2.40), and we have

$$\Pi_{mp}^e = \Pi_p^e - \oint_{\partial V^e} \sigma_c^T \boldsymbol{n}^T \boldsymbol{u}_\lambda dS = \Pi_p^e - \int_{V^e} \sigma_c^T \varepsilon_\lambda \, dV \tag{2.41}$$

It can be verified that for the stationary condition for the new functional (Equation 2.41), $\delta\Pi_{mp}^e = 0$, the PTC for $\boldsymbol{u}_\lambda$ no longer appears. Therefore, based on the revised energy functional Π_{mp}^e, the PTC is no longer required.

Now, consider the element displacements, for which the compatible and incompatible parts are assumed independently by

$$\boldsymbol{u} = \boldsymbol{u}_q + \boldsymbol{u}_\lambda = N_q \boldsymbol{q} + N_\lambda \lambda \tag{2.42}$$

in which $\boldsymbol{N}_q$ is the compatible displacement interpolation function wherein the constant-strain mode and the rigid body mode should be included, and $\boldsymbol{q}$ is the corresponding nodal displacements, while $\boldsymbol{N}_\lambda$ represents the additional incompatible mode with λ as the corresponding displacement parameters. The latter can be statically condensed in the element level. The corresponding element strains are

$$\varepsilon = \boldsymbol{D}(\boldsymbol{u}_q + \boldsymbol{u}_\lambda) = \varepsilon_q + \varepsilon_\lambda = \boldsymbol{B}_q \boldsymbol{q} + B_\lambda \lambda \tag{2.43}$$

where $\boldsymbol{B}_q$ and $\boldsymbol{B}_\lambda$ are, respectively, the compatible strain matrix and incompatible strain matrix.

Introducing Equation 2.42 and Equation 2.43 into the modified potential energy functional (Equation 2.41), and noting that $\sigma = \boldsymbol{C}\varepsilon_c$, $\varepsilon_c = \boldsymbol{B}_c\boldsymbol{q}$, in which ε_c is the constant strain vector, and $\boldsymbol{B}_c$ is the corresponding strain matrix. Thus, the strain energy functional for 2-D/3-D cases can be written as

$$\Pi_{mp}^e = \int_{V^e} \frac{1}{2}\varepsilon^T \boldsymbol{C}\varepsilon\, dV - \int_{V^e} \varepsilon_c^T \boldsymbol{C}\varepsilon_\lambda dV = \frac{1}{2}\begin{Bmatrix} q \\ \lambda \end{Bmatrix}^T \boldsymbol{K}_p^e \begin{Bmatrix} q \\ \lambda \end{Bmatrix} - \frac{1}{2}\begin{Bmatrix} q \\ \lambda \end{Bmatrix}^T \boldsymbol{K}_m^e \begin{Bmatrix} q \\ \lambda \end{Bmatrix} \tag{2.44}$$

The element stiffness matrix is now

$$\boldsymbol{K}_{mp}^e = \boldsymbol{K}_p^e - \boldsymbol{K}_m^e \tag{2.45}$$

$$\boldsymbol{K}_p^e = \int_{V^e} \begin{bmatrix} \boldsymbol{B}_q & \boldsymbol{B}_\lambda \end{bmatrix}^T \boldsymbol{C} \begin{bmatrix} \boldsymbol{B}_q & \boldsymbol{B}_\lambda \end{bmatrix} dV = \begin{bmatrix} \boldsymbol{K}_{qq} & \boldsymbol{K}_{q\lambda} \\ \boldsymbol{K}_{q\lambda}^{\mathrm{T}} & \boldsymbol{K}_{\lambda\lambda} \end{bmatrix} \tag{2.46}$$

$$\boldsymbol{K}_m^e = \begin{bmatrix} 0 & \boldsymbol{G} \\ \boldsymbol{G}^T & 0 \end{bmatrix} \tag{2.47}$$

$$\mathbf{G} = \int_{V^e} B_c^T \boldsymbol{C}\boldsymbol{B}_\lambda dV = B_c^{\mathrm{T}}\boldsymbol{C}\int_{V^e} B_\lambda dV \tag{2.48}$$

Note that the constant strain matrix $\boldsymbol{B}_c$ can be easily obtained because $\boldsymbol{B}_c = \boldsymbol{B}_q|_{(\xi=0,\eta=0)}$ in 2-D problems or $\boldsymbol{B}_c = \boldsymbol{B}_q|_{(\xi=0,\eta=0,\zeta=0)}$ in 3-D problems, where ξ, η, and ζ are element natural coordinates. The R-s approach only results in a simple revision of the initial element stiffness matrix $\boldsymbol{K}_{\mathrm{p}}^e$, so it is quite easily carried out. Some examples of this R-s approach are shown in Reference 34.

The incompatible strain $\varepsilon_\lambda = \boldsymbol{B}_\lambda\lambda$ might undergo a change due to the revision of element stiffness in the R-s approach. Without a loss in generality, let the changed strain be

$$\varepsilon_{m\lambda} = \varepsilon_\lambda + \varepsilon_m \tag{2.49}$$

in which the additional constant strain ε_m can be determined by the PTC as shown by Wilson and Ibrahimbegovic [32]; let

$$\int_{V^e} \varepsilon_{m\lambda} dV = \int_{V^e} (\varepsilon_\lambda + \varepsilon_m)\, dV = 0 \tag{2.50}$$

Hence

$$\varepsilon_m = \frac{1}{V^e}\int_{V^e} \varepsilon_\lambda dV = -\frac{1}{V^e}\int_{V^e} B_\lambda dV\lambda \tag{2.51}$$

Substituting Equation 2.51 into Equation 2.49 leads to

$$\varepsilon_{m\lambda} = \left(\boldsymbol{B}_\lambda - \frac{1}{V^e} \int_{V^e} B_\lambda dV \right) \lambda = B_{m\lambda} \lambda \tag{2.52}$$

In the element stress recovery, the stress solution obtained by using $\boldsymbol{B}_{m\lambda}$ is usually more accurate than that obtained by using $\boldsymbol{B}_\lambda$.

The present R-s approach was initially called the "constant stress multiplier method" [31] when it was used to develop the assumed stress hybrid finite element [35]. In addition, the R-s approach can be extended to the development of multivariable incompatible elements based on the Hellinger–Reissner functional, $\Pi^e_{HR}(\boldsymbol{u},\sigma)$, or the Hu–Washizu functional, $\Pi^e_{HW}(\boldsymbol{u},\sigma,\varepsilon)$, when the assumed displacements are expressed as $\boldsymbol{u} = \boldsymbol{u}_q + \boldsymbol{u}_\lambda$. In order to apply the PTC for the resulting elements, we can replace the functional Π^e_{HR} and Π^e_{HW}, with the respective modified versions shown in the following:

$$\Pi^e_{mHR} = \Pi^e_{HR} - \oint_{\partial V^e} \sigma_c^T \boldsymbol{n}^T \boldsymbol{u}_\lambda dS = \Pi^e_{HR} - \int_{V^e} \sigma_c^T \varepsilon_\lambda dV \tag{2.53}$$

and

$$\Pi^e_{mHW} = \Pi^e_{HW} - \oint_{\partial V^e} \sigma_c^T \boldsymbol{n}^T \boldsymbol{u}_\lambda dS = \Pi^e_{HW} - \int_{V^e} \sigma_c^T \varepsilon_\lambda dV \tag{2.54}$$

2.8 The PTC in Curvilinear Coordinates

Up to this point, the discussions and applications of the convergence condition PTC for incompatible elements have been restricted to Cartesian coordinates. In general, these discussions are not applicable to problems formulated in curvilinear coordinates. Various plane isoparametric elements can be extended directly to problems formulated in curvilinear coordinates. But the same process is no longer applicable for plane incompatible elements. For example, although the QM6 element developed by Taylor et al. [5] can pass the patch test for plane stress and plane strain problems, it cannot pass the patch test for axisymmetric problems [36]. In order to extend the theory and method of incompatible elements to problems in curvilinear coordinates, it is necessary to find the corresponding forms of PTC in various curvilinear coordinate systems.

Consider an infinitesimal parallelepiped (Figure 2.6) in a system of 3-D curvilinear coordinates a_i $(i = 1,2,3)$. Let v_i represent the displacement vector within an element. Then the element strain tensor is

$$e_{ij} = \frac{1}{2}\left(v_i\big|_j + v_j\big|_i \right) = \frac{1}{2}\left(v_{i,j} + v_{j,i} \right) - v_m \Gamma^m_{ij} \tag{2.55}$$

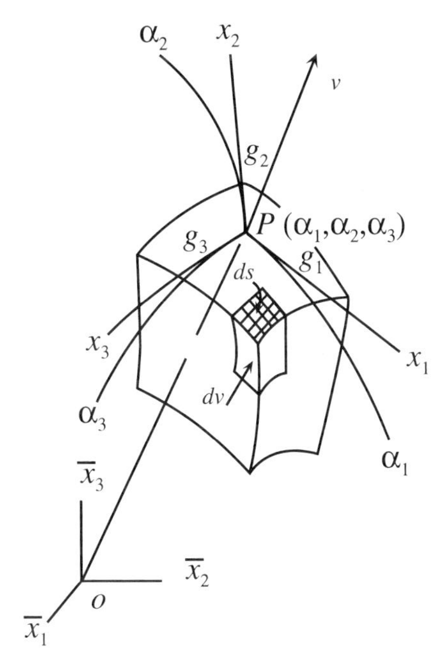

FIGURE 2.6
Infinitesimal parallelepiped with curvilinear coordinates.

where ( ) and ( )$|_i$ are, respectively, the partial derivative and covariant derivative of the quantity ( ) with respect to a_i. Γ_{ij}^{m} is the Christoffel symbol. The strain energy functional of the element is represented by

$$\int\int\int_{V^e} \frac{1}{2} c^{ijkl} e_{kl} e_{ij}\, dV = \int\int\int_{V^e} \frac{1}{2} \tau^{ij} v_i\big|_j\, dV \tag{2.56}$$

where the stress tensor is

$$\tau^{ij} = c^{ijkl} e_{kl} \tag{2.57}$$

For incompatible elements, we choose trial displacement $v_i = v_i^q + v_i^\lambda$, in which v_i^q and v_i^λ are, respectively, the compatible and incompatible parts of the trial function. We define the strain energy functional of the incompatible element system as

$$\begin{aligned}\Pi &= \sum_e \Pi^e(v_i^q, v_i^\lambda) \\ &= \sum_e \left[\int\int\int_{V^e} \frac{1}{2} \tau^{ij}(v_i^q + v_i^\lambda)\big|_j\, dV - \int\int_{S_\sigma^e} \bar{T}^i v_i^q dS\right]\end{aligned} \tag{2.58}$$

From the stationary condition, we can derive the various equilibrium conditions under the curvilinear coordinate system. In addition, the following condition for energy compatibility can be obtained:

$$\sum_e \oint\!\!\oint_{\partial V^e} \tau_c^{ij} \nu_j \partial v^{\lambda} dS = 0 \tag{2.59}$$

in which ν_j is the component of the surface normal ν (Figure 2.6).

In the following we shall discuss only the formulation of the PTC of the element. At the limiting case of element discretization with the size of the element h approaching zero, the element stress tensor τ^{ij} reaches a set of constants τ_c^{ij}. Instead of Equation 2.59, we may simply use

$$\oint\!\!\oint_{\partial V^e} \tau_c^{ij} \nu_j v_i^{\lambda} dS = 0 \tag{2.60}$$

Based on the divergence theorem for curvilinear coordinate systems, the above equation may also be written in volume integral form as follows:

$$\int\!\!\int\!\!\int_{V^e} (\tau_c^{ij} v_i^{\lambda}\big|_j + \tau_c^{ij}\big|_j v_i^{\lambda}) dV = 0 \tag{2.61}$$

in which

$$v_i^{\lambda}\big|_j = v_{i,j}^{\lambda} - v_m^{\lambda} \Gamma_{i,j}^{m}$$

$$\tau_c^{ij}\big|_j = \tau_{c,j}^{ij} - \tau_c^{im} \Gamma_{jm}^{j} + \tau_c^{mj} \Gamma_{jm}^{j} = \tau_c^{im} \Gamma_{jm}^{j} + \tau_c^{mj} \Gamma_{jm}^{j}$$

Equation 2.60 and Equation 2.61 are used for the PTC of general curvilinear coordinate systems. We can substitute the following physical components

$$\sigma_c^{ij} = \sqrt{\frac{g_{jj}}{g^{ii}}}\,\tau_c^{ij}, \quad u_i^{\lambda} = \sqrt{g^{ii}}\,v_i^{\lambda}, \quad n_j = \sqrt{g^{jj}}\,v_j \tag{2.62}$$

in which g^{ii} and g_{jj} are metric tensors. Thus, under curvilinear coordinate systems, the equation in physical components for the PTC can be written as

$$\oint\!\!\oint_{\partial V^e} \frac{1}{\sqrt{g^{jj} g_{jj}}} \sigma_c^{ij} n_j u_i^{\lambda} ds = 0 \tag{2.63}$$

For orthogonal curvilinear coordinate systems, g^{jj} is equal to $1/g_{jj}$; hence Equation 2.64 can be simplified to

$$\oint\!\!\oint_{\partial V^e} \sigma_c^{ij} n_j u_i^{\lambda} ds = 0 \tag{2.64}$$

We can see that the physical component equation (Equation 2.64) for the PTC of orthogonal curvilinear coordinates and the tensorial equation (Equation 2.60) are identical. n_j in Equation 2.64 is the directional cosine of the surface normal ν (Figure 2.6) against the local Cartesian coordinates r to z.

We can apply the above result to an axisymmetric problem, as shown in Figure 2.7. The differential relationship for the cylindrical coordinates (r,θ,z) is $dS = rd\theta dl$. Hence,

$$\oiint_{\partial V^e} (\bullet\bullet\bullet)\,dS = 2\pi \oint_{\partial A^e} (\bullet\bullet\bullet)\,rdl$$

The integral $\oint_{\partial A^e}$ is taken along the circumference of the axisymmetric element. Hence, from Equation 2.60 we can get the PTC of the axisymmetric problem as follows:

$$\oint_{\partial V^e} \sigma_c^T n^T u_\lambda r dl = 0 \tag{2.65}$$

where

$$\sigma_c = \{\sigma_c^{ij}\} = \begin{Bmatrix} \sigma_c^{rr} \\ \sigma_c^{zz} \\ \sigma_c^{\theta\theta} \\ \sigma_c^{zr} \end{Bmatrix}$$

$$\boldsymbol{u}^\lambda = \{u_i^\lambda\} = \begin{Bmatrix} u_r^\lambda \\ u_z^\lambda \end{Bmatrix}$$

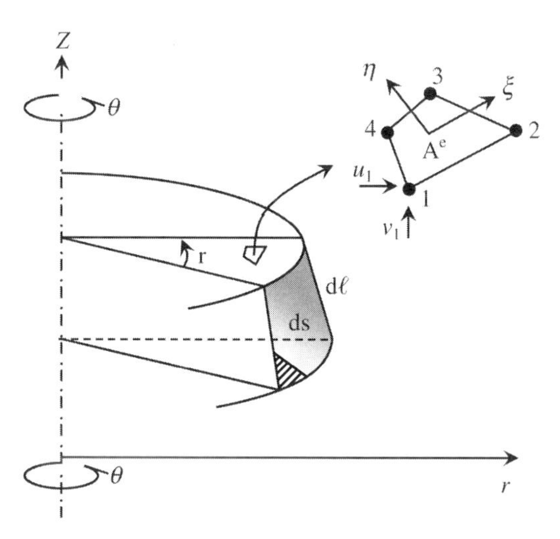

FIGURE 2.7
Axisymmetric element in cylindrical coordinate.

$$n = \{n_i\} = \begin{bmatrix} n_r & 0 & 0 & n_z \\ 0 & n_z & 0 & n_r \end{bmatrix}$$

Because σ_c is constant stress, Equation 2.65 can be replaced by the following more practical relation:

$$\oint_{\partial V^e} \begin{bmatrix} n_r \\ n_z \end{bmatrix} u_i^\lambda r dl = 0 \tag{2.66}$$

We can also formulate the PTC of axisymmetric problems in terms of area integrals by starting from Equation 2.58 in tensor notations. Under cylindrical coordinate systems, $dV = rd\theta dA$, $\Gamma^r_{\theta\theta} = -r = \Gamma^\theta_{\theta r} = \Gamma^\theta_{r\theta} = 1/r$. All other Γ^i_{jk} are zero. By noting the corresponding relations between physical and tensorial components,

$$\sigma_c = \begin{Bmatrix} \tau_c^{rr} \\ \tau_c^{zz} \\ r^2\tau_c^{\theta\theta} \\ \sigma_c^{zr} \end{Bmatrix}, \quad \boldsymbol{u}^\lambda = \begin{Bmatrix} v_r^\lambda \\ v_z^\lambda \end{Bmatrix}$$

we can obtain the following area integral form of the PTC. It is equivalent to Equation 2.60:

$$2\Pi \int \int_{\partial A^e} \sigma_c^T \begin{bmatrix} \frac{\partial}{\partial r} + \frac{1}{r} & 0 \\ 0 & \frac{\partial}{\partial z} \\ 0 & 0 \\ \frac{\partial}{\partial z} & \frac{\partial}{\partial r} + \frac{1}{r} \end{bmatrix} \boldsymbol{u}^\lambda r dA = 0 \tag{2.67}$$

Clearly, Equation 2.67 can be replaced by the following simpler relation:

$$\int \int_{A^e} \begin{bmatrix} \frac{\partial}{\partial r} + \frac{1}{r} \\ \frac{\partial}{\partial z} \end{bmatrix} u_i^\lambda r dA = 0 \tag{2.68}$$

This is the area integral form of the PTC for axisymmetric bodies. It is equivalent to Equation 2.66.

Many incompatible elements such as QM6, QP6 [24], NQ6, and NQ10 [14] can pass the following PTC for plane elastic problems, but they cannot pass the PTC conditions for Equation 2.66 or Equation 2.68. Thus, they are not suitable for the analysis axisymmetric problems.

2.9 Equivalent Nodal Load and Recovery of Stress

The effect of incompatible function N_i should not be included in the determination of the equivalent nodal loads corresponding to known distributed body forces and boundary forces for an incompatible element that contains internal displacement $u_\lambda = \Sigma_i N_{\lambda i}\lambda_i$. Otherwise, the equilibrium condition of the resulting nodal loads and the original distributed loads will not be maintained. This can be shown as follows: Let the distributed body force be $\bar{f}$. The interpolating function corresponding to the displacement field $u_q = \Sigma_i N_i u_i$ should maintain the condition $\Sigma_i N_i = 1$. Thus, the sum of the equivalent nodal load, which is

$$\sum_i F_i = \sum_i \int_{V^e} N_i \bar{f} dV = \int_{V^e} \bar{f} dV \tag{2.69}$$

is in uniform equilibrium with $\bar{f}$.

But with the additional internal displacement, $u = \Sigma_i N_i u_i + \Sigma_j u_{\lambda_j}\lambda_j$, and after static condensation, the result is

$$u = \sum_i (N_i + N_i^\Delta) u_i \tag{2.70}$$

in which N_i^Δ is a nonzero term resulting from the internal displacement. The sum of the equivalent nodal loads corresponding to Equation 2.70 is

$$\sum_i F_i = \sum_i \int_{V^e} (N_i + N_i^\Delta)\bar{f} dV = \int_{V^e} \bar{f} dV + \int_{V^e} \left(\sum_i N_i^\Delta \right) \bar{f} dV \tag{2.71}$$

The existence of the integral term at the right-hand side of Equation 2.71 is a violation of the equilibrium conditions (Equation 2.69).

The following notes should be made about the evaluation of stresses for incompatible elements:

1. When the internal displacements are not included in the determination of element nodal loads, the inclusion of u_λ in the formulation will improve the stress distribution and the accuracy in the elements. Reasonably added u_λ will improve the order of completeness of the originally assigned compatible displacement field u_q; hence it will

also improve the order of completeness of the resulting stress field in the element. For example, for the plane biquadratic eight-node element (Q8), the element interpolation function only contains complete polynomials of the $P_1(x,y)$ type. For irregularly shaped elements, even the stress values at the Gauss integral points are very unreliable (as shown by the example in Figure 8.11 of Reference 20). But if an internal displacement $u_i = (1-\xi^2)(1-\eta^2)\lambda$ is added, the element displacement becomes $u \in P_2(x,y)$. Hence, a reasonable linear stress distribution can result.

2. When the incompatible displacement $u = (u_q + u_\lambda) \in P_n(x,y)$, and does not contain any incomplete terms of an order higher than n, the stress value at any nodal point can directly be evaluated by the element displacement $u(\xi,\eta)$. In this case, even for an irregularly shaped element, the resulting stress distribution can be $\sigma \in P_{n-1}(x.y)$.
3. When the displacement u of an incompatible element contains incomplete higher-order terms under $x-y$ coordinates, the resulting stress distribution is often unreasonable. For example, for the incompatible elements Q6 and QM6, the element displacement $u \in P_2(\xi,\eta)$. Since the Jacobi matrix J of an irregular quadrilateral is a linear function, the resulting stresses will contain second-order terms in ξ and η. This higher-order stress distribution is unreasonable and will result in large errors for the stresses at the nodal points along the element boundary. In this case, the stresses at the nodal point along the element boundary should be determined by extrapolation from the more reliable values at the Gauss points of the element. This is, if the element displacement $u \in P_n(\xi,\eta)$ the corresponding element stress distribution would not be higher than (n – 1)-th. Then, the stresses at the $n \times n$ Gauss points are evaluated and the stresses at the boundary nodes can be determined by (n – 1)-th extrapolation.

For a four-node plane-incompatible element (Figure 2.8) based on a 2×2 Gauss integration, the stresses at the four corner nodal points can be determined by

$$\begin{Bmatrix} \sigma_A \\ \sigma_B \\ \sigma_C \\ \sigma_D \end{Bmatrix} = \frac{1}{4} \begin{bmatrix} 1+\sqrt{3}/2 & -1/2 & 1-\sqrt{3}/2 & -1/2 \\ -1/2 & 1+\sqrt{3}/2 & -1/2 & 1-\sqrt{3}/2 \\ 1-\sqrt{3}/2 & -1/2 & 1+\sqrt{3}/2 & -1/2 \\ -1/2 & 1-\sqrt{3}/2 & -1/2 & 1+\sqrt{3}/2 \end{bmatrix} \begin{Bmatrix} \sigma_1 \\ \sigma_2 \\ \sigma_3 \\ \sigma_4 \end{Bmatrix} \tag{2.72}$$

in which σ_1 to σ_4 represent the resulting stresses (σ_x, σ_y, τ_{xy}) at the four Gauss integration points and σ_A to σ_D are the extrapolated stress values of the four corner nodes.

Similarly, for an eight-node solid incompatible element (Figure 2.9) based on $2 \times 2 \times 2$ Gauss integration, the stresses at the eight corner nodes can be determined by

$$\begin{Bmatrix} \sigma_A \\ \sigma_B \\ \sigma_C \\ \sigma_D \\ \sigma_E \\ \sigma_F \\ \sigma_G \\ \sigma_H \end{Bmatrix} = \frac{3}{8} \begin{bmatrix} K & M & L & M & M & L & N & L \\ & K & M & L & L & M & L & N \\ & & K & M & N & L & M & L \\ & & & K & L & N & L & M \\ & & & & K & M & L & M \\ & & \textit{symmetric} & & & K & M & L \\ & & & & & & K & M \\ & & & & & & & K \end{bmatrix} \begin{Bmatrix} \sigma_1 \\ \sigma_2 \\ \sigma_3 \\ \sigma_4 \\ \sigma_5 \\ \sigma_6 \\ \sigma_7 \\ \sigma_8 \end{Bmatrix} \tag{2.73}$$

where

$$K = (1+\sqrt{3})\left(1+\frac{\sqrt{3}}{3}\right)^2$$

$$L = (1+\sqrt{3})\left(1-\frac{\sqrt{3}}{3}\right)^2$$

$$M = (1-\sqrt{3})\left(1+\frac{\sqrt{3}}{3}\right)^2$$

$$N = (1-\sqrt{3})\left(1-\frac{\sqrt{3}}{3}\right)^2$$

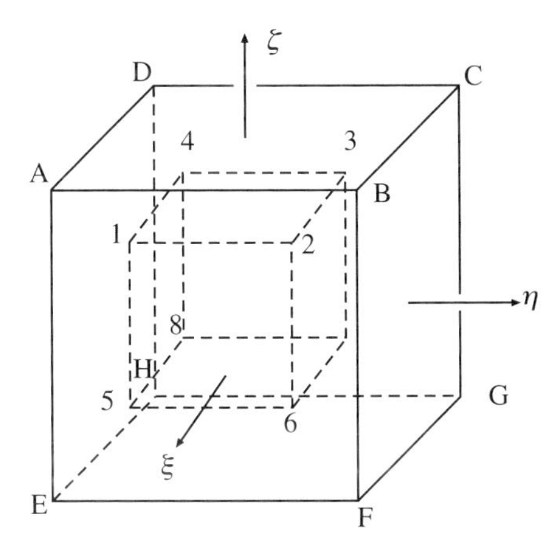

FIGURE 2.8
Linear stress extrapolation of six-faced solid.

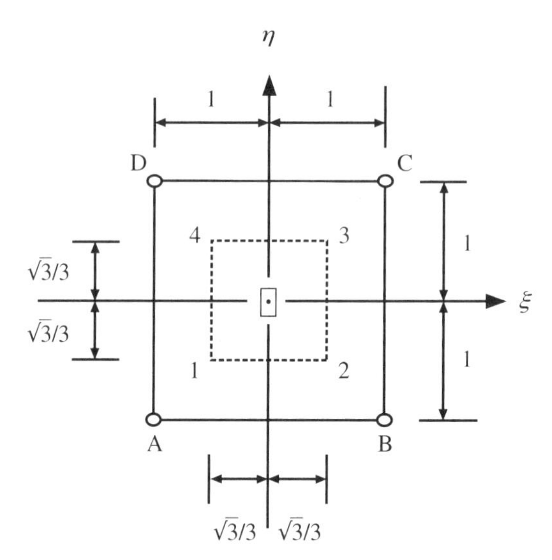

FIGURE 2.9
Bilinear stress extrapolation of quadrilateral element.

References

1. Strang, G. and Fix, G.J., *An Analysis of the Finite Element Method,* Prentice-Hall, Englewood Cliffs, NJ, 1973.
2. Ciarlet, P.G., *The Finite Element for Elliptic Problems,* North-Holland, Amsterdam, 1978.
3. Feng, K., A study of theory of incompatible finite elements (in Chinese), *Comput. Math.,* 4, 378, 1979.
4. Bazeley, G.P., Cheung, Y.K., Irons, B.M., and Zienkiewicz, O.C., Triangular elements in plate bending: conforming and nonconforming solutions, in *Proc. Conf. on Matrix Methods in Structural Mechanics,* Przemieniecki, J.S., Bader, R.M., Bozich, W.F., Johnson, J.R., and Mykytow, W.J., Eds., Wright-Patterson Air Force Base, Dayton, OH, 1965, p. 547.
5. Taylor, R.L., Simo, J.C., Zienkiewicz, O.C., and Chan, A.H.C., The patch test: a condition for assessing FEM convergence, *Int. J. Numer. Methods Eng.,* 22, 39, 1986.
6. Chien, W.Z., *Variational Methods and Finite Elements* (in Chinese), Science Press, Beijing, 1980.
7. Hu, H.C., Convergence requirements for displacements by using Ritz method (in Chinese), *J. Mech.,* 16, 36, 1984.
8. Wu, J.K., Wang, M.Z., and Wang, W., *Introduction to Theory of Elasticity* (in Chinese), Beijing University Press, Beijing, 2001.
9. Mikhlin, S.G., *Variational Methods in Mathematical Physics,* Pergamon Press, New York, 1964.
10. Oden, J.T. and Reddy, J.N., *Variational Methods in Theoretical Mechanics,* Springer-Verlag, Heidelberg, 1983.

11. Fung, Y.C., *Foundations of Solid Mechanics,* Prentice-Hall, Englewood Cliffs, NJ, 1965.
12. Washizu, K. *Variational Methods in Elasticity and Plasticity,* 3rd ed., Pergamon Press, Oxford, 1982.
13. Hu, H.C., *Variational Principles in Theory of Elasticity and its Applications* (in Chinese), Science Press, Beijing, 1981.
14. Wu, C.C. and Cheung, Y.K., Numerical stability and constitution analysis of λ-type incompatible elements, *Int. J. Numer. Methods Eng.,* 31, 1669, 1991.
15. Wu, C.C., Huang, M.G., and Pian, T.H.H., Consistency condition and convergence criteria of incompatible elements: general formulation of incompatible functions and its application, *Comput. Struct.,* 27, 639, 1987.
16. Irons, B.M. and Razzaque, A., Experience with the patch test for convergence of finite element method, in *Mathematical Foundations of the Finite Element Method,* Aziz, A.K., Ed., Academic Press, New York, 1972.
17. Sander, G. and Beckers, P., The influence of the choice of connectors in finite element method, *Int. J. Numer. Methods Eng.,* 11, 1491, 1977.
18. Wu, C.C. and Bufler, H., Multivariable finite elements: consistency and optimization (in Chinese), *Sci. China (A),* 9, 946, 1990.
19. Shi, Z.C., On the convergence properties of the quadrilateral elements of Sander and Beckers, *Math. Comp.,* 42, 493, 1984.
20. Zienkiewicz, O.C. and Taylor, R.L., *The Finite Element Method,* 4th ed., Vol. 1, McGraw-Hill, New York, p. 989.
21. Bergan, P.G. and Mygard, M.K., Finite element with increased freedom in choosing shape functions, *Int. J. Numer. Methods Eng.* 20, 643, 1984.
22. Stummel, F., The limitations of the patch test, *Int. J. Numer. Methods Eng.,* 15, 177, 1980.
23. Shi, Z.C., An explicit analysis of Stummel's patch test examples, *Int. J. Numer. Meth. Eng.,* 1233, 1984.
24. Wachspress, E.L., Incompatible quadrilateral basis functions, *Int. J. Numer. Methods Eng.,* 12, 589, 1978.
25. Li, Y. and Wu, C.C., A new quadrilateral nonconforming model and its convergence, *Appl. Math. Mech.,* 12, 699, 1986.
26. Long, Y.Q., Generalized compatible element (in Chinese), *J. Civil Eng.,* 1, 1, 1987.
27. Specht, B., Modified shape functions for the three-node plate bending element passing the patch test, *Int. J. Numer. Methods Eng.,* 26, 705, 1988.
28. Pian, T.H.H. and Wu, C.C., General formulation of incompatible shape function and an incompatible isoparametric element, in *Proc. International China-American Workshop on Finite Element Methods,* Shen, Y. and Pian, T.H.H., Eds., Beijing Institute of Aeronautics and Astronautics, Beijing, 1986, p. 159.
29. Weissman, S.L. and Taylor, R.L., Resultant fields for mixed plate bending elements, *Comp. Methods Appl. Mech. Eng.,* 79, 321, 1990.
30. Weissman, S.L. and Taylor, R.L., A unified approach to mixed finite element methods: application to in-plane problems, *Comp. Methods Appl. Mech. Eng.,* 58, 127, 1992.
31. Wu, C.C., Huang, Y.Q., and Pian, T.H.H., Constant stress multiplier method and implementation to incompatible models, in *Modeling and Simulation Based Engineering,* Vol. 2, Atluri, S.N. and O'Donoghue P.E., Eds., Tech Science Press, Palmdale, CA, 1998, p. 1873.

32. Wilson, E.L. and Ibrahimbegovic, A., Use of incompatible displacement modes for the calculation of element stiffness or stresses, *Finite Element Anal. Des.*, 7, 229, 1990.
33. Dong, Y.F. and Freitas, J.A.T., A quadrilateral hybrid stress element for Mindlin plates based on incompatible displacements, *Int. J. Numer. Methods Eng.* 37, 279, 1994.
34. Wu, C.C., Huang, Y.Q., and Ramm, E., A further study on incompatible models: revise-stiffness approach and quadratic-completeness mode, *Comp. Methods Appl. Mech. Eng.*, 190, 5923, 2001.
35. Wu, C.C., Di, S.L., and Pian, T.H.H., Optimizing formulation of axisymmetric hybrid stress elements, *Acta Aeronaut. Astronaut. Sinica*, 8, A439, 1987.
36. Taylor, R.L., Beresford, P.J., and Wilson, E.L., A noncompatible element for stress analysis, *Int. J. Num. Meth. Eng.*, 10, 1211–1219, 1976.

3

Incompatible Elements for the Theory of Elasticity

3.1 Introduction

Based on the theory of incompatible elements discussed in the previous chapter, some incompatible elements will be developed for two- and three-dimensional (3-D) problems and axisymmetric problems in this chapter. Two kinds of methodologies will be adopted for model design. They are the Revised-displacement (R-u) approach based on Wu–Huang–Pian (WHP) incompatible formulation and the Revised-stiffness (R-s) approach based on the energy functional without the patch test condition (PTC) as mentioned in Section 2.6. For plate bending problems our discussion will concentrate on the incompatible model based on Reissner–Mindlin plate theory.

3.2 Four-Node Plane Incompatible Elements: NQ6

For plane problems, the four-node element is the simplest and also the best example for illustrating theory and applications. For the convenience of further studies, the four-node plane element of Figure 1.1 is used as an example and some of the useful equations are presented.

For isoparametric elements the same interpolation functions are used for the element coordinates and for the element displacements, that is,

$$\begin{Bmatrix} x \\ y \end{Bmatrix} = \sum_{i=1}^{4} N_i(\xi, \eta) \begin{Bmatrix} x_i \\ y_i \end{Bmatrix} \tag{3.1}$$

$$\begin{Bmatrix} u \\ v \end{Bmatrix} = \sum_{i=1}^{4} N_i(\xi, \eta) \begin{Bmatrix} u_i \\ v_i \end{Bmatrix} \tag{3.2}$$

in which $N_i(\xi, \eta) = \frac{1}{4}(1 + \xi_i \xi)(1 + \eta_i \eta)$ is the bilinear interpolation function. By defining

$$\begin{bmatrix} a_1 & b_1 \\ a_2 & b_2 \\ a_3 & b_3 \\ a_4 & b_4 \end{bmatrix} = \frac{1}{4} \begin{bmatrix} -1 & 1 & 1 & -1 \\ 1 & -1 & -1 & 1 \\ -1 & -1 & 1 & 1 \\ 1 & 1 & 1 & 1 \end{bmatrix} \begin{bmatrix} x_1 & y_1 \\ x_2 & y_2 \\ x_3 & y_3 \\ x_4 & y_4 \end{bmatrix} \tag{3.3}$$

Equation 3.1 can be expressed as

$$\begin{bmatrix} x & y \end{bmatrix} = \begin{bmatrix} \xi & \xi\eta & \eta & 1 \end{bmatrix} \begin{bmatrix} a_1 & b_1 \\ a_2 & b_2 \\ a_3 & b_3 \\ a_4 & b_4 \end{bmatrix} \tag{3.4}$$

Similarly, the element displacements can be expressed as

$$\begin{bmatrix} u & v \end{bmatrix} = \begin{bmatrix} \xi & \xi\eta & \eta & 1 \end{bmatrix} \begin{bmatrix} A_1 & B_1 \\ A_2 & B_2 \\ A_3 & B_3 \\ A_4 & B_4 \end{bmatrix} \tag{3.5}$$

the factors in which are

$$\begin{bmatrix} A_1 & B_1 \\ A_2 & B_2 \\ A_3 & B_3 \\ A_4 & B_4 \end{bmatrix} = \frac{1}{4} \begin{bmatrix} -1 & 1 & 1 & -1 \\ 1 & -1 & 1 & -1 \\ -1 & -1 & 1 & 1 \\ 1 & 1 & 1 & 1 \end{bmatrix} \begin{bmatrix} u_1 & v_1 \\ u_2 & v_2 \\ u_3 & v_3 \\ u_4 & v_4 \end{bmatrix} \tag{3.6}$$

The local coordinates (ξ,η) and the system coordinates (x, y) are related by their derivatives in the following manner:

$$\begin{Bmatrix} \dfrac{\partial(\,)}{\partial \xi} \\ \dfrac{\partial(\,)}{\partial \eta} \end{Bmatrix} = J \begin{Bmatrix} \dfrac{\partial(\,)}{\partial x} \\ \dfrac{\partial(\,)}{\partial y} \end{Bmatrix} \tag{3.7}$$

in which the Jacobi matrix is

$$J = \begin{bmatrix} \dfrac{\partial x}{\partial \xi} & \dfrac{\partial y}{\partial \xi} \\ \dfrac{\partial y}{\partial \eta} & \dfrac{\partial y}{\partial \eta} \end{bmatrix} = \begin{bmatrix} a_1 + a_2\eta & b_1 + b_2\eta \\ a_3 + a_2\xi & b_3 + b_2\xi \end{bmatrix} \tag{3.8}$$

The corresponding determinant is

$$|J| = J_0 + J_1\xi + J_2\eta \tag{3.9}$$

in which

$$J_0 = a_1 b_3 - a_3 b_1 = \frac{1}{4}(\textit{area of element}) > 0$$

$$J_1 = a_1 b_2 - a_2 b_1$$

$$J_2 = a_2 b_3 - a_3 b_2$$

When the element is a rectangle or parallelogram shape, factors $a_2 = b_2 = J_1 = J_2 = 0$, and $|\boldsymbol{J}| = J_0$ is a positive constant.

The inverse transformation corresponding to Equation 3.7 is

$$\begin{Bmatrix} \dfrac{\partial()}{\partial x} \\ \dfrac{\partial()}{\partial y} \end{Bmatrix} = J^{-1} \begin{Bmatrix} \dfrac{\partial()}{\partial \xi} \\ \dfrac{\partial()}{\partial \eta} \end{Bmatrix} \tag{3.10}$$

In many situations it is possible to let $|J| = J_0$ to simplify the computation. In that case Equation 3.10 can be expressed as

$$\begin{Bmatrix} \dfrac{\partial()}{\partial x} \\ \dfrac{\partial()}{\partial y} \end{Bmatrix} = \frac{1}{J_0} \begin{bmatrix} b_3 + b_2\xi & -(b_1 + b_2\eta) \\ -(a_3 + a_2\xi) & a_1 + a_2\eta \end{bmatrix} \begin{Bmatrix} \dfrac{\partial()}{\partial \xi} \\ \dfrac{\partial()}{\partial \eta} \end{Bmatrix} \tag{3.11}$$

We can prove that the use of J_0 to approximate $|\boldsymbol{J}|$ is equivalent to the introduction of a geometric approximation by replacing the original four-side element into a parallelogram of the same length (as shown by the dotted lines in Figure 1.1). If we let the geometric distortion factor of the element $a_2 = b_2 = 0$, we have

$$\begin{Bmatrix} \dfrac{\partial(\ )}{\partial x} \\ \dfrac{\partial(\ )}{\partial y} \end{Bmatrix} = \frac{1}{J_0} \begin{bmatrix} b_3 & -b_1 \\ -a_3 & a_1 \end{bmatrix} \begin{Bmatrix} \dfrac{\partial(\ )}{\partial \xi} \\ \dfrac{\partial(\ )}{\partial \eta} \end{Bmatrix} \tag{3.12}$$

When an element is not a rectangle or parallelogram, the use of Equation 3.12 to replace Equation 3.10 is equivalent to the use of $J(\xi,\eta) \approx J(0,0)$. This may lead to a large error in the solution.

The equations for incompatible elements introduced in the previous chapter will now be applied to plane elasticity problems by the construction of a four-node incompatible element. Referring to Figure 1.1, for any component of the element displacements $\mathbf{u} = \{u, v\}$, take a trial solution, $\mathbf{u} = \mathbf{u}_q + \mathbf{u}_\lambda$, in which $\mathbf{u}_q$ is the ordinary bilinear interpolation function and $\mathbf{u}_\lambda$ is the incompatible displacement defined by

$$u_\lambda = \begin{bmatrix} 1-\xi^2 & 1-\eta^2 \end{bmatrix} \begin{Bmatrix} \lambda_1 \\ \lambda_2 \end{Bmatrix} + \begin{bmatrix} \xi & \eta \end{bmatrix} \begin{Bmatrix} \lambda_1^* \\ \lambda_2^* \end{Bmatrix} = N_\lambda \lambda + N^* \lambda^* \tag{3.13}$$

where $N_\lambda = \begin{bmatrix} 1-\xi^2 & 1-\eta^2 \end{bmatrix}$ is the incompatible basic function for the Wilson element, which was suggested by Wilson et al. [1]. It turns out that the resulting element formulated by using only N_λ cannot pass the constant stress patch test. The function N^* is added to pass the test. The equations that correspond to Equation 2.36 and Equation 2.35 are, respectively,

$$P_\lambda = \oint \begin{Bmatrix} l \\ m \end{Bmatrix} N_\lambda dS = \frac{8}{3} \begin{bmatrix} -b_2 & b_2 \\ a_2 & -a_2 \end{bmatrix} \tag{3.14}$$

and

$$P_* = \oint \begin{Bmatrix} l \\ m \end{Bmatrix} N^* dS = 4 \begin{bmatrix} b_3 & -b_1 \\ -a_3 & a_1 \end{bmatrix} \tag{3.15}$$

The geometric parameters a_i and b_i are determined by Equation 3.3. $|P_*|$ is the area of the element. Hence, P_*^{-1} does exist. Thus, according to the general equation (Equation 2.34) for the generation of incompatible function, the required incompatible internal displacement of the element is

$$u_\lambda^* = (N_\lambda - N^* P_*^{-1} P)\lambda = \begin{bmatrix} N_{\lambda 1}^* & N_{\lambda 2}^* \end{bmatrix} \begin{Bmatrix} \lambda_1 \\ \lambda_2 \end{Bmatrix} \tag{3.16}$$

in which

$$\left.\begin{aligned} N_{\lambda 1}^{*} &= (1-\xi^{2}) + \frac{2}{3}\left(\frac{J_1}{J_0}\xi - \frac{J_2}{J_0}\eta\right) \\ N_{\lambda 2}^{*} &= (1-\eta^{2}) - \frac{2}{3}\left(\frac{J_1}{J_0}\xi - \frac{J_2}{J_0}\eta\right) \end{aligned}\right\} \tag{3.17}$$

The factors in the above equations can be determined by the following definition of the Jacobi determinant:

$$|J| = J_0 + J_1\xi + J_2\eta = (a_1b_3 - a_3b_1) + (a_1b_2 - a_2b_1)\xi + (a_2b_3 - a_3b_2)\eta \tag{3.18}$$

The four-node incompatible element based on the displacement $u = u_q + u_{\lambda}^{*}$ is named NQ6 [2, 3].

Weissman and Taylor [3] suggested a 2-D incompatible mode. It can be stated in the form of Equation 3.17 as

$$\left.\begin{aligned} N_{\lambda 1}^{*} &= \left(1 - \frac{J_2}{J_0}\eta\right)(1-\xi^{2}) + \frac{J_1}{J_0}\xi(1-\eta^{2}) \\ N_{\lambda 2}^{*} &= \left(1 - \frac{J_1}{J_0}\xi\right)(1-\eta^{2}) + \frac{J_2}{J_0}\eta(1-\xi^{2}) \end{aligned}\right\} \tag{3.19}$$

Equation 3.19 contains the $\xi^2\eta$ and $\xi\eta^2$ terms. Hence, in the determination of the element stiffness matrix, a 3×3 numerical integration is required. A comparison of the solution for the plate bending problem in Reference 4 has shown that the use of the incompatible function (Equation 3.17) leads to even better results.

For the four-node plane incompatible element NQ6, the element displacement $u = u_q + u_{\lambda} \in P_2(\xi, \eta)$. This indicates that in natural coordinates, the element displacement is complete in the second order. However, because $u \notin P_2(x, y)$, the displacement in x-y coordinates is not complete in the second order. Thus, for irregularly shaped four-node elements, even if the resulting displacement solution is good, the resulting stress distribution may have large errors. In order to improve the stress solutions, additional treatment is needed such that $u \in P_2(x, y)$.

The following example illustrates the construction of a four-node plane-incompatible element. Starting from an element displacement with completeness in the second order, let

$$u = \begin{bmatrix} 1 & x & y & x^2 & y^2 & xy \end{bmatrix} \begin{Bmatrix} a_1 \\ \vdots \\ a_6 \end{Bmatrix} \tag{3.20}$$

With coordinate transformation by Equation 3.4, this equation can be rewritten as

$$u = u_q + u_\lambda \tag{3.21}$$

in which u_q is a bilinear interpolation function given by

$$u_q = c_1\xi + c_2\xi\eta + c_3\eta + c_4 \tag{3.22}$$

with

$$\begin{Bmatrix} c_1 \\ c_2 \\ c_3 \\ c_4 \end{Bmatrix} = \frac{1}{4}\begin{bmatrix} -1 & 1 & 1 & -1 \\ 1 & -1 & 1 & -1 \\ -1 & -1 & 1 & 1 \\ 1 & 1 & 1 & 1 \end{bmatrix}\begin{Bmatrix} u_1 \\ u_2 \\ u_3 \\ u_4 \end{Bmatrix} \tag{3.23}$$

and

$$u_\lambda = \begin{bmatrix} \xi^2 & \eta^2 & \xi^2\eta & \xi\eta^2 & \xi^2\eta^2 \end{bmatrix}\begin{Bmatrix} \lambda_1 \\ \vdots \\ \lambda_5 \end{Bmatrix} \tag{3.24}$$

The latter is the required additional internal displacement of the element for maintaining $u \in P_2(x, y)$.

We can show that the displacement u_λ in Equation 3.24 does not satisfy the PTC. However, we can modify u_λ by using the WHP equation (Equation 2.34) for incompatible function of the element. The desirable element internal displacement is

$$u_\lambda^* = \left[\xi^2 - \frac{2}{3}\left(\frac{J_1}{J_0}\xi - \frac{J_2}{J_0}\eta\right) \quad \eta^2 + \frac{2}{3}\left(\frac{J_1}{J_0}\xi - \frac{J_2}{J_0}\eta\right) \right.$$
$$\left. \xi^2\eta - \frac{1}{3}\eta \quad \xi\eta^2 - \frac{1}{3}\xi \quad \xi^2\eta^2 \right]\begin{Bmatrix} \lambda_1 \\ \vdots \\ \lambda_5 \end{Bmatrix} \tag{3.25}$$

It should be noted that the element displacement $u = u_q + u_\lambda^* \in P_2(x, y)$ and satisfies the PTC. It contains nine independent parameters. The corresponding incompatible element is named NQ9.

The method for constructing the incompatible element described above is the use of the WHP equation to modify the initially assumed incompatible displacement so that the PTC can be satisfied. Thus, this method can be named the R-u method. This method and the R-s approach presented in Section 2.6

are both basic methods for developing incompatible elements. Some details for the incompatible element new formulations can be seen in Reference 5.

3.3 P_2-Linked Incompatible Methods with the Fewest Degrees of Freedom (DOF)

In comparison with the NQ6 element, each displacement component of the NQ9 element has three additional element internal parameters. Thus, in determining the element stiffness matrix of NQ9, additional computing time is required for statically condensing the internal parameters λ. This section presents the construction of a four-node plane element with completeness in the second order in the displacement field without increasing the element-internal displacement parameters.

In the previously developed four-node incompatible elements such as Q6, QM6, NQ6, etc., the element displacement $u = u_q + u_\lambda \in \boldsymbol{P}_2(\xi,\eta)$ has quadratic completeness with respect to the element-natural coordinates (ξ,η) but not to the system coordinates (x, y), that is, $u \notin P_2(x,y)$.

Therefore, the related element stress, $\sigma \notin P_1(x,y)$, and a rational distribution for the resulting stresses cannot be obtained when irregular meshes are adopted. In order to create a four-node incompatible mode with $u \in P_2(x,y)$, we can select Equation 3.24 as an additional incompatible displacement mode for each element. However, in order for this deformation mode to satisfy the PTC, it should be revised before its implementation for incompatible element formulation. In this way, the incompatible model NQ9 is obtained. This displacement mode contains five internal parameters. Hence, in the construction of the element stiffness matrix, more time must be spent for the elimination of λ_i.

With a rectangular coordinate frame built for an individual element shown in Figure 3.1,

$$\left.\begin{aligned}\bar{x} = x - a_4 = a_1\xi + a_2\xi\eta + a_3\eta \\ \bar{y} = y - b_4 = b_1\xi + b_2\xi\eta + b_3\eta\end{aligned}\right\} \tag{3.26}$$

Thus, a quadratic complete element-displacement model should take the form of

$$u = \alpha_1 + \alpha_2\bar{x} + \alpha_3\bar{y} + \alpha_4\overline{xy} + \lambda_1\bar{x}^2 + \lambda_2\bar{y}^2 \in P_2(\bar{x},\bar{y}) \tag{3.27}$$

Introduction of Equation 3.26 into Equation 3.27 results in

$$\begin{aligned}u =& \alpha_1 + (a_1\alpha_2 + b_1\alpha_3)\xi + (a_3\alpha_2 + b_3\alpha_3)\eta \\ &+ [a_2\alpha_2 + b_2\alpha_3 + (a_1b_3 + a_3b_1)\alpha_4 + 2a_1a_3\lambda_1 + 2b_1b_3\lambda_2]\xi\eta \\ &+ f_3\alpha_4 + f_1\lambda_1 + f_2\lambda_2\end{aligned} \tag{3.28}$$

in which

$$f_1 = \bar{x} - 2a_1a_3\xi\eta$$
$$f_2 = \bar{y} - 2b_1b_3\xi\eta$$
$$f_3 = \overline{xy} - (a_1b_3 + a_3b_1)\xi\eta$$

The sum of the first four terms in Equation 3.28 is the bilinear part of the compatible element displacement, that is, u_q, which was defined in Equation 3.22. Thus, a certain relationship between the nodal displacement parameters c_i and the general parameters (α_I, λ_i) can be obtained:

$$\left.\begin{aligned} &\alpha_1 = c_4 \\ &\alpha_2 = (c_1b_3 - c_3b_1)/J_0 \\ &\alpha_3 = (c_3a_1 - c_1a_3)/J_0 \\ &\alpha_4 = (c_2 - a_2\alpha_2 - b_2\alpha_3 - 2a_1a_3\lambda_1 - 2b_1b_3\lambda_2)/(a_1b_3 + a_3b_1) \end{aligned}\right\} \tag{3.29}$$

Substituting Equation 3.29 into Equation 3.28, we obtain the element displacement

$$u = u_q + u_{q\lambda} + u_{\lambda\lambda} \tag{3.30a}$$

in which u_q is determined by bilinear interpolation, and the last two terms are all incompatible modes. They are

$$u_{q\lambda} = \frac{1}{a_1b_3 + a_3b_1}\left(c_2 - \frac{J_2c_1 + J_1c_3}{J_0}\right)f_3 = \sum_{i=1}^{4} N_{q\lambda}^{i} u_i \tag{3.30b}$$

and

$$u_{\lambda\lambda} = \left(f_1 - \frac{2a_1a_3}{a_1b_3 + a_3b_1}f_3\right)\lambda_1 + \left(f_2 - \frac{2b_1b_3}{a_1b_3 + a_3b_1}f_3\right)\lambda_2 \tag{3.30c}$$

in which

$$N_{q\lambda}^{i} = \frac{1}{4(a_1b_3 + a_3b_1)}\left(\xi_i - \frac{J_2\xi_i + J_1\eta_i}{J_0}\right)f_3, \ (i = 1,2,3,4)$$

$$J_0 = a_1b_3 - a_3b_1, \; J_1 = a_1b_2 - a_2b_1, \; J_2 = a_2b_3 - a_3b_2$$

$N^i_{q\lambda}$ in Equation 3.30b is the incompatible function that corresponds to the nodal parameter u_i, while $u_{\lambda\lambda}$ of Equation 3.30c corresponds to the internal parameters λ_I.

The displacement function given by Equation 3.30a is the desired quadratic-complete incompatible mode with the fewest internal parameters, with dim $(\lambda_i) = 2$. In general, the incompatible modes $u_{q\lambda}$ and $u_{\lambda\lambda}$ do not meet the PTC and need to be revised. To this end, only the incompatible functions f_1, f_2, and f_3 in Equation 3.28 need to be modified. From the WHP formulation, we obtain

$$\left.\begin{aligned} f_1^* &= f_1 - r_1(a_1\xi + a_3\eta) \\ f_2^* &= f_2 - r_2(b_1\xi + b_3\eta) \\ f_3^* &= f_3 - \frac{1}{2}(a_1r_2 + b_1r_1)\xi - \frac{1}{2}(a_3r_2 + b_3r_1)\eta \end{aligned}\right\} \tag{3.31}$$

in which

$$r_1 = \frac{2}{3}\left(a_1\frac{J_1}{J_0} + a_3\frac{J_2}{J_0}\right), \qquad r_2 = \frac{2}{3}\left(b_1\frac{J_1}{J_0} + b_3\frac{J_2}{J_0}\right)$$

The incompatible shape functions for the displacement component v can be obtained in a manner similar to that for u. Based on the assumed element displacement field in Equation 3.30 with quadratic completeness, two incompatible plane elements $\in P_2(x,y)$ with the fewest DOF are developed. One is termed as R-u:Q6 $\in P_2$, for which the incompatible functions are revised as in Equation 3.31; the other is called R-s:Q6 $\in P_2$, for which the R-s approach is adopted.

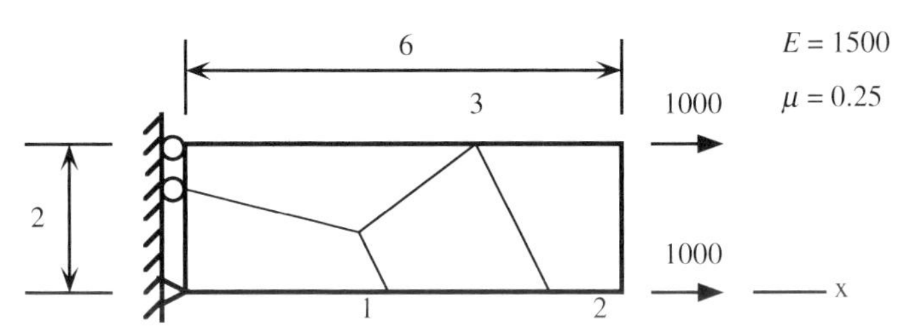

FIGURE 3.1
Patch test: a uniformly stretched specimen (E = 1500, μ = 0.25, t = 1).

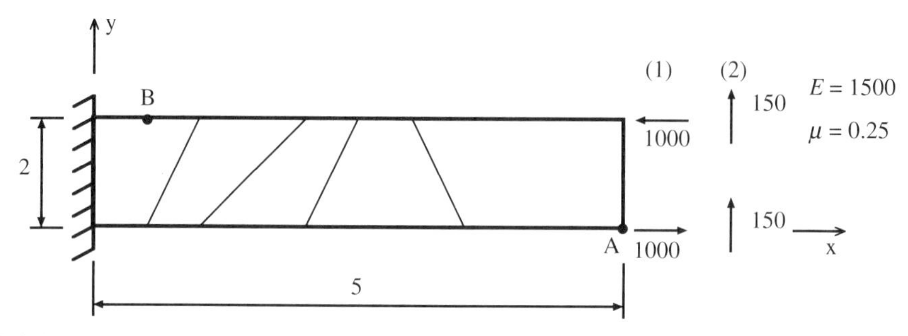

FIGURE 3.2
Plane stress cantilever beam (E = 1500, μ = 0.25).

The following four-node quadrilateral plane elements are employed in the numerical tests:

1. Q4: bilinear isoparametric element with 2 × 2 integration.
2. Q6: Wilson's incompatible element with 2 × 2 integration.
3. QM6: incompatible element by Taylor et al. [4] with 2 × 2 integration.
4. NQ6: incompatible element by Wu et al. [2] with 2 × 2 integration.
5. R-u:Q6 $\in P_2$: present R-u-based plane-incompatible element with $u \in P_2(x,y)$, for which the displacement u is given by Equation 3.30, with f^* given by Equation 3.31 and by using 2 × 2 or 3 × 3 integration.
6. R-s:Q6 $\in P_2$: present R-s-based plane-incompatible element with $u \in P_2(x,y)$, for which the displacement u_λ is given by Equation 3.30 and by using 2 × 2 or 3 × 3 integration.

The following are the numerical tests for these elements:

Test 1. Patch test for plane stress/strain problem (Figure 3.1): This is a uniformly stretched specimen modeled by four irregular elements under constant stress and strain status. Numerical results show that except for Q6, all the other incompatible models in the above list have the ability to represent the constant stress and strain status and can pass the patch test.

Test 2. Bending of cantilever beam (Figure 3.2 and Table 3.1): A cantilever beam subjected to bending and shear loads is chosen here to examine element performance under bending action. Figure 3.2 shows the geometry, applied loads, and material constants of the beam and the element meshes used in the computation. The numerical results given in Table 3.1 indicate that the quadratic-complete incompatible models R-u:Q6 $\in P_2$ and R-s:Q6 $\in P_2$ are able to produce excellent numerical solutions for both displacements and stresses.

TABLE 3.1

Comparison of Results from Four-Node Plane Elements

Element	Load Case 1		Load Case 2	
	v_A	σ_{xB} / σ_{xC}	v_A	σ_{xB} / σ_{xC}
Q4	45.7	−1762/2895	50.7	−2448/4032
Q6	98.4	−2427/3932	100.3	−3325/5514
		(−3002/2998)		(−4414/4225)
QM6	96.1	−2513/4006	98.0	−3390/5479
		(−3014/3002)		(−4074/4112)
NQ6	96.1	−2439/3939	98.0	−3294/5389
		(−3014/3002)		(−4074/4112)
R-u:Q6 ∈ P2	95.7	−2995/3048	97.9	−4108/4261
R-s:Q6 ∈ P2	99.5	−3015/3007	101.8	−4137/4206
Exact	100	−3000/3000	102.6	−4050/4275

( ): Results were figured by extrapolation.

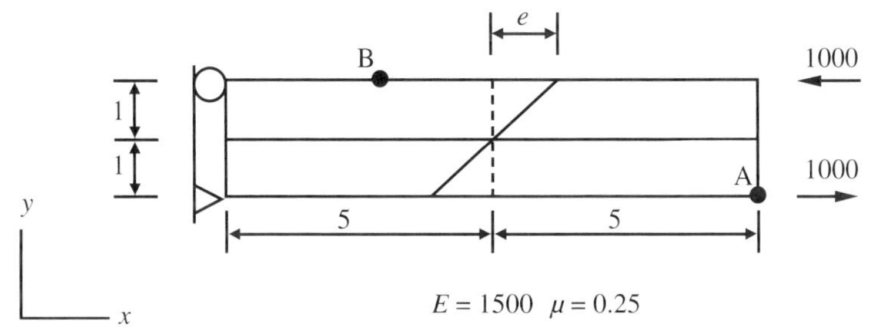

FIGURE 3.3
Distortion test of four-element beam under pure bending (E = 1500, μ = 0.25).

In addition, numerical tests have shown that the present quadratic-complete elements can be implemented by using only low-order 2×2 integration without loss of accuracy. Hence, $P_2(x,y)$-linked lower-order models R-u:Q6 ∈ P_2 and R-s:Q6 ∈ P_2 possess higher computational efficiency and should be strongly recommended.

Test 3. Mesh distortion test for numerical stability (Figure 3.3): A beam under pure bending is modeled by four irregular elements as shown in Figure 3.3. This test is used to examine the instability of numerical solutions caused by mesh distortions. Mesh distortion sensitivity of the tip displacement v_A and the normal stress σ_{xB} at the midpoint of the element edge are shown in Figure 3.4 and Figure 3.5, respectively. The incompatible models QM6 and NQ6 provide fairly stable displacement solutions, while their stress solutions are very sensitive to the mesh distortion factor "e." This occurs because a quadratic-incomplete

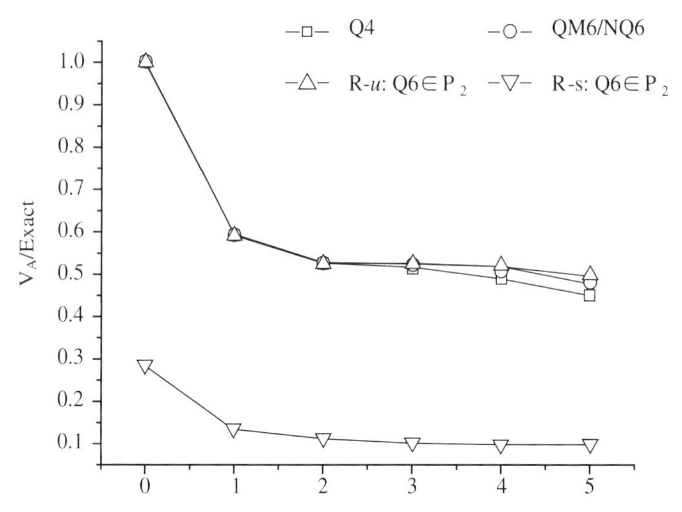

FIGURE 3.4
Mesh distortion sensitivity of the tip displacement (see Figure 3.3).

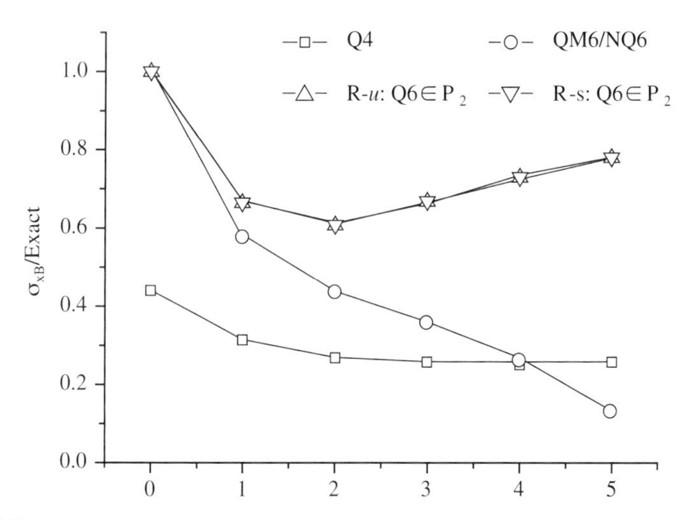

FIGURE 3.5
Mesh distortion sensitivity of the normal stress (see Figure 3.3).

displacement, $u \notin P_2(x,y)$, is employed in the formulation of QM6 and NQ6. All the present P_2(x,y)-linked incompatible models produce stable solutions for both displacements and stresses. These tests indicate that the completeness, with respect to the system coordinates, of the assumed element displacement is very important to ensure a better stress distribution within the element. Otherwise, a stress recovery in terms of the bilinear extrapolation is needed.

3.4 Eight-Node 3-D Solid Incompatible Elements

For the analysis of 3-D solids, the 3-D isoparametric elements are the most convenient to use. However, there are some limitations, such as the appearance of locking phenomena for bending and incompressible problems. Wilson et al. [1] have developed an eight-node hexahedral solid element by adding incompatible displacements to 3-D isoparametric nodes. This element, called H11 in this chapter, can avoid the above-mentioned difficulties. However, irregularly shaped H11 elements may not satisfy the PTC. This section presents a 3-D eight-node hexahedral solid element NH11 that can satisfy the PTC.

By extending the equations for incompatible elements in Section 2.5, we denote the linear interpolation of the displacements of an eight-node six-faced isoparametric element (Figure 3.6) as

$$u_q = \sum_{i=1}^{8} \frac{1}{8}(1+\xi_i\xi)(1+\eta_i\eta)(1+\zeta_i\zeta)u_i \tag{3.32}$$

and the additional incompatible displacements to be added to u_q:

$$u_\lambda = \begin{bmatrix} \xi^2 & \eta^2 & \varsigma^2 \end{bmatrix} \begin{Bmatrix} \lambda_1 \\ \lambda_2 \\ \lambda_3 \end{Bmatrix} = N_\lambda \lambda \tag{3.33}$$

Thus, the element displacement $u = (u_q + u_\lambda) \in P_2(\xi, \eta, \zeta)$ to be complete in the second order with respect to the natural coordinates of the element. But for an irregularly shaped 3-D element, Equation 3.33 does not satisfy the PTC and a modification is required. In accordance with WHP formulation, we define

$$N^* = [\xi \ \eta \ \varsigma] \tag{3.34}$$

and correspondingly,

$$P_* = \oint \begin{Bmatrix} l \\ m \\ n \end{Bmatrix} N^* dS = \int_{V^e} \begin{Bmatrix} \partial/\partial x \\ \partial/\partial y \\ \partial/\partial z \end{Bmatrix} N^* dV = \int_{V^e} J^{-1} dV \tag{3.35}$$

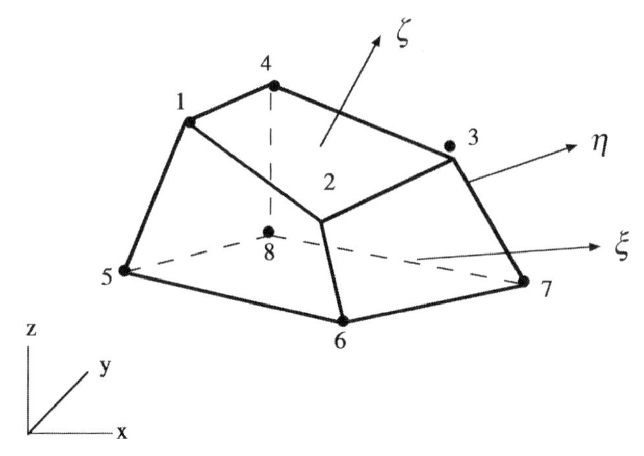

FIGURE 3.6
Eight-node six-faced solid element.

in which the Jacobi matrix J is given by

$$J = \begin{bmatrix} \dfrac{\partial x}{\partial \xi} & \dfrac{\partial y}{\partial \xi} & \dfrac{\partial z}{\partial \xi} \\ \dfrac{\partial x}{\partial \eta} & \dfrac{\partial y}{\partial \eta} & \dfrac{\partial z}{\partial \eta} \\ \dfrac{\partial x}{\partial \zeta} & \dfrac{\partial y}{\partial \zeta} & \dfrac{\partial z}{\partial \zeta} \end{bmatrix} \tag{3.36a}$$

$$= \left[\begin{array}{l|c|c} a_1 + a_4\eta + a_5\zeta + a_7\eta\xi & & \\ a_2 + a_4\xi + a_6\zeta + a_7\xi\zeta & (a_i \rightarrow b_i) & (b_i \rightarrow c_i) \\ a_3 + a_5\xi + a_6\eta + a_7\xi\eta & & \end{array}\right]$$

The factors a_i, b_i, c_i, which are expressed in terms of the element nodal coordinates (x_i, y_i, z_i), are shown as

$$\begin{bmatrix} a_1 & b_1 & c_1 \\ \vdots & \vdots & \vdots \\ a_7 & b_7 & c_7 \end{bmatrix} = \frac{1}{8} \begin{bmatrix} -1 & 1 & 1 & -1 & -1 & 1 & 1 & -1 \\ -1 & -1 & 1 & 1 & -1 & -1 & 1 & 1 \\ 1 & 1 & 1 & 1 & -1 & -1 & -1 & -1 \\ 1 & -1 & 1 & -1 & 1 & -1 & 1 & -1 \\ -1 & 1 & 1 & -1 & 1 & -1 & -1 & 1 \\ -1 & -1 & 1 & 1 & 1 & 1 & -1 & -1 \\ 1 & -1 & 1 & -1 & -1 & 1 & -1 & 1 \end{bmatrix} \begin{bmatrix} x_1 & y_1 & z_1 \\ \vdots & \vdots & \vdots \\ x_8 & y_8 & z_8 \end{bmatrix}$$

(3.36b)

corresponding to $\mathbf{N}_\lambda$ in Equation 3.33, we have

$$\begin{aligned} \boldsymbol{P}_\lambda &= \oint \begin{Bmatrix} l \\ m \\ n \end{Bmatrix} N_\lambda dS = \int_{V^e} \begin{Bmatrix} \partial/\partial x \\ \partial/\partial y \\ \partial/\partial z \end{Bmatrix} N_\lambda dV \\ &= 2\int_{V^e} J^{-1} \begin{bmatrix} \xi & 0 & 0 \\ 0 & \eta & 0 \\ 0 & 0 & \zeta \end{bmatrix} dV \end{aligned} \tag{3.37}$$

Based on Equation 2.34, the required incompatible internal displacement of the element is

$$u_\lambda^* = ([\xi^2\, \eta^2\, \zeta^2] - [\xi\, \eta\, \zeta]\boldsymbol{P}_*^{-1}\boldsymbol{P}_\lambda) \begin{Bmatrix} \lambda_1 \\ \lambda_2 \\ \lambda_3 \end{Bmatrix} = \mathbf{N}_\lambda^* \lambda \tag{3.38}$$

Equations similar to the above are applicable to the other two displacement components, and we will have

$$\boldsymbol{u}_\lambda = \begin{Bmatrix} u_\lambda \\ v_\lambda \\ w_\lambda \end{Bmatrix} = \begin{bmatrix} \mathbf{N}_\lambda^* & 0 & 0 \\ 0 & \mathbf{N}_\lambda^* & 0 \\ 0 & 0 & \mathbf{N}_\lambda^* \end{bmatrix} \begin{Bmatrix} \lambda_1 \\ \vdots \\ \lambda_9 \end{Bmatrix} \tag{3.39}$$

The explicit expressions for the matrices P_* and P_λ are given in the following:

$$\begin{aligned} \boldsymbol{P}_* &= \int_{V^e} \begin{bmatrix} \partial/\partial x \\ \partial/\partial y \\ \partial/\partial z \end{bmatrix} N^* dV = \int_{V^e} J^{-1} dV = \int_{-1}^{1}\int_{-1}^{1}\int_{-1}^{1} J^{-1}\,|J|\,d\xi d\eta d\zeta \\ &= \frac{8}{3}\begin{pmatrix} 3(b_2c_3 - b_3c_2) + b_4c_5 - b_5c_4 & 3(b_3c_1 - b_1c_3) + b_6c_4 - b_4c_6 \\ 3(c_2a_3 - c_3a_2) + c_4a_5 - c_5a_4 & 3(c_3a_1 - c_1a_3) + c_6a_4 - c_4a_6 \\ 3(a_2b_3 - a_3b_2) + a_4b_5 - a_5b_4 & 3(a_3b_1 - a_1b_3) + a_6b_4 - a_4b_6 \end{pmatrix. \\ &\qquad \left. \begin{matrix} 3(b_1c_2 - b_2c_1) + b_5c_6 - b_6c_5 \\ 3(c_1a_2 - c_2a_1) + c_5a_6 - c_6a_5 \\ 3(a_1b_2 - a_2b_1) + a_5b_6 - a_6b_5 \end{matrix} \right) \end{aligned}$$

$$P_\lambda = \int_{V^e} \begin{bmatrix} \partial/\partial x \\ \partial/\partial y \\ \partial/\partial z \end{bmatrix} N_\lambda dV = 2\int_{-1}^{1}\int_{-1}^{1}\int_{-1}^{1} J^{-1}\begin{pmatrix} \xi & 0 & 0 \\ 0 & \eta & 0 \\ 0 & 0 & \zeta \end{pmatrix} |J|\, d\xi d\eta d\zeta$$

$$= \frac{16}{3}\begin{pmatrix} b_2c_5 - b_5c_2 + b_4c_3 - b_3c_4 & b_3c_4 - b_4c_3 + b_6c_1 - b_1c_6 \\ c_2a_5 - c_5a_2 + c_4a_3 - c_3a_4 & c_3a_4 - c_4a_3 + c_6a_1 - c_1a_6 \\ a_2b_5 - a_5b_2 + a_4b_3 - a_3b_4 & a_3b_4 - a_4b_3 + a_6b_1 - a_1b_6 \end{pmatrix.$$

$$\begin{matrix} b_1c_6 - b_6c_1 + b_5c_2 - b_2c_5 \\ c_1a_6 - c_6a_1 + c_5a_2 - c_2a_5 \\ a_1b_6 - a_6b_1 + a_5b_2 - a_2b_5 \end{matrix}\Bigg)$$

With these explicit expressions it is no longer required to obtain P_* and P_λ by numerical integration. This will not only reduce the computing time required for the construction of the element stiffness matrices. It will also reduce the computing time required for the determination of stresses later on.

A cantilever beam shown in Figure 3.7 illustrates the ability of the NH11 element for bending analysis and the computing efficiency of the above method based on an explicit treatment. The beam is modeled by five irregularly shaped elements. The vertical displacement v_A at A and the stress σ_{xB} at B are determined by using the isoparametric element H8, and the incompatible elements H11 and NH11 are given in Table 3.2. The improvement of elements H11 and NH11 over element H8 in the calculation of both displacement and stress is clearly indicated. Similar to 2-D incompatible element problems, because the H11 and NH11 elements used are off irregular shape, the displacement u $\notin$ P_2 (x,y,z). In such cases, the use of the linear extrapolation equation (Equation 2.67) can improve the solution of element stress distribution. The improvement from using the explicit form of N_λ^* in the construction of NH11, the reduction of time for the determination of element stiffness matrix, is 25% and the reduction of time for the determination element stresses is 22%.

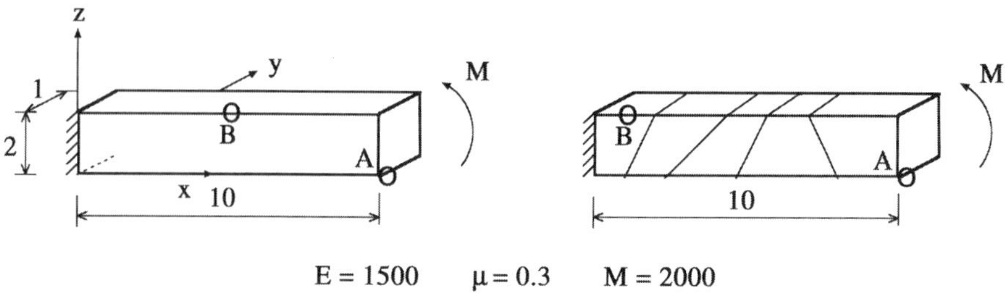

FIGURE 3.7
Bending of a cantilever beam.

TABLE 3.2

Comparison of Results from Eight-Node Six-Faced Solid Elements

Element	One Element		Five Elements	
	v_A	σ_{xB}	v_A	σ_{xB}
H8	9.0	−297.0	44.6	−1744
H11	100	−3000	97.6	−2511 (−2958)
NH11	100	−3000	95.8	−2483 (−3021)
Analytical solution	100	−3000	100	−3000

The results in the parentheses were figured by extrapolation.

3.5 Axisymmetric Incompatible Elements

3.5.1 Patch Test Condition

The PTC for a general 3-D incompatible element defined in orthogonal coordinates is of the form

$$\oiint_{\partial V^e} \sigma_c^{ij} n_j u_i^{\lambda} dS = 0 \tag{3.40}$$

in which u_i^{λ} is the incompatible part of the element displacement u_i, and σ_c^{ij} and n_j are, respectively, the constant stress tensor and the unit normal tensor to the element boundary ∂V^e. The above consistency condition can be interpreted such that the work done by the incompatible displacement on the element boundary should vanish when a constant stress field is considered. For plane and 3-D elements the PTC (Equation 3.40) can be written as

$$\int_{V^e} \varepsilon_{ij}^{\lambda} dV = 0 \quad or \quad \int_{V^e} B_{\lambda} dV = 0 \tag{3.41}$$

in which $\varepsilon_{ij}^{\lambda}$ and B_{λ} are, respectively, the element incompatible strain tensor and the element strain matrix within the element area V^e. This condition (Equation 3.41) is widely used to develop incompatible 2-D/3-D elements due to its simplicity and convenience.

However, Equation 3.41 is not suitable for the axisymmetric problems with cylindrical coordinates (r, θ, z) shown in Figure 3.8. Since dS = rdθ dl, the PTC (Equation 3.40) for the axisymmetric problems takes the form of

$$2\pi\oint_{\partial A^e} \sigma_c^{ij} n_j u_i^\lambda r dl = 0 \tag{3.42}$$

Here the curvilinear integral is carried out along the element edges in the r-z section of the element. By noting that for an axisymmetric problem, the shear stresses $\sigma^{r\theta} = \sigma^{z\theta} = 0$ and the hoop stress $\sigma^{\theta\theta}$ contribute nothing to the energy integral in Equation 3.42, the PTC condition for the axisymmetric problem can be formulated as

$$2\pi\oint_{\partial A^e} \sigma_c^T n^T u_\lambda r dl = 0 \tag{3.43}$$

where the constant stresses, the incompatible displacements, and the direction cosine matrix are, respectively,

$$\sigma_c = \begin{Bmatrix} \sigma_c^{rr} \\ \sigma_c^{zz} \\ \sigma_c^{rz} \\ \sigma_c^{\theta\theta} \end{Bmatrix}, \qquad u_\lambda = \begin{Bmatrix} u_\lambda \\ v_\lambda \end{Bmatrix} \quad \text{and} \quad u = \begin{bmatrix} n_r & 0 & n_z & 0 \\ 0 & n_z & n_r & 0 \end{bmatrix}$$

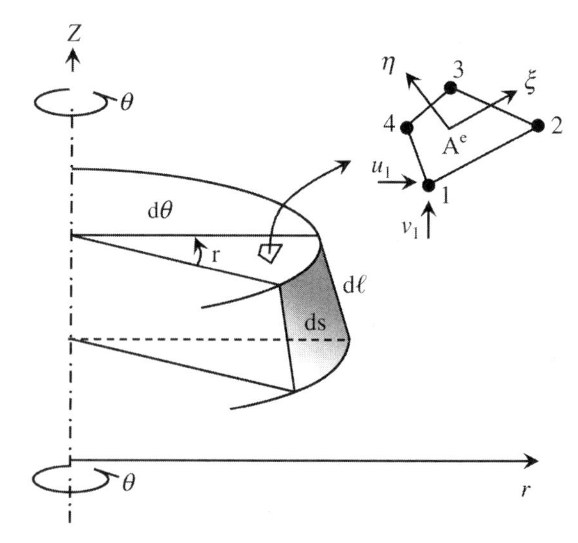

FIGURE 3.8
Four-node axisymmetric element in cylindrical coordinates (r,z,θ).

Noting that σ_c = constant, the PTC (Equation 3.43) can be replaced by the following more practical form:

$$\oint_{\partial A^e} \begin{bmatrix} n_r \\ n_z \end{bmatrix} u_\lambda r dl = 0 \quad \text{(for } u_\lambda \text{ too)} \tag{3.44}$$

with a similar relation for v_λ.

Corresponding to the PTC in Equation 3.43 and Equation 3.44 in the form of element linear integrals, there are also two area integrals within the axisymmetric element section A^e, that is,

$$2\pi \iint_{A^e} \sigma_c^T \begin{bmatrix} \frac{\partial}{\partial r}+\frac{1}{r} & 0 \\ 0 & \frac{\partial}{\partial z} \\ \frac{\partial}{\partial z} & \frac{\partial}{\partial r}+\frac{1}{r} \\ 0 & 0 \end{bmatrix} u_\lambda r \, dA = 0 \tag{3.45}$$

or equivalently

$$\iint_{A^e} \begin{bmatrix} \frac{\partial}{\partial r}+\frac{1}{r} \\ \frac{\partial}{\partial z} \end{bmatrix} u_\lambda r dA = 0 \quad \text{(for } u_\lambda \text{ too)} \tag{3.46}$$

and the corresponding equation for v_λ. The condition in Equation 3.46 can be satisfied more strictly as

$$\iint_{A^e} \begin{bmatrix} \frac{\partial}{\partial r} \\ \frac{\partial}{\partial z} \\ \frac{1}{r} \end{bmatrix} u_\lambda r dA = 0 \quad \text{(for } v_\lambda \text{ too)} \tag{3.47}$$

and the corresponding equation for v_λ. Similar to the PTC in the 2-D/3-D case, the axisymmetric PTC can also be formulated as

$$\int_{V^e} \varepsilon_{ij}^\lambda \, r dV = 0 \quad \text{or} \quad \int_{V^e} B_\lambda \, r dV = 0 \tag{3.48}$$

where $\varepsilon_{ij}^{\lambda}$ is the incompatible strain tensor, while B_{λ} is the strain-displacement matrix for the axisymmetric element.

As an application of the above axisymmetric PTC, we consider Wilson's incompatible function

$$u_{\lambda} = (1-\xi^2)\lambda_1 + (1-\eta^2)\lambda_2 = N_{\lambda}\lambda \tag{3.49}$$

The axisymmetric elements based on Equation 3.49 cannot pass the axisymmetric patch test even for the rectangular finite element meshes. In terms of the WHP formulation of incompatible function, the mode (Equation 3.49) can be revised as

$$\overset{*}{u_{\lambda}} = (N_{\lambda} - N_{*}P_{*}^{-1}P_{\lambda})\lambda \equiv \overset{*}{N_{\lambda}}\lambda \tag{3.50a}$$

in which $N_{*} = \begin{bmatrix} \xi & \eta \end{bmatrix}$, and the P-matrices consistent with the PTC (Equation 3.46) are

$$P_{*} = \iint_{A^e} \begin{bmatrix} \dfrac{\partial}{\partial r} + \dfrac{1}{r} \\ \dfrac{\partial}{\partial z} \end{bmatrix} N_{*}\, r\, dA, \quad P_{\lambda} = \iint_{A^e} \begin{bmatrix} \dfrac{\partial}{\partial r} + \dfrac{1}{r} \\ \dfrac{\partial}{\partial z} \end{bmatrix} N_{\lambda}\, r\, dA \tag{3.50b}$$

The incompatible element based on the mode in Equation 3.50 is termed R-u: Axi-Q6 $\in P_1$, where "$\in P_1$" means that the element displacement $(u = u_q + \overset{*}{u_{\lambda}})$ is linearly complete in respect to the rectangular coordinates (r, z) for irregular meshes. *R-u* means that revised-displacement is needed to pass the PTC.

3.5.2 Revised-Stiffness Approach

As an alternative way to create an incompatible model, the R-s approach is presented here. Only a simple revision to the current stiffness matrix is needed, while the initial element displacement does not need to be changed.

For a given incompatible element with the assumed displacement $u = u_q + u_{\lambda}$ where u_q is the compatible part and u_{λ} the incompatible one, the element potential energy can be formulated as

$$\Pi_p^e = \int_{V^e} \frac{1}{2} \varepsilon^T C \varepsilon \, dV \tag{3.51}$$

in which the element strain $\varepsilon = \mathbf{D}(u_q + u_{\lambda})$, and $\mathbf{C}$ is the elasticity matrix. The stationary condition for the functional (Equation 3.51), $\delta\Pi_p^e(u_q, u_{\lambda}) = 0$, leads to an additional energy consistency constraint, that is, the PTC (Equation 3.40) for the 2-D/3-D problem or the PTC (Equation 3.42) for the axisymmetric one. For the purpose of releasing the energy constraint for u_λ, a modified functional is obtained by deleting the nonzero energy integral (Equation 3.40 or Equation 3.42) from the functional (Equation 3.51), and we have, in the 2-D/3-D case,

$$\Pi_{mp}^e = \Pi_p^e - \oint_{\partial A^e} \sigma_c^T n^T u_\lambda ds = \Pi_p^e - \int_{V^e} \sigma_c^T \varepsilon_\lambda dV \tag{3.52a}$$

In the axisymmetric case

$$\Pi_{mp}^e = \Pi_p^e - 2\pi \oint_{\partial A^e} \sigma_c^T n^T u_\lambda r ds$$

$$= \Pi_p^e - 2\pi \iint_{A^e} \sigma_c^T \begin{bmatrix} \frac{\partial}{\partial r} + \frac{1}{r} & 0 \\ 0 & \frac{\partial}{\partial z} \\ \frac{\partial}{\partial z} & \frac{\partial}{\partial r} + \frac{1}{r} \\ 0 & 0 \end{bmatrix} u_\lambda r \, dA \tag{3.52b}$$

Note that, σ_c, n, and u_λ in the functional (Equation 3.52) have different meanings for 2-D/3-D and axisymmetric problems. It can be verified that as a stationary condition for the new functional (Equation 3.52), $\delta \Pi_{mp}^e = 0$, the PTC for u_λ no longer appears. Therefore, the PTC is free for the element based on the revised energy functional Π_{mp}^e. Implementation of this approach will be given below for 2-D/3-D and axisymmetric problems.

Let us consider an element displacement model with the compatible and incompatible parts assumed independently in the form

$$u = u_q + u_\lambda = N_q q + N_\lambda \lambda \tag{3.53}$$

in which N_q is the compatible displacement interpolation function wherein the constant-strain mode and the rigid body mode should be included, N_λ is the additional incompatible mode, and q and λ are element nodal displacement vector and internal displacement parameters. The latter can be statically eliminated within an element.

The element strain can be obtained using strain-displacement relations:

$$\varepsilon = \mathbf{D}u = \varepsilon_q + \varepsilon_\lambda = B_q q + B_\lambda \lambda \tag{3.54}$$

in which B_q and B_λ are the compatible strain matrix and the incompatible strain matrix, respectively.

Introducing Equation 3.53 and Equation 3.54 into the modified functional (Equation 3.52a) and noting that $\sigma_c = C\varepsilon_c$, $\varepsilon_c = B_c q$, in which ε_c and B_c are the constant strain vector and related strain matrix, the functional for the 2-D/3-D case can be written as

$$\Pi_{mp}^e = \int_{V^e} \frac{1}{2}\varepsilon^T C\varepsilon \, dV - \int_{V^e} \varepsilon_c^T C\varepsilon_\lambda dV = \frac{1}{2}\begin{Bmatrix} q \\ \lambda \end{Bmatrix}^T K_p^e \begin{Bmatrix} q \\ \lambda \end{Bmatrix} - \frac{1}{2}\begin{Bmatrix} q \\ \lambda \end{Bmatrix}^T K_m^e \begin{Bmatrix} q \\ \lambda \end{Bmatrix} \tag{3.55}$$

Therefore, the element stiffness matrix is

$$K_{mp}^e = K_p^e - K_m^e \tag{3.56}$$

where

$$K_p^e = \int_{V^e} \begin{bmatrix} B_q & B_\lambda \end{bmatrix}^T C \begin{bmatrix} B_q & B_\lambda \end{bmatrix} dV = \begin{bmatrix} K_{qq} & K_{q\lambda} \\ K_{q\lambda}^T & K_{\lambda\lambda} \end{bmatrix} \tag{3.57}$$

$$K_m^e = \begin{bmatrix} 0 & G \\ G^T & 0 \end{bmatrix}, \qquad G = \int_{V^e} B_c^T C B_\lambda dV = B_c^T C \int_{V^e} B_\lambda dV \tag{3.58}$$

Note that the constant strain matrix B_c can easily be obtained by the fact that $B_c = B_q\big|_{(\xi=0,\ \eta=0)}$ for 2-D problems or $B_c = B_q\big|_{(\xi=0,\ \eta=0,\ \zeta=0)}$ for the 3-D case, where ξ, η, and ζ are natural coordinates. This posttreatment approach only results in a simple revision to the initial element stiffness matrix K_p^e, so it is easy to carry out.

The incompatible strain $\varepsilon_\lambda = B_\lambda \lambda$ might undergo a change due to the revision of element stiffness. Therefore, a rational modified strain is necessary to calculate the stresses when the elemental displacement is available. To this end, we can directly adopt Wilson and Ibrahimbegovic's formulation [6] and obtain the modified incompatible strain matrix

$$B_{m\lambda} = B_\lambda - \frac{1}{V^e}\int_{V^e} B_\lambda dV \tag{3.59}$$

where V^e is the element area or volume. We find that the stress solution obtained by using $B_{m\lambda}$ is usually more accurate than that obtained by using B_λ.

For the axisymmetric case, we can obtain the formulations similar to the 2-D/3-D case, and the element stiffness matrix has the similar form in Equation 3.56, except that

$$K_p^e = 2\pi \int_{A^e} [B_q \quad B_\lambda]^T C \; [B_q \quad B_\lambda] \, r dA \tag{3.60}$$

$$G = 2\pi \iint_{A^e} B_c^T C \begin{bmatrix} \frac{\partial}{\partial r} + \frac{1}{r} & 0 \\ 0 & \frac{\partial}{\partial z} \\ \frac{\partial}{\partial z} & \frac{\partial}{\partial r} + \frac{1}{r} \\ 0 & 0 \end{bmatrix} N_{\lambda} r \, dA \tag{3.61}$$

Again, we can use a modified incompatible strain matrix to compute the element stresses as mentioned in the 2-D/3-D case, which takes the form of

$$B_{m\lambda} = B_{\lambda} - \frac{1}{\int_{A^e} r dA} \int_{A^e} B_{\lambda} r dA \tag{3.62}$$

Using the above formulations, an axisymmetric incompatible element based on Equation 3.49 is developed and termed as R-s: Axi-Q6 $\in$ P_1.

3.5.3 Numerical Test: Axisymmetric Stress Analysis and Incompressible Calculation

Figure 3.9 shows an infinitely long cylinder with uniformly distributed inner pressure p. A unit section of this cylinder is modeled by five elements. The axisymmetric isoparametric element Axi-Q4 and the axisymmetric incompatible elements R-u: AxiQ6 $\in$ P_1 and R-s: Axi-Q6 $\in$ P_1 are considered in the analysis. The numerical solutions listed in Table 3.3 show that, for the case of an incompressible material, the problems of displacement locking and stress instability cannot be avoided by Axi-Q4. However, the incompatible models R-u: AxiQ6 $\in$ P_1 and R-s: Axi-Q6 $\in$ P_1 can provide excellent solutions for the case of an incompressible material, and no numerical problems are encountered when the Poisson's ratio μ approaches 0.5.

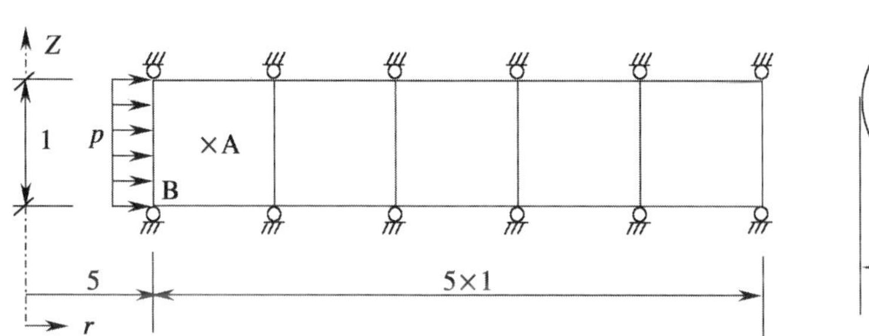

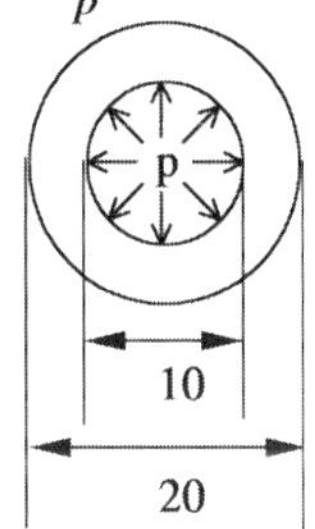

FIGURE 3.9
Thick cylinder under internal pressure ($E = 1$, $p = 10/\pi$).

TABLE 3.3
Solutions of a Cylinder under Inner Pressure (Figure 3.9)

μ	Element	u_B	σ_{rrA}	σ_{zzA}	$\sigma_{\theta\theta A}$
0.3	Axi-Q4	30.15	–1.91	1.17	5.81
	R-u: Axi-Q6$\in P_1$	30.57	–2.48	0.63	4.58
	R-s: Axi-Q6$\in P_1$	30.19	–2.44	0.65	4.60
	Exact	30.35	–2.45	0.64	4.57
0.49	Axi-Q4	28.79	13.7	16.9	20.7
	R-u: Axi-Q6$\in P_1$	32.03	–2.48	1.03	4.58
	R-s: Axi-Q6$\in P_1$	31.53	–2.42	1.07	4.60
	Exact	31.78	–2.45	1.04	4.57
0.4999	Axi-Q4	28.36	159.0	160.0	160.0
	R-u: Axi-Q6$\in P_1$	32.09	–2.48	1.05	4.58
	R-s: Axi-Q6$\in P_1$	31.57	–2.42	1.09	4.60
	Exact	31.83	–2.45	1.06	4.57

3.6 Hermite Type Incompatible Plate Elements

This section extends the method of incompatible elements to plate bending problems. The shear deformation of a Reissner–Mindlin (R–M) plate will be considered. In early attempts to construct R–M plate elements, when the midplane displacement w and rotations θ_x and θ_y were approximated by interpolation functions of the same order, the resulting elements were suitable for plates with relatively large thickness-span ratios (t/l). However, for thin plates, when $t/l \to 0$, the bending deformation may be constrained. This is referred to as the locking phenomena. Use of appropriate reduced and selective integration in the element formulation process may eliminate such locking phenomena. For example, in the construction of the very successful four-node R–M plate element S-1 by Hughes et al. [9], applied 2×2 integration for the construction of the stiffness matrix in bending was used. But the stiffness matrix in shear was constructed using only one-point integration. However, the use of reduced integration in the construction of element stiffness matrices may lead to an unstable system because of the existence of the zero-energy deformation mode. For example, the S-1 element has two zero-energy deformation modes. When the elements are irregularly arranged, the S-1 element cannot pass the constant curvature patch test.

This section discusses a four-node R–M plate element that is formulated by applying Kirchhoff constraint at the element nodes and by using a bicubic displacement field and a corresponding incompatible rotation field. For this element, locking phenomena can be avoided without using any reduced integration, and stable numerical solutions can be maintained.

3.6.1 Hermite Type Lateral Displacement Field

The displacement field of the midplane of a quadrilateral plate element can be considered as the modification and extension of the bicubic displacement field w that was used by Melosh [7]. The following is the C^o type bicubic interpolation function for the standard element shown in Figure 3.10:

1. Standard element and its nodal coordinates
2. Quadrilateral plate element in rectangular coordinates

$$w = \sum_{i=1}^{4} N_i w_i + \sum_{i=1}^{4} N_{\xi i}\left(\frac{\partial w}{\partial \xi}\right)_i + \sum_{i=1}^{4} N_{\eta i}\left(\frac{\partial w}{\partial \eta}\right)_i \tag{3.63}$$

in which

$$N_i = \frac{1}{8}(1+\xi_i\xi)(1+\eta_i\eta)(2+\xi_i\xi+\eta_i\eta)-\xi^2-\eta^2)$$

$$N_{\xi i} = -\frac{1}{8}\xi_i(1+\xi_i\xi)(1+\eta_i\eta)(1-\xi^2)$$

$$N_{\eta i} = -\frac{1}{8}\eta_i(1+\xi_i\xi)(1+\eta_i\eta)(1-\eta^2)$$

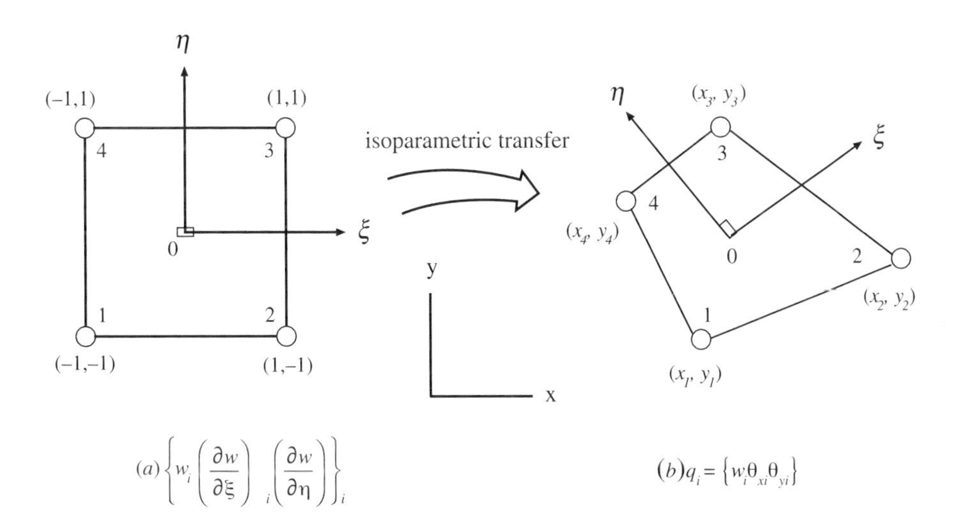

FIGURE 3.10
Four-node Reissner–Mindlin plate bending element.

At the nodal points i (= 1, 2, 3, 4) of the element let the displacements w satisfy the discretized Kirchhoff constraint condition (i.e., let the shearing strain at all points i be zero):

$$\begin{Bmatrix} \gamma_{xz} \\ \gamma_{yz} \end{Bmatrix}_i = \begin{Bmatrix} \dfrac{\partial w}{\partial x} \\ \dfrac{\partial w}{\partial y} \end{Bmatrix}_i - \begin{Bmatrix} \theta_x \\ \theta_y \end{Bmatrix}_i = 0 \tag{3.64}$$

By applying the isoparametric transformation (Equation 3.10), the relation between the types of rotational parameters in Figure 3.1 can be obtained:

$$\begin{Bmatrix} \dfrac{\partial w}{\partial \xi} \\ \dfrac{\partial w}{\partial \eta} \end{Bmatrix}_i = J(\xi_i, \eta_i) \begin{Bmatrix} \theta_x \\ \theta_y \end{Bmatrix}_i \tag{3.65}$$

in which the Jacobi matrix at the i-th node is

$$J(\xi_i, \eta_i) = \begin{bmatrix} a_1 + a_2\eta_i & b_1 + b_2\eta_i \\ a_3 + a_2\xi_i & b_3 + b_2\xi_i \end{bmatrix} \tag{3.66}$$

Substituting Equation 3.65 into Equation 3.63, we can obtain the following bicubic interpolation functions with $q_i = \{w_i \; \theta_{xi} \; \theta_{yi}\}$ as the nodal parameters:

$$w_q = [N_1 \; N_2 \; N_3 \; N_4] \begin{Bmatrix} q_1 \\ \vdots \\ q_4 \end{Bmatrix} = \mathbf{N}\, q \tag{3.67a}$$

$$N_i^T = \begin{Bmatrix} N_i \\ N_{xi} \\ N_{yi} \end{Bmatrix} = \begin{bmatrix} 1 & 0 \\ 0 & J^T(\xi_i, \eta_i) \end{bmatrix} \begin{Bmatrix} \mathbf{N}_i \\ \mathbf{N}_{\xi i} \\ \mathbf{N}_{\eta i} \end{Bmatrix} \tag{3.67b}$$

We can show that the displacement given by Equation 3.67 is continuous with that of the neighboring elements, that is, $w_q \in C^0$. We can also show that the displacement field consists of the following 12 independent basic functions:

$$\begin{array}{c} 1 \\ \xi \quad \eta \\ \xi^2 \quad \xi\eta \quad \eta^2 \\ \xi^3 \quad \xi^2\eta \quad \xi\eta^2 \quad \eta^3 \\ \xi^3\eta \qquad \xi\eta^3 \end{array}$$

This means that $w_q \in P_3(\xi,\eta)$. But for an arbitrary quadrilateral element, w is complete only in linear terms. One method for increasing the order of completeness of the displacement field is to add a bubble function term as an internal displacement. The resulting displacement is

$$w = N_q q + (1-\xi^2)(1-\eta^2)\lambda_0 \tag{3.68}$$

in which λ_0 is an internal parameter of the element. For Equation 3.68, $w \in C^0$ *and* $\in P_2(x,y)$.

This is very important in order for the plate element to pass the constant bending patch test and for the improvement in the resulting stress distribution.

3.6.2 Incompatible Rotation Field

The rotation field of the plate element is defined by a four-node bilinear interpolation function with an added second-order incompatible internal displacement, that is,

$$\begin{aligned} \theta &= \begin{Bmatrix} \theta_x \\ \theta_y \end{Bmatrix} = \theta_q + \theta_\lambda \\ &= \sum_i^4 \frac{1}{4}(1+\xi_i\xi)(1+\eta_i\eta)\begin{Bmatrix} \theta_{xi} \\ \theta_{yi} \end{Bmatrix} + \begin{bmatrix} 1-\xi^2 & 1-\eta^2 & 0 & 0 \\ 0 & 0 & 1-\xi^2 & 1-\eta^2 \end{bmatrix}\begin{Bmatrix} \lambda_1 \\ \vdots \\ \lambda_4 \end{Bmatrix} \\ &= \boldsymbol{N}_{\theta q}\boldsymbol{q} + \boldsymbol{N}_{\theta\lambda}\lambda \end{aligned} \tag{3.69}$$

where $\theta \in P_2(\xi,\eta)$. Under a natural coordinate system, this is a good match with the third-order displacement field $w \in P_3(\xi,\eta)$ or with Equation 3.68. This is the reason for preventing the constraint caused by the appearance of zero θ energy as mentioned in Reference 7. Hence, the possibility of any displacement locking can be avoided in numerical solutions of thin plates.

Similar to the PTC in the solution of the plane elasticity problem, the following constant curvature PTC also exists for the R–M plate element.

$$\oint_{\partial A^e} \begin{Bmatrix} l \\ m \end{Bmatrix} \theta_\lambda dS = \mathbf{0} \quad \text{or equivalently} \int_{A^e} \begin{Bmatrix} \dfrac{\partial}{\partial x} \\ \dfrac{\partial}{\partial y} \end{Bmatrix} \theta_\lambda dA = \mathbf{0} \tag{3.70}$$

The basic functions of the incompatible rotational fields $(1 - \xi^2)$ and $(1 - \eta^2)$ in Equation 3.69 do not satisfy the above PTC; hence, the two incompatible basic functions given in Equation 3.17 should be used instead. Thus, the incompatible internal displacement θ_λ should be replaced by

$$\theta_\lambda^* = N_{\theta\lambda}^* \lambda = \begin{bmatrix} N_{\lambda 1}^* & N_{\lambda 2}^* & 0 & 0 \\ 0 & 0 & N_{\lambda 1}^* & N_{\lambda 2}^* \end{bmatrix} \begin{Bmatrix} \lambda_1 \\ \vdots \\ \lambda_4 \end{Bmatrix} \tag{3.71}$$

in which $N_{\lambda 1}^*$ and $N_{\lambda 2}^*$ are the same as those in Equation 3.17.

3.6.3 Element Construction

Let the displacement vector at the nodes of a four-node R–M plate element be

$$d = \{w_1 \; \theta_{x1} \; \theta_{y1} \cdots w_4 \; \theta_{x4} \; \theta_{y4} \; \lambda_0 \; \lambda_1 \; \lambda_2 \; \lambda_3 \; \lambda_4\} \tag{3.72}$$

The changes in curvature are

$$\mathbf{k} = \begin{Bmatrix} \kappa_x \\ \kappa_y \\ \kappa_{xy} \end{Bmatrix} = \begin{bmatrix} -\dfrac{\partial}{\partial x} & 0 \\ 0 & -\dfrac{\partial}{\partial y} \\ -\dfrac{\partial}{\partial y} & -\dfrac{\partial}{\partial x} \end{bmatrix} \begin{Bmatrix} \theta_x \\ \theta_y \end{Bmatrix} = \boldsymbol{B}_b \boldsymbol{d} \tag{3.73}$$

The shear changes in transverse shear are

$$\gamma = \begin{Bmatrix} \gamma_{xz} \\ \gamma_{yz} \end{Bmatrix} = \begin{Bmatrix} \dfrac{\partial w}{\partial x} \\ \dfrac{\partial w}{\partial y} \end{Bmatrix} - \begin{Bmatrix} \theta_x \\ \theta_y \end{Bmatrix} = \boldsymbol{B}_s \boldsymbol{d} \tag{3.74}$$

The material matrices of the element are

$$C_b = \frac{Et^3}{12(1-\mu^2)}\begin{bmatrix} 1 & \mu & 0 \\ \mu & 1 & 0 \\ 0 & 0 & (1-\mu)/2 \end{bmatrix}$$

$$C_s = \frac{kEt}{2(1+\mu)}\begin{bmatrix} 1 & 0 \\ 0 & 1 \end{bmatrix}$$

with shearing coefficient $k = 5/6$. For a plate element with unit thickness, the bending deformation energy is

$$U_b = \iint_{A^e} \frac{1}{2}\kappa^T C_b \kappa dA = \frac{1}{2}\boldsymbol{d}^T \boldsymbol{K}_b \boldsymbol{d} \tag{3.75}$$

The shear deformation energy is

$$U_s = \iint_{A^e} \frac{1}{2}\gamma^T C_s \gamma dA = \frac{1}{2}\boldsymbol{d}^T \boldsymbol{K}_s \boldsymbol{d} \tag{3.76}$$

The stiffness matrix and shearing stiffness matrix are, respectively,

$$K_b = \iint_{A^e} \boldsymbol{B}_b^T \boldsymbol{C}_b \boldsymbol{B}_b dA,\ \boldsymbol{K}_s = \iint_{A^e} \boldsymbol{B}_s^T \boldsymbol{C}_s \boldsymbol{B}_s dA, \tag{3.77}$$

The total strain energy of the element is

$$U = U_b + U_s = \frac{1}{2}\boldsymbol{d}^T \boldsymbol{K}\boldsymbol{d} = \frac{1}{2}\boldsymbol{d}^T(\boldsymbol{K}_b + \boldsymbol{K}_s)\boldsymbol{d} \tag{3.78}$$

In computing the element stiffness matrix for K_b it is acceptable to use 2×2 or 3×3 points integration. However, for $\boldsymbol{K}_s$ it is necessary to use 3×3 points integration, so that the contribution of the complete second-order terms in w to the element strain energy can be taken fully into account.

3.6.4 Numerical Examples

3.6.4.1 Patch Test

The square plate shown in Figure 3.11 contains five irregularly shaped elements. Under constant bending and constant twisting conditions, the four-node R–M plate element can pass the patch tests, while the S1 element cannot pass such tests. Figure 3.12 is a square plate supported at three corners and loaded at the fourth corner. In this case the twisting moment M_{xy} is a constant equal to $P/2$. Hence, this is a constant curvature patch test. The present element can pass this test. However, because of the appearance of the zero-energy deformation mode, it is not possible to find the solution for the S1 element.

3.6.4.2 Bending Circular Plates

For the problem of bending circular plates supported along the boundary, we can take only one quarter of the plate and adopt element meshes as shown in Figure 3.13. Figure 3.14 to Figure 3.19 present comparisons of the

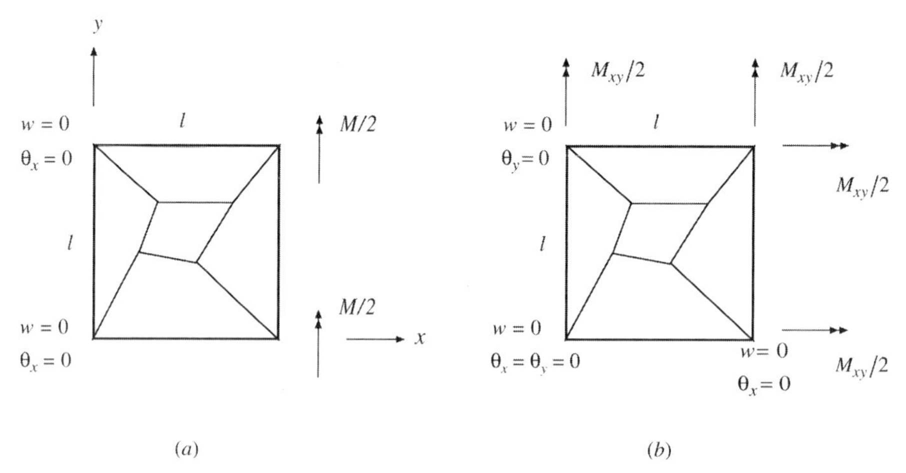

FIGURE 3.11
Constant bending moment and constant twisting moment patch test.

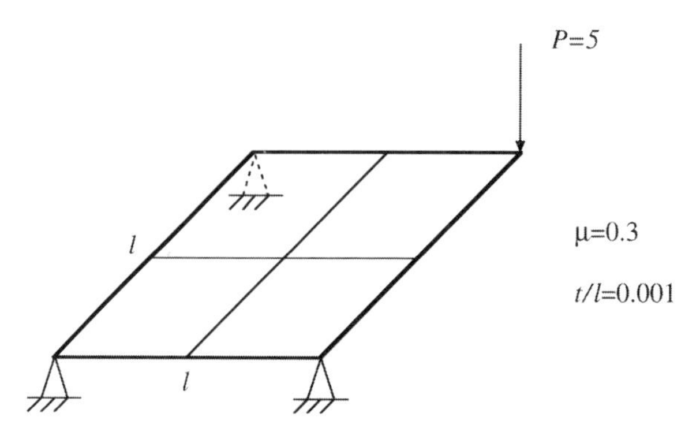

FIGURE 3.12
Square plate under pure twisting condition.

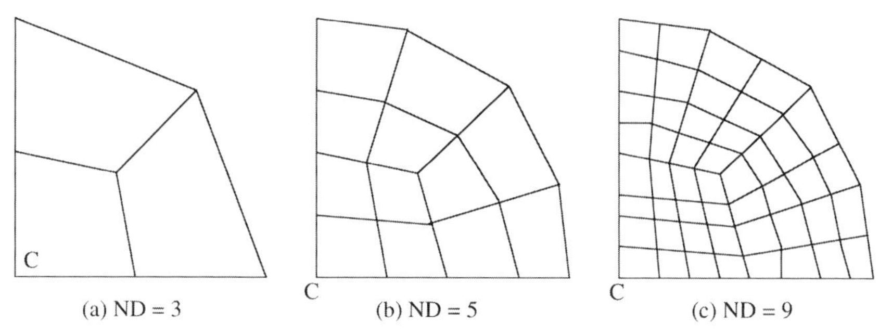

FIGURE 3.13
Discretized meshes for calculation of the circular plate (1/4).

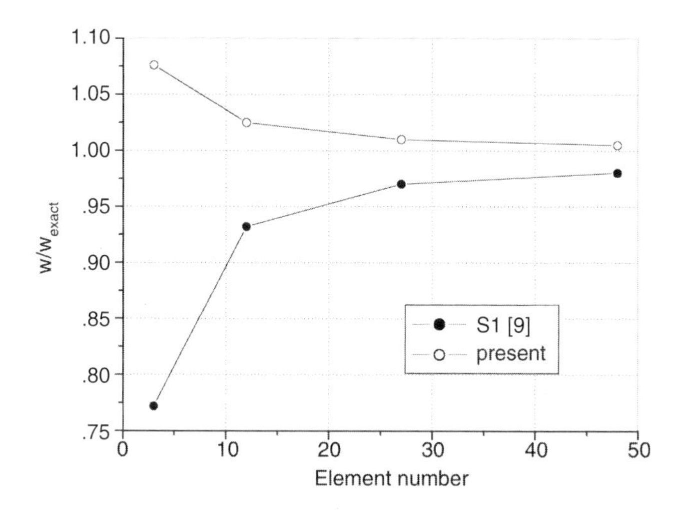

FIGURE 3.14
Convergence of lateral displacement at the center of a simply supported circular plate.

results from the present methods and those using the S1 element.[9] For these solutions, $E = 10.92$ and $\mu = 0.3$. All results are for thin plates with $t/d = 0.01$.

3.7 Bending Model under Reasonable w-θ Constraint

This section presents the general formulation of the R–M type plate bending element that contains an nth-order displacement field and rotational field. By relaxing of the zero θ constraint due to the Kirchhoff condition, the self-locking problem for thin plate problems can be avoided. A new concept and method for the w-θ constraint will be applied in order to guarantee that the resulting plate element passes the patch test. This method will be applied to formulate a set of R–M elements with different numbers of nodal points and to obtain some numerical results for comparison.

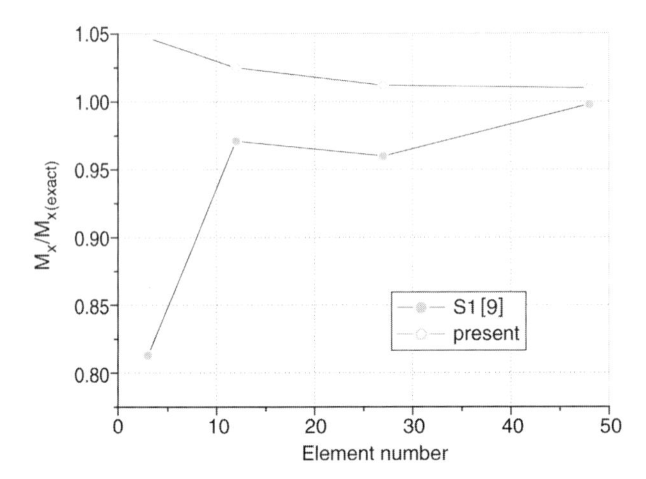

FIGURE 3.15
Convergence of a bending moment at the center of a simply supported circular plate.

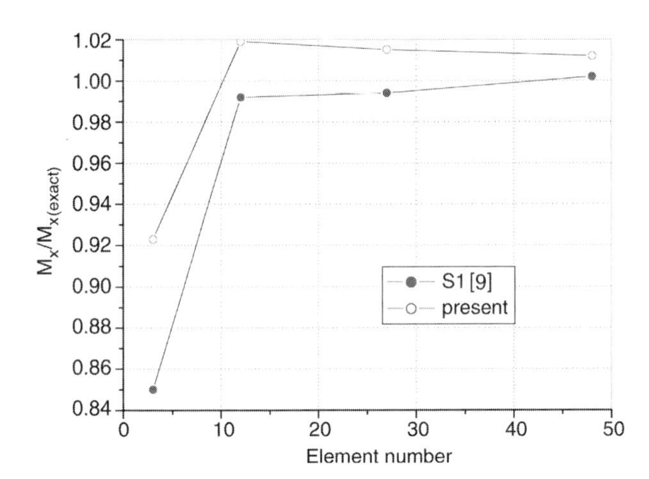

FIGURE 3.16
Convergence of a bending moment at the center of a circular plate with a fixed boundary.

3.7.1 General Formulation of R–M Plate Elements

A set of three R–M plate elements with four, eight, and nine nodal points is shown in Figure 3.20. For these elements the n-th order isoparametric interpolation of the lateral displacement w and the rotations θ_ξ and θ_η defined by the natural coordinates ξ-η are as follows [9]:

$$w = \sum_{i=1}^{n} N_i w_i = \begin{bmatrix} 1 & \xi & \eta & \cdots \end{bmatrix} \begin{Bmatrix} c_1 \\ \vdots \\ c_n \end{Bmatrix} = P_n C_n \tag{3.79a}$$

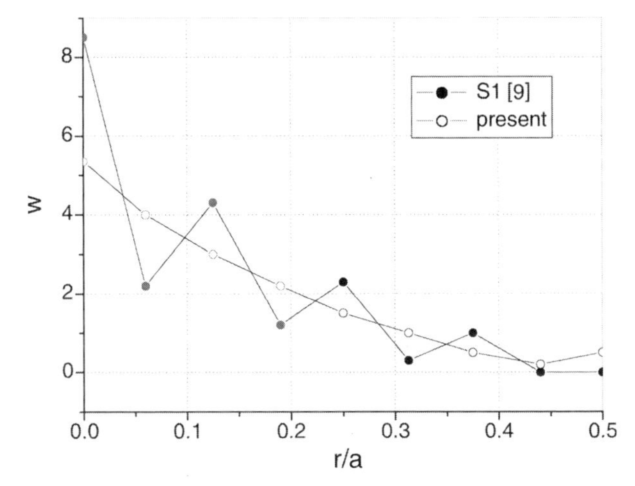

FIGURE 3.17
Radial distribution of lateral displacement for a circular plate with a fixed boundary.

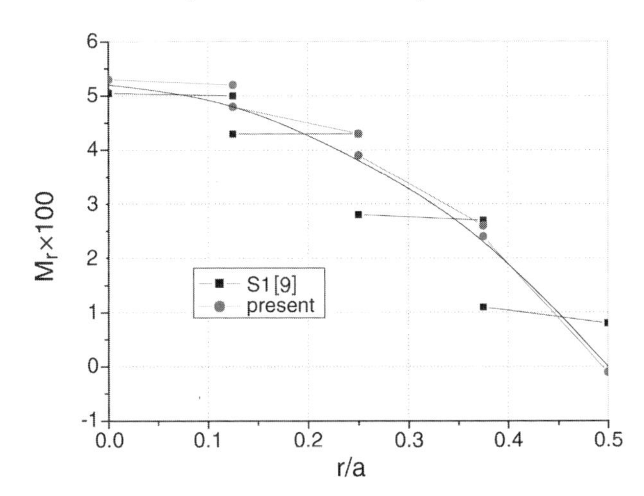

FIGURE 3.18
Radial distribution of moment M_r for a circular plate with a simply supported boundary.

$$\theta_\xi = \sum_{i=1}^{n} N_i \theta_{\xi i} = \begin{bmatrix} 1 & \xi & \eta & \cdots \end{bmatrix} \begin{Bmatrix} \alpha_1 \\ \vdots \\ \alpha_n \end{Bmatrix} = P_n \alpha_n \tag{3.79b}$$

$$\theta_\eta = \sum_{i=1}^{n} N_i \theta_{\eta i} = \begin{bmatrix} 1 & \xi & \eta & \cdots \end{bmatrix} \begin{Bmatrix} \beta_1 \\ \vdots \\ \beta_n \end{Bmatrix} = P_n \beta_n \tag{3.79c}$$

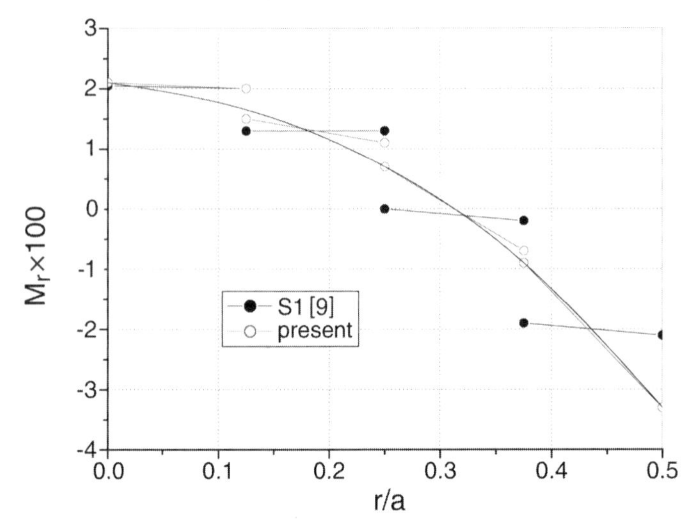

FIGURE 3.19
Radial distribution of moment M_r for a circular plate with a fixed boundary.

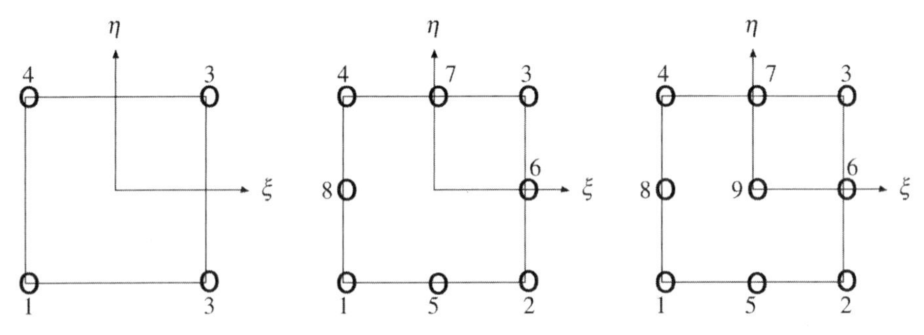

FIGURE 3.20
Plate bending elements with four, eight, and nine nodal points.

The interpolation functions N_i are the same as that for a standard plane isoparametric element. Corresponding to the three kinds of generalized functions, C_n, α_n, and β_n, the basic functions of the four-, eight-, and nine-node point elements are listed according to the orders as

$$\left.\begin{aligned} P_4 &= \begin{bmatrix} 1 & \xi & \eta & \xi\eta \end{bmatrix} \\ P_8 &= \begin{bmatrix} 1 & \xi & \eta & \xi\eta & \xi^2 & \eta^2 & \xi^2\eta & \xi\eta^2 \end{bmatrix} \\ P_9 &= \begin{bmatrix} 1 & \xi & \eta & \xi\eta & \xi^2 & \eta^2 & \xi^2\eta & \xi\eta^2 & \xi^2\eta^2 \end{bmatrix} \end{aligned}\right\} \quad (3.80)$$

When the displacements at the nodal points of the element are indicated in matrix form as

$$Q_n = \begin{bmatrix} w_1 & \theta_{\xi 1} & \theta_{\eta 1} \\ \vdots & \vdots & \vdots \\ w_n & \theta_{\xi n} & \theta_{\eta n} \end{bmatrix} \tag{3.81}$$

the following relation exists between $\boldsymbol{Q}_n$ and displacement parameters:

$$\begin{bmatrix} C_n & \alpha_n & \beta_n \end{bmatrix} = T_n Q_n \tag{3.82}$$

Corresponding to the elements with four, eight, and nine nodal points, the transformation matrices of Equation 3.82 are, respectively,

$$T_4 = \frac{1}{4}\begin{bmatrix} 1 & 1 & 1 & 1 \\ -1 & 1 & 1 & -1 \\ -1 & -1 & 1 & 1 \\ 1 & -1 & 1 & -1 \end{bmatrix} \tag{3.83a}$$

$$T_8 = \frac{1}{4}\begin{bmatrix} -1 & -1 & -1 & -1 & 2 & 2 & 2 & 2 \\ 0 & 0 & 0 & 0 & 0 & 2 & 0 & -2 \\ 0 & 0 & 0 & 0 & -2 & 0 & 2 & 0 \\ 1 & -1 & 1 & -1 & 0 & 0 & 0 & 0 \\ 1 & 1 & 1 & 1 & -2 & 0 & -2 & 0 \\ 1 & 1 & 1 & 1 & 0 & -2 & 0 & -2 \\ -1 & -1 & 1 & 1 & 2 & 0 & -2 & 0 \\ -1 & 1 & 1 & -1 & 0 & -2 & 0 & 2 \end{bmatrix} \tag{3.83b}$$

$$T_9 = \frac{1}{4}\begin{bmatrix} 0 & 0 & 0 & 0 & 0 & 0 & 0 & 0 & 4 \\ 0 & 0 & 0 & 0 & 0 & 2 & 0 & -2 & 0 \\ 0 & 0 & 0 & 0 & -2 & 0 & 2 & 0 & 0 \\ 1 & -1 & 1 & -1 & 0 & 0 & 0 & 0 & 0 \\ 0 & 0 & 0 & 0 & 0 & 2 & 0 & 2 & -4 \\ 0 & 0 & 0 & 0 & 2 & 0 & 2 & 0 & -4 \\ -1 & -1 & 1 & 1 & 2 & 0 & -2 & 0 & 0 \\ -1 & 1 & 1 & -1 & 0 & -2 & 0 & 2 & 0 \\ 1 & 1 & 1 & 1 & -2 & -2 & -2 & -2 & 4 \end{bmatrix} \tag{3.83c}$$

Let the displacement vector of the nodal points of the element be

$$q = \{w_1 \ \theta_{x1} \ \theta_{y1} \cdots w_n \ \theta_{xn} \ \theta_{yn}\} \tag{3.84}$$

in which θ_{xi} and θ_{yi} are the rotation of the element nodal point in the x-y coordinate system. Then, if the thickness of the plate is unity, the strain energy of the element is given by

$$U = U_b + U_s = \frac{1}{2} q^T K q \tag{3.85}$$

$$U_b = \frac{1}{2}\iint_A k^T S_b k dA = \frac{1}{2} q^T K_b q \qquad \text{(with } \boldsymbol{S}_b \text{ changed to } \boldsymbol{C}_b\text{)} \tag{3.86}$$

$$U_s = \frac{1}{2}\iint_A \gamma^T S_s \gamma \, dA = \frac{1}{2} q^T K_s q \qquad \text{(with } \boldsymbol{S}_s \text{ changed to } \boldsymbol{C}_s\text{)} \tag{3.87}$$

in which $\boldsymbol{K} = \boldsymbol{K}_b + \boldsymbol{K}_s$ is the element stiffness matrix of the element, and $\boldsymbol{K}_b$ and $\boldsymbol{K}_s$ are, respectively, the bending and shearing stiffness matrices of the element. The curvature κ and the transverse shear γ, as well as the material matrices $\boldsymbol{C}_b$ and $\boldsymbol{C}_s$, are defined in the same way as those in Section 3.5.

For the rotation and the shear deformation of the element the following Jacobi transformation relation exists:

$$\theta = \begin{Bmatrix} \theta_x \\ \theta_y \end{Bmatrix} = J^{-1} \begin{Bmatrix} \theta_\xi \\ \theta_\eta \end{Bmatrix} \tag{3.88}$$

$$\gamma = \begin{Bmatrix} \gamma_x \\ \gamma_y \end{Bmatrix} = J^{-1} \begin{Bmatrix} \gamma_\xi \\ \gamma_\eta \end{Bmatrix} \tag{3.89}$$

in which γ_ξ and γ_η are the shear deformation in the $\xi - \eta$ coordinate system. Under thin plate conditions, the following Kirchhoff condition holds:

$$\gamma = J^{-1} \begin{Bmatrix} \gamma_\xi \\ \gamma_\eta \end{Bmatrix} = 0 \tag{3.90}$$

or the equivalent equation

$$\begin{Bmatrix} \gamma_\xi \\ \gamma_\eta \end{Bmatrix} = \begin{Bmatrix} \dfrac{\partial w}{\partial \xi} \\ \dfrac{\partial w}{\partial \eta} \end{Bmatrix} - \begin{Bmatrix} \theta_\xi \\ \theta_\eta \end{Bmatrix} = 0 \tag{3.91}$$

By substituting the element displacements defined by Equation 3.79abc into Equation 3.91, two kinds of constraint conditions for the element displacements can be obtained. They are (a) zero θ constraint:

$$\alpha_I = 0, \ \beta_I = 0 \tag{3.92}$$

and (b) w-θ constraint:

$$\alpha_{II} = C_\alpha, \ \beta_{II} = C_\beta \tag{3.93}$$

in which, for a four-node element,

$$\alpha_I = \{\alpha_2 \ \alpha_4\}, \quad \beta_I = \{\beta_6 \ \beta_7 \ \beta_8\}$$

$$\alpha_{II} = \{\alpha_1 \ \alpha_3\}, \quad \beta_{II} = \{\beta_1 \ \beta_2\}$$

$$C_\alpha = \{c_2 \ c_4\}, \quad C_\beta = \{c_3 \ c_4\}$$

for an eight-node element,

$$\alpha_I = \{\alpha_5 \ \alpha_7 \ \alpha_8\}, \quad \beta_I = \{\beta_6 \ \beta_7 \ \beta_8\}$$

$$\alpha_{II} = \{\alpha_1 \ \alpha_2 \ \alpha_3 \ \alpha_4 \ \alpha_6\}, \quad \beta_{II} = \{\beta_1 \ \beta_2 \ \beta_3 \ \beta_4 \ \beta_5\}$$

$$C_\alpha = \{c_2 \ 2c_5 \ c_4 \ 2c_7 \ c_8\}, \quad C_\beta = \{c_1 \ c_3 \ 2c_6 \ 2c_8 \ c_7\}$$

and for a nine-node element,

$$\alpha_I = \{\alpha_5 \ \alpha_7 \ \alpha_9\}, \quad \beta_I = \{\beta_6 \ \beta_8 \ \beta_9\}$$

$$\alpha_{II} = \{\alpha_1 \ \alpha_2 \ \alpha_3 \ \alpha_4 \ \alpha_6 \ \alpha_8\}, \quad \beta_{II} = \{\beta_1 \ \beta_2 \ \beta_3 \ \beta_4 \ \beta_5 \ \beta_7\}$$

$$C_\alpha = \{c_2 \ 2c_5 \ c_4 \ 2c_7 \ c_8 \ 2c_9\}, \quad C_\beta = \{c_3 \ c_4 \ 2c_6 2c_8 \ c_7 \ 2c_9\}$$

Because of the existence of the zero θ constraint condition for the element, the bending deformation mode that is related to the parameters α_I and β_I also disappears. This is the cause of locking phenomena in the bending of thin plates. Thus, the zero θ constraint should be relaxed when the shearing strain energy of the element is determined. However, the existence of the w-θ constraint condition will strictly define the relation between the rotational degrees of freedom and the lateral deformation degrees of freedom. Under a certain condition this may prevent the appearance of the constant stress condition, and hence the resulting element cannot pass the patch test. However, in such case, there exist two possible solutions: (a) use of the weak form of the Kirchhoff condition through the numerical integration process and (b) the appropriate choice of the w field and θ field, such that $w \in P_3(\xi,\eta)$ and $\theta \in P_2(\xi,\eta)$ [8].

References

1. Wilson, E.L., Taylor, R.L., Doherty, W.P., and Ghaboussi, J., Incompatible displacement models, in *Numerical and Computer Methods in Structural Mechanics*, Fenves, S.J. et al., Eds., Academic Press, New York, 1973, p. 43.
2. Wu, C.C., Huang, M.G., and Pian, T.H.H., Consistency condition and convergence criteria of incompatible elements: general formulation of incompatible functions and its application, *Comput. Struct.*, 27, 639, 1987.
3. Weissman, S.L. and Taylor, R.L., Resultant fields for mixed plate bending elements, *Comput. Methods Appl. Mech. Eng.*, 79, 321, 1990.
4. Taylor, R.L., Beresford, P.J., and Wilson, E.L., A nonconforming element for stress analysis, *Int. J. Numer. Methods Eng.*, 10, 1211, 1976.
5. Wu, C.C., Huang, Y.Q., and Ramm, E., A further study on incompatible models: revise-stiffness approach and completeness of trial functions, *Comput. Methods Appl. Mech. Eng.*, 190, 5923, 2001.
6. Wilson, E.L. and Ibrahimbegovic, A., Use of incompatible displacement modes for the calculation of element stiffnesses or stresses, *Finite Element Anal. Des.*, 7, 229, 1990.
7. Melosh, R.J., Basis for derivation of matrices for the direct stiffness method, *AIAA J.*, 1, 1631, 1963.
8. Cheung, Y.K., Jiao, Z.P., and Wu, C.C., General formulation of C^0 bending models based on the rational $w - \theta$ constraint, *Comput. Mech.*, 16, 53, 1995.
9. Hughes, T.J.R., Cohen, M., and Haroun, M., Reduced and selective integration techniques in the finite element analysis of plates, *Nucl. Eng. Design*, 46, 203–222, 1978.

4

Foundation in Mechanics of Hybrid Stress Elements

4.1 Introduction

The introduction of incompatible displacement functions can improve numerical solutions by compatible displacement finite elements. Similarly, solutions by hybrid finite elements can also be improved by introducing incompatible displacements [1–4]. Pian and Sumihara [2] have presented a rational method for the formulation of the hybrid stress finite-element method. The introduction of incompatible displacements applies a constraining condition to the assumed stresses and results in a 5β plane hybrid element of superior properties. However, the formulation process requires a geometric perturbation of element geometry. A new method for the optimal design of hybrid stress finite elements is presented in References 5 and 6. This chapter presents this method of optimal design of hybrid stress finite elements based on an energy consistency requirement. In addition, this chapter also presents a penalty-equilibrium method for improving the numerical accuracy of the hybrid stress finite-element method.

4.2 Energy Consistency Analysis for Incompatible Hybrid Elements

We begin by considering the Hellinger–Reissner variational principle for nonlinear elasticity. When displacement boundary conditions are satisfied, the energy functions of a discretized system are

$$
\begin{aligned}
\Pi_{HR} &= \sum_e \Pi_{HR}^e \\
&= \sum_e \left\{ \int_{V^e} \frac{1}{2}(u_{i,j} + u_{j,i} + u_{m,i}u_{m,j})\sigma_{ij} - B(\sigma_{ij}) \right\} dV - \int_{S_\sigma^e} \bar{T}_i u_i dS
\end{aligned}
\tag{4.1}
$$

in which the complementary energy functional $B(\sigma_{ij}) = \frac{1}{2} S_{ijkl}\sigma_{ij}\sigma_{kl}$ and $\sigma_{ij} = \sigma_{ji}$ is the second Piola–Kirchhoff stress tensor.

Ordinarily, the trial solution of displacements u_i is compatible within the entire region of the system $V = \bigcup_e V^e$; that is, $u_i \in C^0(V)$, and the stress trial solution σ_{ij} is allowed to be piecewise continuous. The question now is, What is the situation when $u_i \notin C^0(V)$? As for the method for analyzing incompatible elements [7, 8], we can ignore the effect of displacement discontinuity between elements and define the energy functional of the system as the sum of the individual elements:

$$\Pi = \sum_e \left\{ \int_{V^e} \left[\frac{1}{2}(u_{i,j} + u_{j,i} + u_{m,i}u_{m,j})\sigma_{ij} - B(\sigma_{ij}) \right] dV - \int_{S_\sigma^e} \bar{T}_i u_i^0 dS \right\} \tag{4.2}$$

in which u_i^0 is the interpolation function based on element nodal displacements. It is the compatible part of the element displacement. Our purpose is to find a condition from which the finite element solution for (σ_{ij}, u_i) will converge to the correct solution in the variational sense. To be stationary, this required condition must be the same as that for the functional Π and hence the virtual work equation $\delta\Pi = 0$ can be established.

Let the element displacement u_i be divided into the compatible part u_i^0 and the incompatible part u_i^Δ, that is,

$$u_i = u_i^0 + u_i^\Delta \tag{4.3}$$

in which u_i^0 and u_i^Δ are linearly independent. Substituting Equation 4.3 into Equation 4.2 and carrying out variation of the functional Π, we have

$$\begin{aligned} \delta\Pi(\sigma_{ij}, u_i^0, u_i^\Delta) = \sum_e \int_{V^e} \Bigg\{ \Bigg[& \frac{1}{2}(u_{i,j} + u_{j,i} + u_{m,i}u_{m,j}) \\ & - \frac{\partial}{\partial \sigma_{ij}} B(\sigma_{ij}) \Bigg] \delta\sigma_{ij} - (\sigma_{mi} + \sigma_{ij}u_{mj})_{,i}(\delta u_m^0 + \delta u_m^\Delta) \Bigg\} dV \\ & + \sum_e \left(\int_{\partial V^e} T_i \delta u_i^0 dS + \int_{\partial V^e} T_i \delta u_i^\Delta dS - \int_{S_\sigma^e} \bar{T}_i \delta u_i^0 dS \right) \end{aligned} \tag{4.4}$$

Here, the element boundary force is

$$T_i = n_j \sigma_{mj}(\delta_{im} + u_{i,m}) \quad \text{on } \partial V^e \tag{4.5}$$

We note that for an individual element, $\partial V^e = S_\sigma^e \cup S_u^e \cup S_{ab}$, in which S_u^e is the element boundary with prescribed displacement, and S_b is the boundary connecting elements a and b. The prescribed boundary displacement can be maintained only in the compatible part of u_i. This means $u_i^0 = \bar{u}_i$ or $\delta u_i^0 = 0$ on S_u^e. With the above relations, Equation 4.4 can be rewritten as

$$\begin{aligned}\delta\Pi = \sum_e \int_{V^e} &\left\{ \begin{array}{c} \left[\frac{1}{2}(u_{i,j} + u_{j,i} + u_{m,i}u_{m,j}) - S_{ijkl}\sigma_{kl}\right]\delta\sigma_{ij} \\ -[\sigma_{ij}(\delta_{mj} + u_{m,j})]_{,i}(\delta u_m^0 + \delta u_m^{\Delta}) \end{array} \right\} dV \\ &+ \sum_{ab} \int_{S_{ab}} (T_i^{(a)} + T_i^{(b)})\delta u_i^0 dS \\ &+ \sum_e \int_{S_\sigma^e} (T_i - \bar{T}_i)\delta u_i^0 dS + \sum_e \oint_{\partial V^e} T_i \delta u_i^{\Delta} dS \end{aligned} \tag{4.6}$$

in which $T_i^{(a)}$ and $T_i^{(b)}$ are a pair of surface forces along the interelement boundary. From the stationary condition $\delta\Pi = 0$, we can get the equilibrium equations in V^e and along S_{ab} and the $u_i - \sigma_{ij}$ relation and the mechanical boundary condition $T_i = \bar{T}_i$ along S_σ^e. We can also get another constraining condition,

$$\sum_e \oint_{\partial V^e} T_i \delta u_i^{\Delta} dS = \sum_e \oint_{\partial V^e} n_j \sigma_{mj}(\delta_{im} + u_{i,m})\,\delta u_i^{\Delta} dS = 0 \tag{4.7}$$

This means the total virtual work done by the discontinuous virtual displacement over each element boundary ∂V^e must be equal to zero. For an incompatible discretized system, Equation 4.7 is the necessary condition for the functional Π to be stationary. Hence, it is also the necessary condition for the approximate solution to converge to the exact solution. Thus, we can call it an energy consistency condition for an incompatible hybrid finite element. We can also show that for three variable finite elements based on the nonlinear Hu–Washizu variational principle, if incompatible displacements are used, the condition (Equation 4.7) must also be satisfied.

4.3 Patch Test and Element Optimization Condition (OPC)

If we are considering only the convergence of finite element solutions, the energy consistency condition described above can be changed to a simplified form. When the element meshes are refined continuously until the size of mesh $h \to 0$, the stress σ_{ij} in each element reaches a constant state σ_{ij}^c. Accordingly, the limiting form for the condition (Equation 4.7) becomes

$$\sum_e \oint_{\partial V^e} n_j \sigma_{mj}^c (\delta_{im} + u_{i,m})\, \delta u_i^\Delta dS \to 0 \quad \text{when } h \to 0 \tag{4.8}$$

For practical application, the following criterion for solution convergence can be used:

$$\sum_e \oint_{\partial V^e} n_j \sigma_{mj}^c (\delta_{im} + u_{i,m})\, \delta u_i^\Delta dS = 0 \tag{4.9}$$

Under linear elastic conditions Equation 4.9 is simplified to

$$\sum_e \oint_{\partial V^e} n_j \sigma_{ij}^c \delta u_i^\Delta dS = 0 \tag{4.10}$$

Equation 4.10 is the mathematical expression for the patch test condition (PTC) as shown by Irons [7], while Equation 4.9 is its nonlinear form. The PTCs (Equation 4.9 and Equation 4.10) can be applied to individual elements as strong forms. For example, corresponding to Equation 4.10 we have

$$\oint_{\partial V^e} n_j \sigma_{ij}^c \delta u_i^\Delta dS = 0 \tag{4.11a}$$

or an equivalent equation:

$$\oint_{\partial V^e} n_j \delta u_i^\Delta dS = 0 \tag{4.11b}$$

Now we consider a more practical situation: that the discretized mesh h is finite. The energy consistency condition (Equation 4.7) cannot be satisfied by the convergence criterion (Equation 4.9). Let the higher-order stress of the element be

$$\sigma_{ij}^h = \sigma_{ij} - \sigma_{ij}^c \tag{4.12}$$

Then an additional condition for the satisfaction of Equation 4.7 is

$$\sum_e \oint_{\partial V^e} n_j \sigma_{mj}^h (\delta_{im} + u_{i,m})\, \delta u_i^\Delta dS = 0 \tag{4.13}$$

For an individual element it is

$$\int_{\partial V^e} n_j \sigma_{mj}^h (\delta_{im} + u_{i,m})\, \delta u_i^\Delta dS = 0 \tag{4.14}$$

For any element that already satisfies the PTC, the satisfaction of Equation 4.14 also satisfies, rigorously, the energy consistency condition (Equation 4.7). Hence, the optimal property of the element can be obtained. Thus, the constraint by Equation 4.14 can be considered the OPC for a hybrid stress element [9]. For a linearized element it can be simplified into

$$\oint_{\partial V^e} n_j \sigma_{ij}^h \delta u_i^\Delta dS = 0 \tag{4.15}$$

Based on the discussion in Section 2.7 we can conclude that the PTC and OPC conditions established earlier can be applied to Cartesian coordinates and also to a different kind of orthogonal curvilinear coordinates.

4.4 Optimization Method for Hybrid Stress Finite Elements

We apply the convergence condition PTC (Equation 4.11) and the OPC (4.15) to the optimal design of linear-elastic hybrid stress finite elements. We define the element displacements as

$$\boldsymbol{u} = \boldsymbol{u}_q + \boldsymbol{u}_\lambda = \boldsymbol{N}_q \boldsymbol{q} + \boldsymbol{N}_\lambda \boldsymbol{\lambda} \tag{4.16}$$

where $\boldsymbol{q}$ represents nodal parameters of compatible displacements, and λ represents internal parameters for defining the incompatible displacements $\boldsymbol{u}_\lambda$. The internal displacements $\boldsymbol{u}_\lambda = \boldsymbol{N}_\lambda \lambda$ do not pass the PTC. But based on the Wu–Huang–Pian (WHP) general formulation for incompatible functions obtained in Section 2.5, we can change it to a form that can satisfy the PTC. Let this form be

$$\boldsymbol{u}_\lambda^* = \boldsymbol{N}_\lambda^* \boldsymbol{\lambda} \tag{4.17}$$

For the OPC (Equation 4.15), we can choose assumed stresses σ^* that have already been optimized. Let the assumed stresses be

$$\sigma = \varphi\beta \tag{4.18}$$

where β represents the stress parameters. When σ is the sum of constant stresses σ_c and higher-order stresses σ_h, we have

$$\sigma = \sigma_c + \sigma_h = \phi_c\beta_c + \phi_h\beta_h = \beta_c + \begin{bmatrix} \phi_I & \phi_{II} \end{bmatrix} \begin{Bmatrix} \beta_I \\ \beta_{II} \end{Bmatrix} \tag{4.19}$$

in which we assign dim (β_{II}) = dim (λ). With these assumed displacements and stresses, the expression for the OPC (Equation 4.15) is

$$\oint_{\partial V^e} \delta u_i^{\Delta} n_j \sigma_{ij}^h dS = \int_{\partial V^e} \delta \boldsymbol{u}_{\lambda}^{*T} \boldsymbol{n} \sigma_h dS = \delta\lambda^T \boldsymbol{M}\beta_h = 0 \tag{4.20a}$$

where

$$\boldsymbol{M} = \oint_{\partial V^e} \boldsymbol{N}_{\lambda}^{*T} \boldsymbol{n} \begin{bmatrix} \phi_I & \phi_{II} \end{bmatrix} dS = \begin{bmatrix} \boldsymbol{M}_I & \boldsymbol{M}_{II} \end{bmatrix} \tag{4.20b}$$

The above equations (4.20a and 4.20b) are a set of constraining equations for β_h, that is,

$$\boldsymbol{M}\beta_h = \begin{bmatrix} \boldsymbol{M}_I & \boldsymbol{M}_{II} \end{bmatrix} \begin{Bmatrix} \beta_I \\ \beta_{II} \end{Bmatrix} = \boldsymbol{0} \tag{4.20c}$$

If the determinant $|\boldsymbol{M}_{II}|$ does not vanish, β_{II} can be expressed in terms of β_I and hence can be eliminated from Equation 4.19. The optimal form of assumed stresses thus is

$$\sigma^* = \sigma_c + \sigma_h^* = \begin{bmatrix} \boldsymbol{I} & \phi_h^* \end{bmatrix} \begin{Bmatrix} \beta_c \\ \beta_I \end{Bmatrix} = \phi^*\beta^* \tag{4.21}$$

where

$$\phi_h^* = \phi_I - \phi_{II}\, \boldsymbol{M}_{II}^{-1}\boldsymbol{M}_I$$

If external applied loads are not included, the element energy functional $\boldsymbol{u}_2$, can be expressed as

$$\begin{aligned} \Pi^e(\sigma^*, \boldsymbol{u}_q, \boldsymbol{u}_{\lambda}^*) &= \int_{V^e} \left[\sigma^{*T}(\boldsymbol{D}\boldsymbol{u}_q + \boldsymbol{D}\boldsymbol{u}_{\lambda}^*) - \frac{1}{2}\sigma^{*T}\boldsymbol{S}\sigma^* \right] dV \\ &= \int_{V^e} \left[\sigma^{*T}(\boldsymbol{D}\boldsymbol{u}_q) + \sigma_h^{*T}(\boldsymbol{D}\boldsymbol{u}_{\lambda}^*) - \frac{1}{2}\sigma^{*T}\boldsymbol{S}\sigma^* \right] dV \end{aligned} \tag{4.22}$$

in which $\boldsymbol{Du}$ = strains and $\boldsymbol{S}$ = elastic constant matrix.

In Equation 4.22 the order of the internal displacements u_λ^* is higher than that of the interpolation function $\boldsymbol{u}_q$, and σ_h^* are the higher parts of σ*. Thus, in Π^e of Equation 4.22, the integral term involving the incompatible displacements is of higher-order smaller term which can be omitted. Thus, the following simplified functional can be formulated as

$$\Pi^e(\sigma^*, \boldsymbol{u}_q) = \int_{V^e} \left[-\frac{1}{2}\sigma^{*T} \boldsymbol{S} \sigma^* + \sigma^{*T}(\boldsymbol{D}\boldsymbol{u}_q) \right] dV \tag{4.23}$$

Here, the element internal displacements u_λ^* no longer appear. The optimal design of hybrid stress finite elements thus is reduced to a simple issue: how to establish the optimal stress modes σ*.

The above method for optimal design of hybrid stress finite elements has already been used successfully for several two- and three-dimensional (2D/3D) elasticity problems [5, 9], axisymmetric analysis [10, 11], and plate and shell analyses [12–14]. Further discussions of these problems will be given in the following chapter.

The OPC (Equation 4.15) can be expressed in terms of displacement vector and stress tensor as shown in the following equivalent form:

$$\oint_{\partial V^e} (\boldsymbol{n}\sigma_h)^T \boldsymbol{u}_\lambda \, dS = 0$$

Under a Cartesian coordinate system, we have the following relation:

$$\oint_{\partial V^e} (\boldsymbol{n}\sigma_h)^T \boldsymbol{u}_\lambda \, dS = \int_{V^e} (\boldsymbol{D}\sigma_h)^T \boldsymbol{u}_\lambda \, dV + \int_{V^e} \sigma_h^T (\boldsymbol{D}\boldsymbol{u}_\lambda) dV$$

Thus, the OPC can be replaced by the following pair of equations of constraint:

$$\int_{V^e} (\boldsymbol{D}^T \sigma_h)^T \boldsymbol{u}_\lambda \, dV = \int_{V^e} (\boldsymbol{D}^T \sigma)^T \boldsymbol{u}_\lambda \, dV = 0 \tag{4.24}$$

and

$$\int_{V^e} \sigma_h^T (\boldsymbol{D}\boldsymbol{u}_\lambda) dV = \int_{V^e} \sigma_h^T \varepsilon_\lambda \, dV = 0 \tag{4.25}$$

Equation 4.24 was first presented by Pian and Chen [1]. Pian and Sumihara [2] applied this constraining condition to the initially assumed stress field with the aid of a perturbation of the element geometry and obtained a four-node plane element with superior properties. This element is called the P-S

5β element. Equation 4.25, which was suggested by Pian and Tong [15], is an orthogonal constraining condition of σ_h and ε_λ. Reference 16 presents some examples of the use of this constraining condition for assumed stresses. Pian and Wu et al. [3, 5, 6, 9–14] obtained the stress pattern of the P-S 5β element by using the OPC (Equation 4.15) directly. Zienkiewicz and Taylor [17] applied the constraint conditions (Equation 4.24 and Equation 4.25) and obtained the same result.

The constraint conditions (Equation 4.24 and Equation 4.25) cannot be applied to problems in curvilinear coordinates. Thus, they cannot be used for the optimal formulations of axisymmetric bodies and curved shells. When they are used in such problems the constant and higher-order stresses will be coupled, and, hence, the resulting element cannot pass the constant stress patch test. However, the constraint conditions (Equation 4.15 and Equation 4.15a) can be used directly for the selection of optimal stresses in the construction of hybrid stress elements of all kinds of orthogonal curvilinear coordinate systems [10, 12].

4.5 Matching Multivariable Parameters

Pian and Tong [15] have pointed out that the number of stress parameters for a hybrid finite element should satisfy the following condition:

$$n_\beta \geq \dim(\boldsymbol{q}) - \text{(element rigid body degrees of freedom)}$$

Here, we make a systematic presentation of the matching of element parameters.

Consider a typical element, for which the rigid body degrees of freedom have been eliminated from the displacements $\boldsymbol{q}$. The remaining nodal displacements are represented by $\boldsymbol{q}^*$. For this element without rigid body movement, the various expressions of element energy functionals can be presented in the following uniform expression [1–6, 9–11]:

$$\Pi^e(\beta, \boldsymbol{d}) = \beta^T \boldsymbol{G}\boldsymbol{d} - \frac{1}{2}\beta^T \boldsymbol{H}\beta \tag{4.26}$$

in which the deformed degrees of freedom of the element are

$$\boldsymbol{d} = \begin{Bmatrix} \boldsymbol{q}^* \\ \lambda \end{Bmatrix} \quad \text{(For Equation 4.23, take } \lambda = 0\text{)}$$

The $\boldsymbol{H}$ matrix in Equation 4.26 is from the deformation complementary energy of the element, and hence is a positive definite matrix. For the time being we can consider the $\boldsymbol{G}$ matrix as a full one, that is,

$$Rank\ (\boldsymbol{G}) = \min\ (n_\beta, n_d)$$

in which $n_\beta = \dim\ (\beta)$ and $n_d = \dim\ (\boldsymbol{d})$.

From the stationary condition of the functional (Equation 4.26), $\delta\Pi^e = 0$, the following virtual work equations can be obtained:

$$\delta\beta^T(\boldsymbol{G}\boldsymbol{d} - \boldsymbol{H}\beta) = 0 \tag{4.27a}$$

$$\delta\boldsymbol{d}^T\ (\boldsymbol{G}^T\beta) = 0 \tag{4.27b}$$

Against the virtual stress factor $\delta\beta$, Equation 4.27a yields the element geometric equation

$$\boldsymbol{C}\boldsymbol{d} = \beta \tag{4.28a}$$

in which $\boldsymbol{C} = \boldsymbol{H}^{-1}\boldsymbol{G}$ is a n_β by a n_d matrix of complete order. If the condition (Equation 4.28a) provides too much constraint to the deformation degree of freedom, some deformation of the element may not appear, and hence the stiffness of the element is increased. Conversely, if the number of the constraint equation on $\boldsymbol{d}$ in Equation 4.28a is less than n_d, the stiffness of the element is reduced. However, from Equation 4.28a, we can obtain the following element equilibrium equation that corresponds to virtual displacement $\delta\boldsymbol{d}$:

$$\boldsymbol{G}^T\beta = \boldsymbol{0} \tag{4.28b}$$

If the above equation provides too many βs in the equilibrium relation, certain constraints of the deformation may be removed, and the flexibility of the element will increase. Otherwise, the flexibility of the element will decrease $(n_\beta > n_d)$.

Now, we consider the following three situations.

4.5.1 Situation I

$$n_\beta > n_d \tag{4.29}$$

In this case, Equation 4.28a provides a strong geometric constraint to the displacement factor $\boldsymbol{d}$. But Equation 4.28b provides a weak equilibrium constraint to the stress parameter β. Thus, the hybrid element is stiffened simultaneously from two directions. When $n_\beta >> n_d$, the resulting element is so stiffened that self-locking phenomena may occur.

4.5.2 Situation II

$$n_\beta < n_d \tag{4.30}$$

In this case, Equation 4.28a is not sufficient to provide the geometric constraint of $\boldsymbol{d}$, and Equation 4.28b provides an extra equilibrium constraint of the stress parameter β. Thus, the hybrid element can be made more flexible simultaneously from two directions. A hybrid stress finite element based on Equation 4.30 may contain a $n_d - n_\beta$ zero energy displacement mode.

4.5.3 Situation III (Optimal Parameter Matching)

$$n_\beta = n_d \tag{4.31}$$

Under this condition, the two element factors of $\boldsymbol{d}$ and β can be determined, respectively, by Equation 4.28a and Equation 4.28b. Equation 4.31 is the best pair for the arrangement. Under this condition, the geometric equation and the equilibrium equation of the element can be satisfied simultaneously. Hence, the property of the resulting element is very desirable. Many very good hybrid and mixed finite elements are constructed under the arrangement of Equation 4.31. See [9] for details.

Under the best arrangement of stress factors described above, the trial set (σ $\boldsymbol{u}$) can satisfy Equation 4.27a and Equation 4.27b simultaneously. This means the virtual force principle and the virtual displacement principle can be fulfilled at the same time. This is one of the examples of rational application of the generalized variational principle to approximate solutions. In the optimization of the hybrid element, we should try to satisfy, or at least to nearly satisfy, the balanced arrangement given by Equation 4.31. The condition given by Equation 4.31 is necessary, but it may not be sufficient.

References

1. Pian, T.H.H. and Chen, D.P., Alternative ways for formulation of hybrid stress elements, *Int. J. Numer. Methods Eng.*, 18, 1679, 1982.
2. Pian, T.H.H. and Sumihara, K., Rational approach for assumed stress finite elements, *Int. J. Numer. Methods Eng.*, 20, 1685, 1984.
3. Pian, T.H.H. and Wu, C.C., Use of additional incompatible displacements for finite element formulation, in *Nonlinear Computational Mechanics*, Wriggers, P. and Wagner, W., Eds., Springer-Verlag, Heidelberg, 1991, 255–266.
4. Wolf, J.P., Alernative hybrid stress finite element models, *Int. J. Numer. Methods Eng.*, 9, 601, 1975.
5. Wu, C.C., Di, S.L., and Huang, M.G., The optimization design of hybrid element, *China Sci. Bull.*, 32, 1236, 1987.

6. Pian, T.H.H. and Wu, C.C., A rational approach for choosing stress term of hybrid finite element formulations, *Int. J. Numer. Methods Eng.*, 26, 2331, 1988.
7. Strang, G. and Fix, G.J., *An Analysis of the Finite Element Method*, Prentice Hall, Englewood Cliffs, NJ, 1973.
8. Ciarlet, P.G., *The Finite Element for Elliptic Problems*, North-Holland, Amsterdam, 1978.
9. Wu, C.C. and Bufler, H., Multivariable finite elements: consistency and optimization, *Sci. China (A)*, 34, 284, 1991.
10. Wu, C.C., Di, S.L., and Pian, T.H.H., On optimization method for hybrid axisymmetric element, *Acta Aeronaut. Astronaut. Sin.*, 8, 439, 1987 (in Chinese).
11. Wu, C.C., Dong, Y.F., and Huang, M.G., Optimization of stress models of hybrid elements and its application to torsion of shafts, *Proc. 2nd Int. Conf. EPMESC*, Guangzhou, China, 1987, pp. 224–226.
12. Di, S.L., Wu, C.C., and Song, Q.G., Optimizing formulation of hybrid model for thin and moderately thick plates, *Appl. Math. Mech.*, 11, 149, 1990.
13. Di, S.L., Wu, C.C., and Song, Q.G., Model optimization of hybrid stress general shell element, *Acta Mech. Sin.*, 2, 1, 1989.
14. Dong, Y.F., Wu, C.C., and Texeira de Freiras, A., The hybrid stress model for Mindlin-Reissner plate based on a stress optimization condition, *Comput. Struct.*, 46, 877, 1993.
15. Pian, T.H.H. and Tong, P., Basis of finite element methods for solid continue, *Int. J. Numer. Methods Eng.*, 1, 3, 1986.
16. Sze, K.Y., Chow, C.L., and Chen, W.J., A rational formulation of isoparametric hybrid stress elements for three-dimensional stress analysis, *Finite Element Anal. Des.*, 7, 61, 1990.
17. Zienkiewicz, O.C. and Taylor, R.L., *The Finite Element Method*, 4th ed., Vol. 1, McGraw-Hill, New York, 1989.

5

Optimization of Hybrid Stress Finite Elements

5.1 Four-Node Plane Hybrid Element

In 1984 Pian and Sumihara [1] presented a high-performance four-node plane stress–strain hybrid element. It is referred to here as the P-S element. The energy functional of this element is the $\Pi^e(\sigma,\boldsymbol{u})$ defined in Equation 4.25. The determination of the optimal element stress σ requires geometric perturbation of the element geometry. Hence, the method is very complicated. This chapter applies the optimization method of the hybrid stress element suggested in Section 4.3 to determine the optimal stress pattern σ^*.

Referring to the four-node plane isoparametric element shown in Figure 1.1, the interpolation for the coordinate of the element is

$$\begin{Bmatrix} x \\ y \end{Bmatrix} = \sum_{i=1}^{4} N_i(\xi,\eta) \begin{Bmatrix} x_i \\ y_i \end{Bmatrix} \tag{5.1}$$

in which $N_i(\xi,\eta) = \frac{1}{4}(1+\xi_i\xi)(1+\eta_i\eta)$ is the bilinear interpolation function.

The geometric coefficient of the element is defined as

$$\begin{bmatrix} a_1 & b_1 \\ a_2 & b_2 \\ a_3 & b_3 \\ a_4 & b_4 \end{bmatrix} = \frac{1}{4} \begin{bmatrix} -1 & 1 & 1 & -1 \\ 1 & -1 & 1 & -1 \\ -1 & -1 & 1 & 1 \\ 1 & 1 & 1 & 1 \end{bmatrix} \begin{bmatrix} x_1 & y_1 \\ x_2 & y_2 \\ x_3 & y_3 \\ x_4 & y_4 \end{bmatrix} \tag{5.2}$$

The following transformation relation exists between the coordinates (ξ,η) and (x,y):

$$\begin{Bmatrix} \dfrac{\partial(\ )}{\partial \xi} \\ \dfrac{\partial(\ )}{\partial \eta} \end{Bmatrix} = \boldsymbol{J} \begin{Bmatrix} \dfrac{\partial(\ )}{\partial x} \\ \dfrac{\partial(\ )}{\partial y} \end{Bmatrix} \tag{5.3}$$

in which the Jacobi is

$$\boldsymbol{J} = \begin{bmatrix} \dfrac{\partial x}{\partial \xi} & \dfrac{\partial y}{\partial \xi} \\ \dfrac{\partial x}{\partial \eta} & \dfrac{\partial y}{\partial \eta} \end{bmatrix} = \begin{bmatrix} a_1 + a_2\eta & b_1 + b_2\eta \\ a_3 + a_2\xi & b_3 + b_2\xi \end{bmatrix} \tag{5.4}$$

The corresponding determinant is

$$|\boldsymbol{J}| = J_0 + J_1\xi + J_2\eta \tag{5.5}$$

$$J_0 = a_1b_3 - a_3b_1 = (1/4)(\text{element area}) > 0$$

$$J_1 = a_1b_2 - a_2b_1$$

$$J_2 = a_2b_3 - a_3b_2$$

When the element is a rectangle or parallelogram, the factors $a_2 = b_2 = J_1 = J_2 = 0$, but $|\boldsymbol{J}| = J_0$ is a positive constant.

Corresponding to Equation 5.3, there is an inverse transformation relation:

$$\begin{Bmatrix} \dfrac{\partial(\ )}{\partial x} \\ \dfrac{\partial(\ )}{\partial y} \end{Bmatrix} = \boldsymbol{J}^{-1} \begin{Bmatrix} \dfrac{\partial(\ )}{\partial \xi} \\ \dfrac{\partial(\ )}{\partial \eta} \end{Bmatrix} \tag{5.6}$$

In any situation it is possible to take $|\boldsymbol{J}| = J_0$ in order to simplify the computation. In that case Equation 5.6 can be expressed as

$$\begin{Bmatrix} \dfrac{\partial(\ )}{\partial x} \\ \dfrac{\partial(\ )}{\partial y} \end{Bmatrix} = \frac{1}{J_0} \begin{bmatrix} b_3 + b_2\xi & -(b_1 + b_2\eta) \\ -(a_3 + a_2\xi) & a_1 + a_2\eta \end{bmatrix} \begin{Bmatrix} \dfrac{\partial(\ )}{\partial \xi} \\ \dfrac{\partial(\ )}{\partial \eta} \end{Bmatrix} \tag{5.7}$$

For this four-node plane element, the bilinear interpolation function for displacement is

$$\boldsymbol{u}_q = \begin{Bmatrix} u_q \\ v_q \end{Bmatrix} = \mathbf{N}_q \boldsymbol{q} \tag{5.8}$$

To this equation we add the incompatible function of Wilson et al. [2]:

$$\boldsymbol{u}_\lambda = \begin{Bmatrix} u_\lambda \\ v_\lambda \end{Bmatrix} = \begin{bmatrix} 1-\xi^2 & 1-\eta^2 & 0 & 0 \\ 0 & 0 & 1-\xi^2 & 1-\eta^2 \end{bmatrix} \begin{Bmatrix} \lambda_1 \\ \vdots \\ \lambda_4 \end{Bmatrix} = \mathbf{N}_\lambda \lambda \tag{5.9}$$

Hence, the element displacement is

$$\mathbf{u} = \mathbf{u}_q + \mathbf{u}_\lambda \in P_2(\xi, \eta) \tag{5.10}$$

and the element stress field is

$$\begin{aligned} \sigma = \begin{Bmatrix} \sigma_x \\ \sigma_y \\ \tau_{xy} \end{Bmatrix} &= \left[\begin{array}{ccc|ccc|ccc} 1 & & & \eta & 0 & \xi & 0 & 0 & 0 \\ & 1 & & 0 & \xi & 0 & \eta & 0 & 0 \\ & & 1 & 0 & 0 & 0 & 0 & \xi & \eta \end{array}\right] \begin{Bmatrix} \beta_1 \\ \vdots \\ \beta_9 \end{Bmatrix} \\ &= \beta_c + \left[\phi_{\mathrm{I}} \quad \middle| \phi_{\mathrm{II}}\right] \begin{Bmatrix} \beta_{\mathrm{I}} \\ \beta_{\mathrm{II}} \end{Bmatrix} \end{aligned} \tag{5.11}$$

From Equation 5.9 and Equation 5.11, the $\boldsymbol{M}$ matrix in Equation 4.20b can be determined:

$$\begin{aligned} \mathbf{M} = \oint_{\partial A^e} \mathbf{N}_\lambda^T \mathbf{n} \left[\phi_{\mathrm{I}} \quad \middle| \phi_{II}\right] dS &= \frac{8}{3} \left[\begin{array}{cc|cccc} -b_1 & 0 & 0 & 0 & 0 & a_1 \\ 0 & 0 & b_3 & 0 & -a_3 & 0 \\ 0 & 0 & 0 & a_1 & 0 & -b_1 \\ 0 & -a_1 & 0 & 0 & b_3 & 0 \end{array}\right] \\ &= \left[\mathbf{M}_{\mathrm{I}} \quad \middle| \mathbf{M}_{\mathrm{II}}\right] \end{aligned} \tag{5.12}$$

The factors a_1, a_2, b_1, and b_3 are given by Equation 5.2.

The corresponding equation of constraint for the optimization condition (OPC) can be written as

$$\left[\mathbf{M}_{\mathrm{I}} \quad \middle| \mathbf{M}_{\mathrm{II}}\right] \begin{Bmatrix} \beta_{\mathrm{I}} \\ \beta_{\mathrm{II}} \end{Bmatrix} = \mathbf{0} \tag{5.13}$$

M_{II} is usually a 4×4 full rank square matrix. Hence, we have

$$\beta_{II} = -M_{II}^{-1} M_{I} \beta_{I}$$

By substituting the above equation into Equation 5.11, we can obtain the optimizing form of the stresses of the element:

$$\sigma^* = \left[\begin{array}{ccc|cccc} 1 & & & & \eta & (a_3/b_3)^2 & \xi \\ & 1 & & (b_1/a_1)^2 & \eta & & \xi \\ & & 1 & (b_1/a_1) & \eta & (a_3/b_3) & \xi \end{array}\right] \begin{Bmatrix} \beta_1 \\ \vdots \\ \beta_5 \end{Bmatrix} \tag{5.14}$$

This can be written in an equivalent form:

$$\sigma^* = \left[\begin{array}{ccc|cc} 1 & & & a_1^2\eta & a_3^2\xi \\ & 1 & & b_1^2\eta & b_3^2\xi \\ & & 1 & a_1 b_1 \eta & a_3 b_3 \xi \end{array}\right] \begin{Bmatrix} \beta_1 \\ \vdots \\ \beta_5 \end{Bmatrix} \tag{5.15}$$

Equation 5.15 is the element stress field given in Reference 1. It not only satisfies the OPC, but also has the best number of β factors ($n_\beta = 5$). Thus, the P-S element has achieved great success.

When the element is a parallelogram that is parallel to the x-y axes, the factors $a_3 = b_1 = 0$. In this case, Equation 5.15 may be expressed equivalently as

$$\sigma^* = \left[\begin{array}{ccc|cc} 1 & & & y & 0 \\ & 1 & & 0 & x \\ & & 1 & 0 & 0 \end{array}\right] \begin{Bmatrix} \beta_1 \\ \vdots \\ \beta_5 \end{Bmatrix} \tag{5.16}$$

This is the first stress pattern given in Reference 3. This is an equilibrium stress field.

Using the method suggested in Reference 4, it is easy to determine the inverse of the $\boldsymbol{H}$ matrix. This will reduce considerably the computing time needed to determine the element stiffness matrix and element stresses. By denoting Equation 5.8 as the sum of constant and higher-order stresses,

$$\sigma^* = \sigma_c + \sigma_h = \beta_c + \phi_h \beta_h = \left[\mathbf{I} \quad | \phi_h\right] \begin{Bmatrix} \beta_c \\ \beta_h \end{Bmatrix} \tag{5.17}$$

The $\boldsymbol{H}$ matrix of Equation 1.44 can be expressed as

$$\boldsymbol{H} = \int\int_{A^e} [\boldsymbol{I} \quad \phi_h]^T \mathbf{S} [\boldsymbol{I} \quad \phi_h] t dA = \begin{bmatrix} \boldsymbol{H}_{cc} & \boldsymbol{H}_{ch} \\ \boldsymbol{H}_{ch}^T & \boldsymbol{H}_{hh} \end{bmatrix} \tag{5.18}$$

In Equation 5.18, if $\boldsymbol{H}_{ch} = 0$, the calculation of $\boldsymbol{H}^{-1}$ will be very simple. To achieve this objective, we only need to replace the stress pattern in Equation 5.8 with the following equation:

$$\sigma^* = \begin{bmatrix} 1 & & & a_1^2\bar{\eta} & a_3^2\bar{\xi} \\ & 1 & & b_1^2\bar{\eta} & b_3^2\bar{\xi} \\ & & 1 & a_1 b_1 \bar{\eta} & a_3 b_3 \bar{\xi} \end{bmatrix} \begin{Bmatrix} \beta_1 \\ \vdots \\ \beta_5 \end{Bmatrix} \tag{5.19}$$

in which

$$\left. \begin{aligned} \bar{\xi} &= \xi - \frac{J_1}{3J_0} \\ \bar{\eta} &= \eta - \frac{J_2}{3J_0} \end{aligned} \right\} \tag{5.20}$$

The definition of the constants J_0, J_1, and J_2 are given in Equation 5.5. The coordinate transformation (Equation 5.20) indicates the parallel movement of the natural coordinates. Hence, it does not affect the characteristics of the stress pattern σ^*. That is, the stress pattern in Equation 5.19 and the stress pattern in Equation 5.15 are entirely equivalent. After taking the stress pattern in Equation 5.19, $\boldsymbol{H}$ in Equation 5.18 is reduced to the form of a diagonal block. Hence, it is easy to write the explicit expression of the inverse matrix as

$$\boldsymbol{H}^{-1} = \begin{bmatrix} \dfrac{1}{4J_0 t} S^{-1} & \boldsymbol{0} \\ \boldsymbol{0} & \boldsymbol{H}_{hh}^{-1} \end{bmatrix} \tag{5.21}$$

in which the determinant for material is

$$\mathbf{S}^{-1} = \mathbf{C} = \frac{E_0}{1-\mu_0^2} \begin{bmatrix} 1 & \mu_0 & 0 \\ \mu_0 & 1 & 0 \\ 0 & 0 & (1-\mu_0)/2 \end{bmatrix} \tag{5.22}$$

For plane stress problems, $E_0 = E$, $\mu_0 = \mu$.
For plane strain problems, $E_0 = E/(1-\mu^2)$, $\mu_0 = \mu/(1-\mu)$.

$$\mathbf{H}_{hh}^{-1} = \frac{9E_0J_0}{4t(m_1m_2 - m_3^2)}\begin{bmatrix} m_1 & m_3 \\ m_3 & m_2 \end{bmatrix} \tag{5.23}$$

$$m_1 = (a_3^2 + b_3^2)^2 (3J_0^2 - J_1^2)$$

$$m_2 = (a_1^2 + b_1^2)^2 (3J_0^2 - J_2^2)$$

$$m_3 = \left[(a_1a_3 + b_1b_3)^2 - \mu_0 J_0^2 \right] J_1 J_3$$

5.2 Penalty Equilibrium Hybrid Element P-S(α)

In comparison with the four-node plane-isoparametric element Q4, the advantages of P-S elements are the high accuracy in stresses and their insensitivity to geometric distortions. In comparison with the four-node plane-hybrid elements developed earlier, the P-S element has no zero energy mode or directional sensitivity. Its numerical solution is stable and it has only five β parameters. Thus, P-S elements have been widely recognized. The disadvantage of the P-S element, false shearing stress, may occur in the solution of bending problems and result in shear locking phenomena.

Many writers [5–9] have tried to improve the P-S element by using different variational principles or different methods for formulation. But the results are very similar to those for the P-S element. Now we apply the penalty equilibrium optimization method to improve the P-S element to avoid the shear locking phenomenon.

When the penalty equilibrium term is added, the element stiffness matrix of the hybrid element can be expressed as [5]

$$\mathbf{K}^e = \mathbf{G}^T \left(\mathbf{H} + \frac{\alpha}{E} \mathbf{H}_p \right)^{-1} \mathbf{G} \tag{5.24}$$

For the P-S element or the corresponding penalty equilibrium element, we have

$$\mathbf{G} = \int\int_{A^e} \phi^{*T} (\mathbf{D}\mathbf{N}_q) t dA \tag{5.25}$$

$$\mathbf{H} = \int\int_{A^e} \phi^{*T} \mathbf{S} \phi^* t dA \tag{5.26}$$

Here, $\mathbf{N}_q$ is determined from Equation 5.1, and ϕ^* is determined by Equation 5.15 and Equation 5.19. The element penalty equilibrium determinant in Equation 5.21 is

$$\boldsymbol{H}_p = \int\int_{A^e} (\partial\phi^*)^T(\partial\phi^*)tdA \tag{5.27}$$

in which the equilibrium operator is

$$\partial = \mathbf{D}^T = \begin{bmatrix} \frac{\partial}{\partial x} & 0 & \frac{\partial}{\partial y} \\ 0 & \frac{\partial}{\partial y} & \frac{\partial}{\partial x} \end{bmatrix}$$

The following transformation exists for four-node plane elements:

$$\begin{Bmatrix} \frac{\partial}{\partial x} \\ \frac{\partial}{\partial y} \end{Bmatrix} = \frac{1}{|\boldsymbol{J}|}\begin{bmatrix} b_3 + b_2\xi & -b_1 - b_2\eta \\ -a_3 - a_2\xi & a_1 + a_2\eta \end{bmatrix}\begin{Bmatrix} \frac{\partial}{\partial \xi} \\ \frac{\partial}{\partial \eta} \end{Bmatrix} \tag{5.28}$$

Substituting Equation 5.28 into Equation 5.27, the penalty equilibrium determinant can be determined. However, in order to determine the expression of $\mathbf{H}_p$, we can set the Jacobi of Equation 5.21 as a constant, that is, to let $|\boldsymbol{J}| \approx |\mathbf{J}(0,0)| = J_0$. Thus, we can determine the penalty equilibrium determinant of the P-S element by

$$\boldsymbol{H}_P = \frac{4t}{3J_0}\begin{bmatrix} \mathbf{0} & & \mathbf{0} \\ & (a_1^2 + b_1^2)J_1^2 & \mathbf{0} \\ \mathbf{0} & \mathbf{0} & (a_3^2 + b_3^2)J_2^2 \end{bmatrix} \tag{5.29}$$

We now present several examples to compare the numerical solutions of penalty equilibrium elements (indicated as P-S[α]), and the Q4 isoparametric elements and to find out whether the penalty equilibrium is effective to overcome the shear locking phenomena that often exist for four-node membrane elements. The numerical solutions have shown that when the penalty factor $\alpha > 10^3$, the solution by the penalty equilibrium element is already approaching a stable condition. In the following numerical analysis, we took $\alpha = 10^4$. We have already shown that the P-S(α) element and the P-S element all have no zero energy modes and can pass all kinds of patch tests for stresses.

The first test is the long beam test suggested by MacNeal and Harder [10]. Figure 5.1 shows a set of cantilever beams with different forms of discretization, acted by pure bending and pure shearing forces. The numerical solutions are listed in Table 5.1. The solutions from using Q4 elements always involve locking phenomena. However, when rectangular elements are used, the P-S element and P-S(α) element lead to the same results. Both elements can pass this test. However, when irregular meshes such as (Figure 5.1b) and (Figure 5.1c) are used, the solutions from P-S elements become worse, especially under the ladder shape mesh (5.1b). The displacement solutions indicate severe locking phenomena. The numerical solutions indicate that

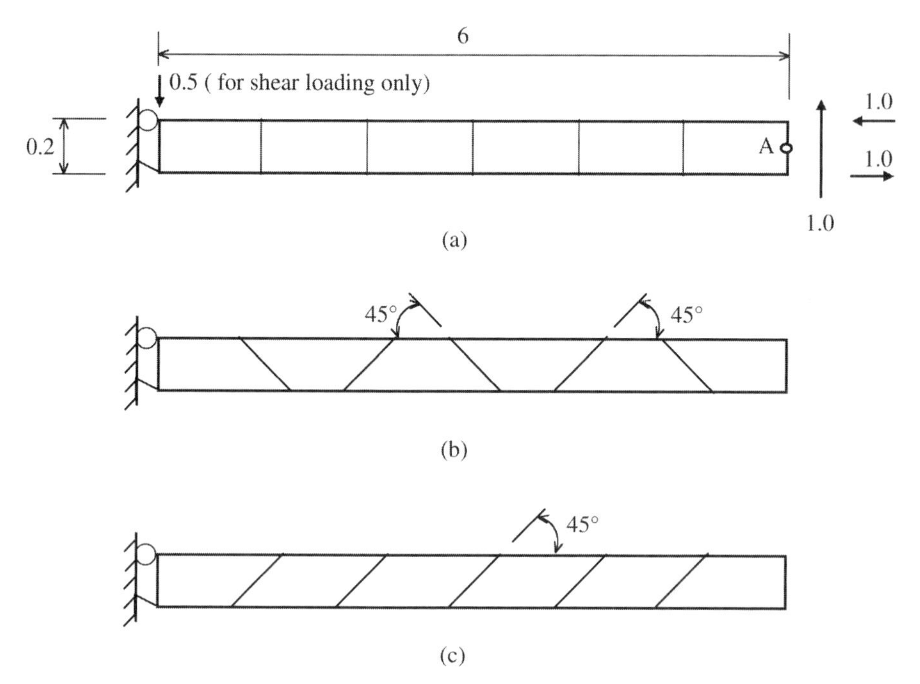

FIGURE 5.1
MacNeal's slim cantilever beam test ($E = 10^7$, $\mu = 0.3$, $t = 0.1$). (a) Rectangular mesh; (b) trapezoidal mesh; (c) parallelogram mesh.

TABLE 5.1
Tip Deflections v_A of Cantilever Beams for MacNeal Tests (Figure 5.1)

Load Status	Pure Bending Load			End Shear Load		
Mesh	(a)	(b)	(c)	(a)	(b)	(c)
Q4 Element	0.0005	0.00012	0.00017	0.0101	0.0003	0.0038
P-S Element [1]	0.0054	0.0009	0.0046	0.1073	0.0239	0.0863
P-S (α) Element [5]	0.0054	0.0054	0.0054	0.1073	0.1073	0.1078
Exact	0.0054	0.0054	0.0054	0.1081	0.1081	0.1081

only the penalty equilibrating element P-S(α) can pass the MacNeal test, independent of the mesh pattern and the loading condition.

The second example is a beam under pure bending, and with two discretizing meshes as shown in Figure 5.2 and Figure 5.3. We can see that the P-S element can effectively improve the bending solution of the Q4 isoparametric element but cannot avoid the shear locking phenomena. Figure 5.3

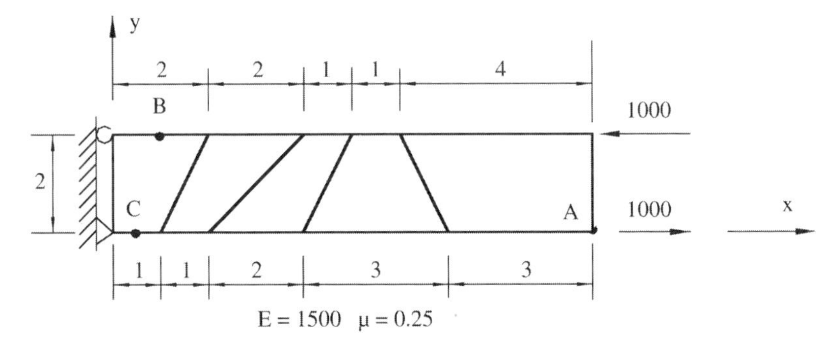

Element	v_A	σ_{xB}/σ_{xC}	τ_{xyB}/τ_{xyC}
Q4	45.65	-1762/2777	427.2/-347.7
P-S	96.18	-3014/3007	9.37/9.37
P-S(α)	99.88	-3000/3000	0.20/0.20
exact	100	-3000/3000	0.00/0.00

FIGURE 5.2
Pure bending solutions of a plane stress beam with five elements.

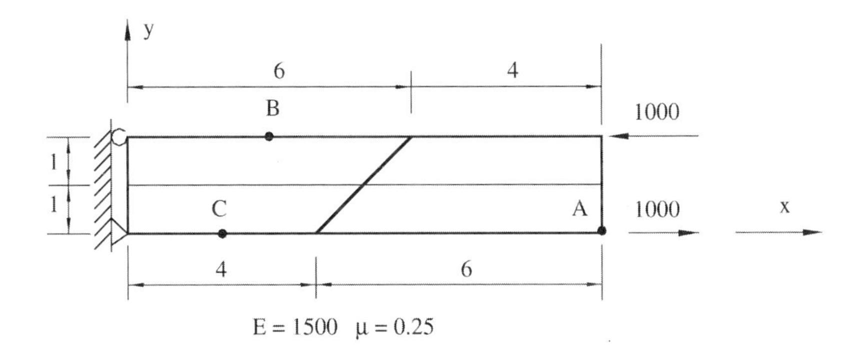

Element	v_A	σ_{xB}/σ_{xC}	τ_{xyB}/τ_{xyC}
Q4	14.73	-412.9/575.8	234/92.36
P-S	59.32	-1741/2174	259.8/317.6
P-S(α)	89.75	-2990/2993	2.14/2.61
Exact	100	-3000/3000	0.00/0.00

FIGURE 5.3
Pure bending solutions of a plane stress beam with four elements.

shows that the P-S element yields serious false shearing stress (τ_{xyB} and τ_{xyC}). Conversely, for the two situations of Figure 5.2 and Figure 5.3 the penalty equilibrating hybrid element P-S(α) can give pure bending solutions and avoid the shear locking phenomena.

The last example is to determine the stability of the numerical solutions and the effect of change in the mesh pattern on the finite element solution. Figure 5.4 and Figure 5.5 are two cantilever beams under different loading conditions. From the resulting displacement and stress curves, we can see

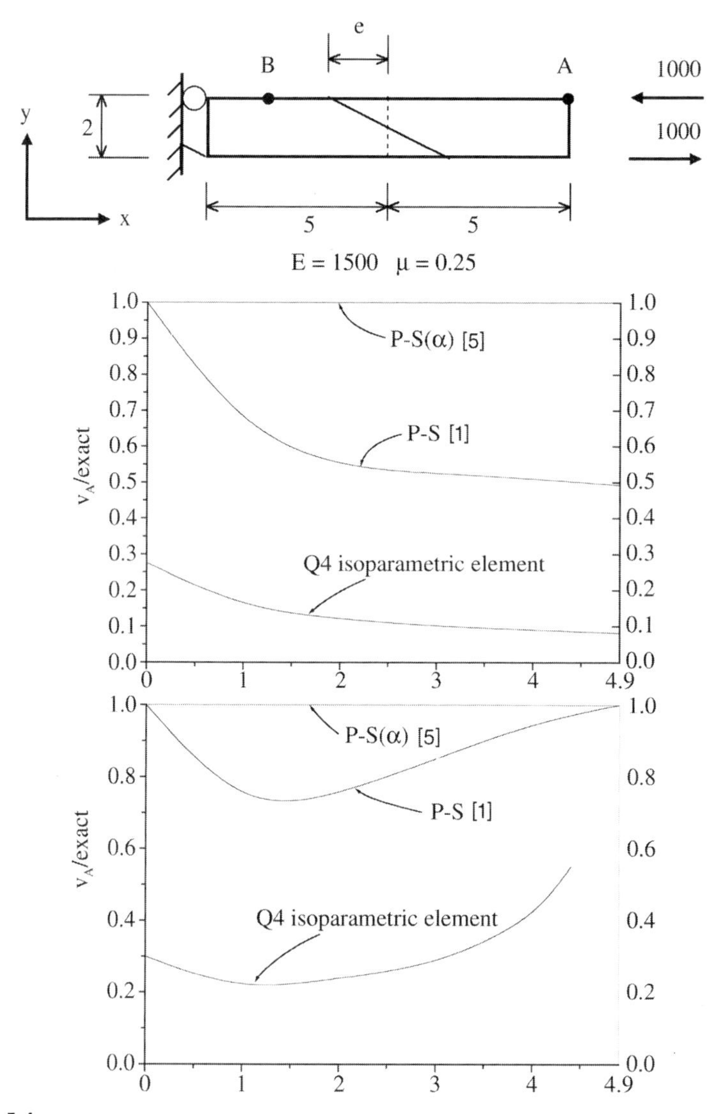

FIGURE 5.4
Pure bending solutions of a beam with two distorted elements.

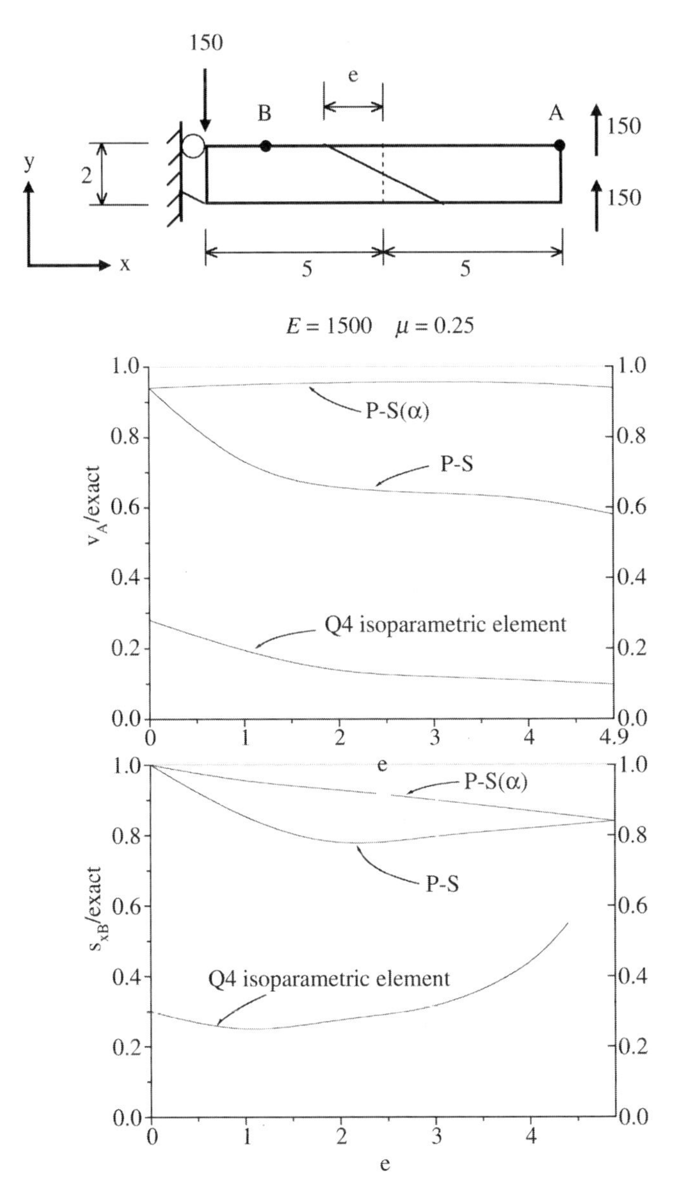

FIGURE 5.5
Bending beam with two distorted elements under an end shear force.

that in comparison to the Q4 element and P-S element, P-S(α) elements provide stable numerical solutions for v_A and σ_{xB}, even when distorted mesh patterns are used.

From the above numerical tests, we can conclude that the penalty equilibrium method is effective for the improvement of hybrid elements for bending analysis.

5.3 Three-Dimensional Body 18β–Optimization Hybrid Element

Now we will apply the method of optimization to three-dimensional (3-D) elasticity problems to determine the stress optimization pattern of an element with 18 stress parameters. Referring to the eight-node six-faced solid in Figure 3.6 and using the 3-D trilinear isoparametric interpolation function,

$$\mathbf{u}_q = \begin{Bmatrix} u_q \\ v_q \\ w_q \end{Bmatrix} = \sum_{i=1}^{8} \frac{1}{8}(1+\xi_i\xi)(1+\eta_i\eta)(1+\zeta_i\zeta)\begin{Bmatrix} u_i \\ v_i \\ w_i \end{Bmatrix} = \mathbf{N}_q\mathbf{q} \tag{5.30}$$

and the additional internal displacement of the element is

$$u_\lambda = \begin{Bmatrix} u_\lambda \\ v_\lambda \\ w_\lambda \end{Bmatrix} = \begin{bmatrix} \xi^2\eta^2\zeta^2 & & \\ & \xi^2\eta^2\zeta^2 & \\ & & \xi^2\eta^2\zeta^2 \end{bmatrix}\begin{Bmatrix} \lambda_1 \\ \vdots \\ \lambda_9 \end{Bmatrix} = \mathbf{N}_\lambda\lambda \tag{5.31}$$

We assume that the initial trial solutions of stresses are of linear field:

$$\sigma = \begin{Bmatrix} \sigma_x \\ \sigma_y \\ \sigma_z \\ \tau_{yz} \\ \tau_{zx} \\ \tau_{xy} \end{Bmatrix} = \sigma_c + \sigma_h = \beta_c + \begin{bmatrix} \phi_{\mathrm{I}} & | \phi_{\mathrm{II}} \end{bmatrix}\begin{Bmatrix} \beta_{\mathrm{I}} \\ \hline \beta_{\mathrm{II}} \end{Bmatrix} \tag{5.32}$$

in which

$$\beta_c = \begin{Bmatrix} \beta_1 \\ \vdots \\ \beta_6 \end{Bmatrix} \quad \beta_{\mathrm{I}} = \begin{Bmatrix} \beta_7 \\ \vdots \\ \beta_{18} \end{Bmatrix} \quad \beta_{\mathrm{II}} = \begin{Bmatrix} \beta_{19} \\ \vdots \\ \beta_{27} \end{Bmatrix}$$

let

$$
[\phi_{\mathrm{I}} \quad | \phi_{\mathrm{II}}] = \left[\begin{array}{ccccccccc|ccc}
\eta & \zeta & \eta\zeta & 0 & 0 & 0 & 0 & 0 & 0 & 0 & 0 & 0 \\
0 & 0 & 0 & \zeta & \xi & \zeta\xi & 0 & 0 & 0 & 0 & 0 & 0 \\
0 & 0 & 0 & 0 & 0 & 0 & \xi & \eta & \xi\eta & 0 & 0 & 0 \\
0 & 0 & 0 & 0 & 0 & 0 & 0 & 0 & 0 & 0 & \xi & 0 \\
0 & 0 & 0 & 0 & 0 & 0 & 0 & 0 & 0 & 0 & 0 & \eta \\
0 & 0 & 0 & 0 & 0 & 0 & 0 & 0 & 0 & \xi & 0 & 0
\end{array}\right.
$$

$$
\left.\begin{array}{ccccccccc}
\xi & 0 & 0 & 0 & 0 & 0 & 0 & 0 & 0 \\
0 & 0 & 0 & 0 & \eta & 0 & 0 & 0 & 0 \\
0 & 0 & 0 & 0 & 0 & 0 & 0 & 0 & \zeta \\
0 & 0 & 0 & 0 & 0 & \zeta & 0 & \eta & 0 \\
0 & 0 & \zeta & 0 & 0 & 0 & \xi & 0 & 0 \\
0 & \eta & 0 & \xi & 0 & 0 & 0 & 0 & 0
\end{array}\right] \tag{5.33}
$$

The purpose of adding the three terms $\eta\zeta$, $\zeta\xi$, and $\xi\eta$ is to prevent the occurrence of the zero energy mode. Corresponding to Equation 5.30, the isoparametric transformation is

$$
\begin{Bmatrix} x \\ y \\ z \end{Bmatrix} = \sum_{i=1}^{8} \frac{1}{8}(1+\xi_i\xi)(1+\eta_i\eta)(1+\zeta_i\zeta)\begin{Bmatrix} x_i \\ y_i \\ z_i \end{Bmatrix} \tag{5.34}
$$

The corresponding expression of Jacobi determinants has already been given in Equation 3.36a and Equation 3.36b.

For convenience in integration, the $\boldsymbol{M}$ matrix can be expressed in terms of the volume integral, which is

$$
\begin{aligned}
\boldsymbol{M} &= \int_{V^e} \left[(\boldsymbol{D}N_\lambda)^T \left[\phi_{\mathrm{I}} \quad | \phi_{\mathrm{II}} \right] + N_\lambda^T \left[\boldsymbol{D}^T \phi_{\mathrm{I}} \quad | \boldsymbol{D}^T \phi_{\mathrm{II}} \right] \right] dV \\
&= \left[\boldsymbol{M}_{\mathrm{I}} \quad | \boldsymbol{M}_{\mathrm{II}} \right]
\end{aligned} \tag{5.35}
$$

The $\boldsymbol{M}_{\mathrm{II}}$ matrix is not zero. The optimal form σ^* can be written in standard form (Equation 4.21). But the integration process is still complicated, and it is difficult to determine the explicit expression of σ^*. If for the Jacobi matrix in (Equation 3.36a), only the values at the centers of the elements are used, we will have

$$
\boldsymbol{J}(\xi,\eta,\zeta) \approx J_c(0,0,0) = \begin{bmatrix} a_1 & b_1 & c_1 \\ a_2 & b_2 & c_2 \\ a_3 & b_3 & c_3 \end{bmatrix} \tag{5.36}
$$

which is a matrix with constant values. The corresponding inversion is

$$J_c^{-1} = \frac{1}{|J_c|}\begin{bmatrix} j_{11} & j_{12} & j_{13} \\ j_{21} & j_{22} & j_{23} \\ j_{31} & j_{32} & j_{33} \end{bmatrix} = \frac{1}{|J_c|}\begin{bmatrix} b_2c_3 - b_3c_2 & b_3c_1 - b_1c_3 & b_1c_2 - b_2c_1 \\ c_2a_3 - c_3a_2 & c_3a_1 - c_1a_3 & c_1a_2 - c_2a_1 \\ a_2b_3 - a_3b_2 & a_3b_1 - a_1b_3 & a_1b_2 - a_2b_1 \end{bmatrix} \tag{5.37}$$

At this point the explicit expressions of $\boldsymbol{M}_{\mathrm{I}}$ and $\boldsymbol{M}_{\mathrm{II}}$ can be determined. We can then apply Equation 4.21 to write the desirable optimal stress pattern of the 3-D 18β hybrid element as follows

$$\sigma^* = \phi^*\beta^* =$$

$$\left[\begin{matrix} I_{6\times6} & j_{22}^2\eta & j_{33}^2\xi & \eta\xi & 0 & j_{21}^2\xi & 0 \\ & j_{12}^2\eta & 0 & 0 & j_{33}^2\xi & j_{11}^2\xi & \zeta\xi \\ & 0 & j_{13}^2\zeta & 0 & j_{23}^2\zeta & 0 & 0 \\ & 0 & 0 & 0 & -j_{23}j_{33}\zeta & 0 & 0 \\ & 0 & -j_{13}j_{33}\xi & 0 & 0 & 0 & 0 \\ & -j_{12}j_{22}\eta & 0 & 0 & 0 & -j_{21}j_{11}\xi & 0 \end{matrix}\right. \tag{5.38}$$

$$\left.\begin{matrix} j_{31}^2\xi & 0 & 0 & 0 & 2j_{21}j_{31}\xi & 0 \\ 0 & j_{32}^2\eta & 0 & 0 & 0 & 2j_{12}j_{32}\eta \\ j_{11}^2\xi & j_{22}^2\eta & \xi\eta & 2j_{13}j_{23}\zeta & 0 & 0 \\ 0 & -j_{32}j_{22}\eta & 0 & -j_{13}j_{33}\zeta & j_{11}^2\xi & -j_{12}j_{22}\eta \\ -j_{31}j_{11}\xi & 0 & 0 & -j_{23}j_{33}\zeta & -j_{21}j_{11}\xi & j_{22}^2\eta \\ 0 & 0 & 0 & j_{33}^2\zeta & -j_{31}j_{11}\xi & -j_{32}j_{22}\eta \end{matrix}\right] \begin{Bmatrix} \beta_1 \\ \vdots \\ \vdots \\ \vdots \\ \vdots \\ \beta_{18} \end{Bmatrix}$$

When an element is a parallel six-faced solid, the coordinate system (ξ-η-ζ) is parallel to the coordinate system (x-y-z), and the coefficient $j_{kl} = 0$ (k is unequal to l), the element stress (Equation 5.38) will be reduced to the results of Loikkanen and Irons [12].

By substituting the trial solution (Equation 5.30) of the element displacement and the optimal stress pattern into the energy functional (Equation 4.23), we can determine the special property determinant of the hybrid element. The six-faced optimal hybrid element will have 18β parameters. We will name it the OPH-18β element. This element can satisfy the best condition for the coefficient arrangement (Equation 4.31).

Here we have two numerical examples. Figure 5.6 is a cantilever beam of uniform cross-section acted by two different loading conditions. Five irregular elements are used. Figure 5.7 is a circular ball with a thick wall under the action of uniform internal pressure q. For symmetric conditions, only three elements are used. The results for the optimal hybrid element OPH-18β are much better than the results for the eight-node isoparametric displacement element.

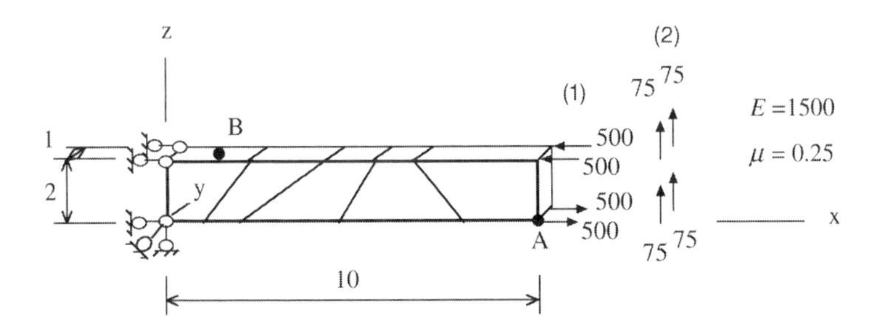

Element	Loading (1)		Loading (2)	
	w_A	σ_{xB}	w_A	σ_{xB}
H8	43.24	-1732	48.43	-2412
OPH-18	96.05	-3016	98.01	-4076
exact	100	-3000	102.6	-4050

FIGURE 5.6
Solution of deflection and stress for a cantilever beam.

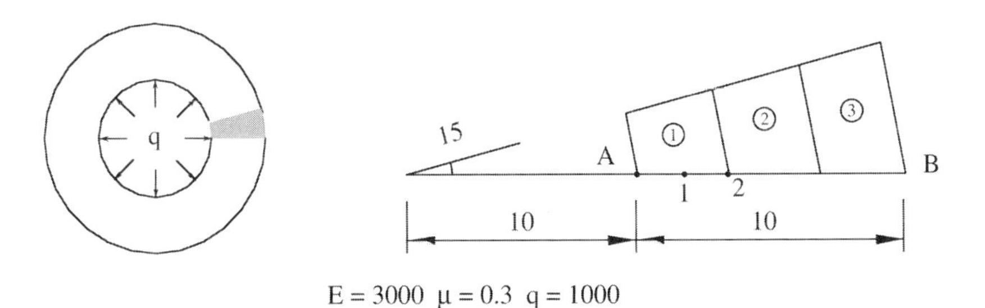

Element	u_{rA}	u_{rB}	σ_{r1}	$\sigma_{\phi 1} = \sigma_{\theta 1}$	σ_{r2}	$\sigma_{\phi 2} = \sigma_{\theta 2}$
H8	25.80	9.88	-521.6	587.6	-425.4	334.9
OPH-18	26.27	9.99	-545.4	528.0	-365.0	329.3
exact	26.67	10	-576.8	502.7	-339.3	383.9

FIGURE 5.7
Axial directional displacements and stress for a thick circular shell.

For the next step, we shall try to apply the penalty equilibrium method to the OPH-18β element. For the present 3-D condition, the penalty equilibrium matrix can be determined by

$$H_p = \int_{-1}^{1}\int_{-1}^{1}\int_{-1}^{1} (\partial\phi^*)^T(\partial\phi^*)|J|\,d\xi\, d\eta\, d\zeta \tag{5.39}$$

in which ϕ^* is determined by Equation 5.31. We denote the corresponding element as OPH-18β(α).

As a typical example of a 3-D problem, we consider a cantilever beam composed of two irregular elements, as shown in Figure 5.8 and Figure 5.9. At the tip of the element two types of loading are applied. In the comparison of numerical solutions five different eight-node, six-faced elements are used. They include the familiar assumed-displacement isoparametric element **H**8, a pair of hybrid elements LO8: 7-APC and LO8: 7-APR, determined through group theory [13], and the optimal hybrid element OPH-18β and its penalty equilibrium form OPH-18β(α) (the penalty factor $\alpha = 10^4$) developed in this section. By comparing the curves in these figures we can conclude that both OPH-18β and OPH-18β(α) can provide stable and reliable solutions for displacement v_A and stress σ_{xB} even if severely distorted meshes are used. A comparison of the results indicates that the numerical approach of the penalty equilibrium hybrid element OPH-18β(α) is much better than that of the original OPH-18β element.

5.4 Axisymmetric 8β–Optimization Hybrid Element

Referring to the four-node axisymmetric element in Figure 3.8, we can write the trial solution of element displacement $\boldsymbol{u}$ as

$$\boldsymbol{u} = \begin{Bmatrix} u \\ v \end{Bmatrix} = \boldsymbol{N}_q \boldsymbol{q}, \quad \boldsymbol{q} = \{u_1 v_1 \quad \cdots \quad u_4 v_4\} \tag{5.40}$$

and the trial solution of element stress as

$$\sigma = \begin{Bmatrix} \sigma_{rr} \\ \sigma_{zz} \\ \tau_{rz} \\ \tau_{\theta\theta} \end{Bmatrix} = \left[\begin{array}{cccc|cccc} 1 & & & & r & z & 0 & 0 \\ & 1 & & & 0 & 0 & 0 & 0 \\ & & 1 & & 0 & 0 & 0 & 0 \\ & & & 1 & 0 & 0 & r & z \end{array}\right] \begin{Bmatrix} \beta_1 \\ \vdots \\ \vdots \\ \beta 8 \end{Bmatrix} = \phi\beta \tag{5.41}$$

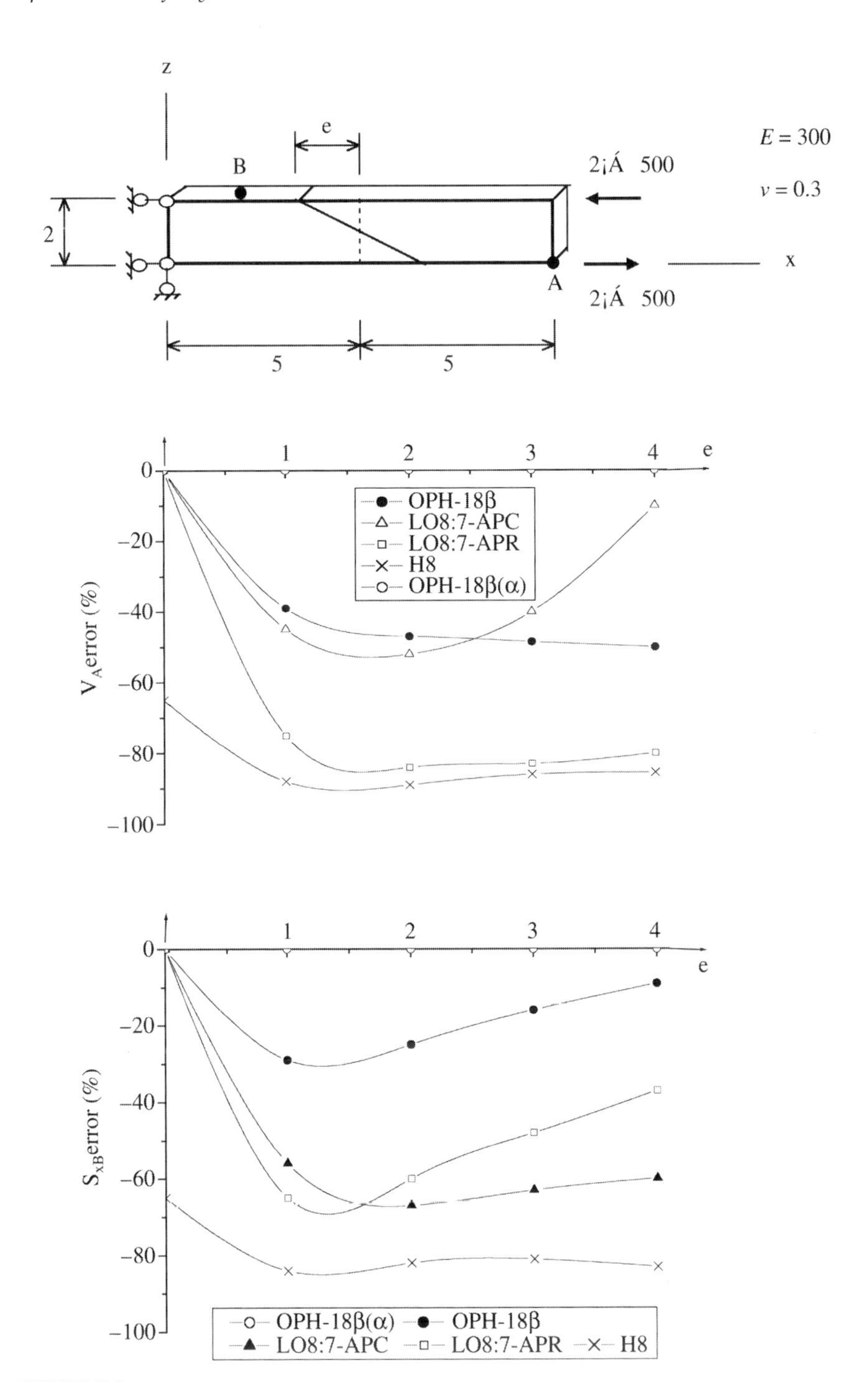

FIGURE 5.8
Distortion experiment I of a 3-D solid (pure bending load).

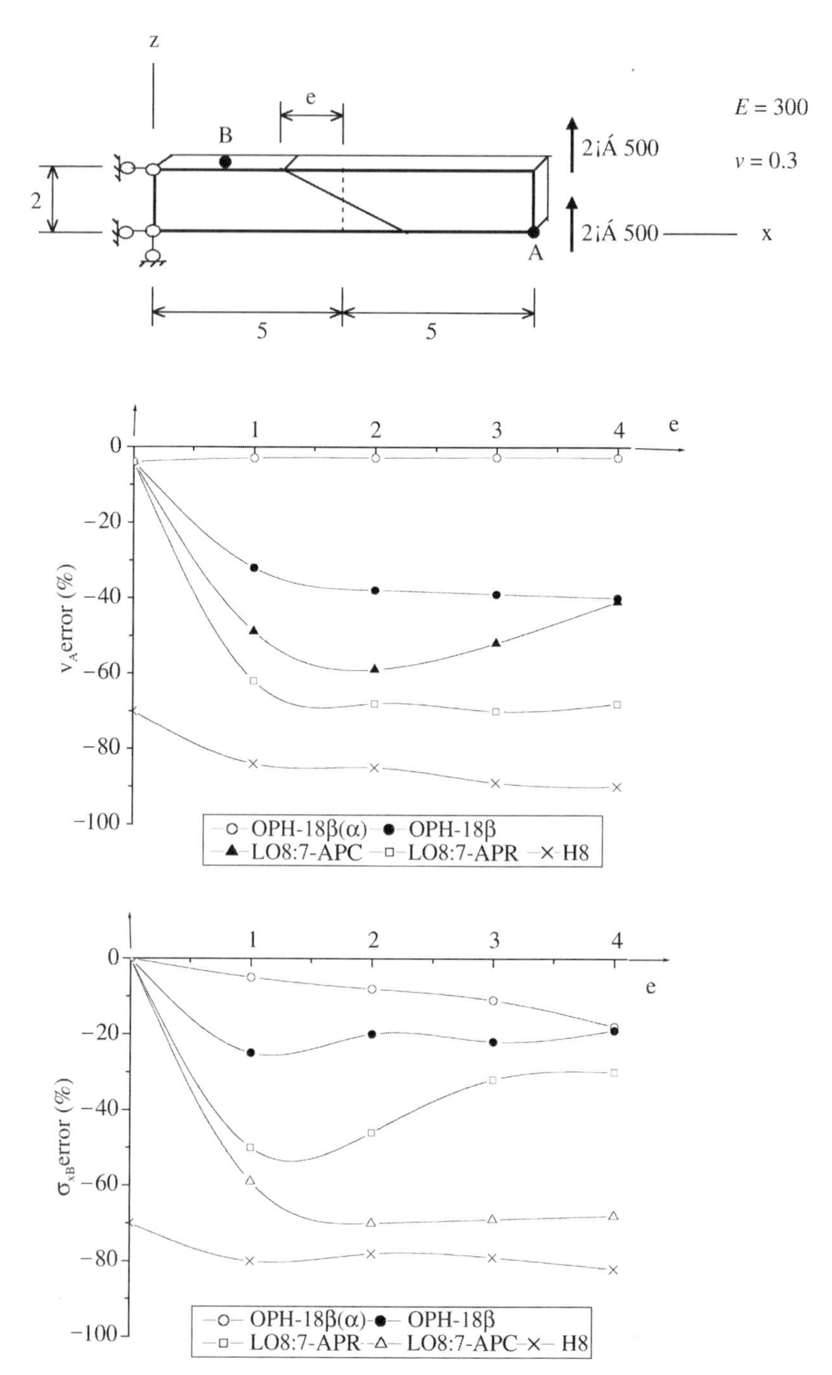

FIGURE 5.9
Distortion experiment II of a 3-D solid (crosswise shearing load).

With the element geometric coefficient a_i, b_i in Equation 5.3 we have

$$\begin{Bmatrix} r = a_1\xi + a_2\xi\eta + a_3\eta + a_4 \\ z = b_1\xi + b_2\xi\eta + b_3\eta + b_4 \end{Bmatrix} \tag{5.42}$$

In Equation 5.34 $\sigma_{rr}, \sigma_{\theta\theta} \in \boldsymbol{P}_1(r, z)$ are complete in linear terms. This is desirable for obtaining a reasonable axial and circumferential stress distribution. We can take σ_{zz} and τ_{rz} as constants. In this way, the following required condition for arranging the element parameters of a hybrid element can be completely satisfied, with smaller numbers of stress parameters ($n_\beta = 8$):

$$n_\beta \geq n_q - n_0 \tag{5.43}$$

Substituting the test solutions u and σ into Equation 4.77, we obtain

$$\Pi_{R^*}^e = \beta^T \boldsymbol{G}\boldsymbol{q} - \frac{1}{2}\beta^T \boldsymbol{H}\beta - \alpha\beta^T \boldsymbol{H}_p\beta \tag{5.44}$$

in which

$$\mathbf{G} = 2\pi\int_{-1}^{1}\int_{-1}^{1} \phi^T(\mathbf{DN}_q) r|\mathbf{J}|\, d\xi\, d\eta \tag{5.45}$$

$$\mathbf{H} = 2\pi\int_{-1}^{1}\int_{-1}^{1} \phi^T \mathbf{S}\phi\, r|\mathbf{J}|\, d\xi\, d\eta \tag{5.46}$$

$$\boldsymbol{H}_p = 2\pi\int_{-1}^{1}\int_{-1}^{1} (\partial\phi)^T(\partial\phi) r|\boldsymbol{J}|\, d\xi\, d\eta \tag{5.47}$$

The equilibrium operator in Equation 5.47 is

$$\partial = \begin{bmatrix} \left(\dfrac{\partial}{\partial r} + \dfrac{1}{r}\right) & 0 & \dfrac{\partial}{\partial z} & -\dfrac{1}{r} \\ 0 & \dfrac{\partial}{\partial z} & \left(\dfrac{\partial}{\partial r} + \dfrac{1}{r}\right) & 0 \end{bmatrix} \tag{5.48}$$

A penalty equilibrium constraint may be applied only at the center point. Thus, we can write the explicit expression of the penalty equilibrium determinant $\boldsymbol{H}_p$. Note that at the center point (r_0, z_0) of the four-sided element,

$$|J| = J_0 = a_1 b_3 - a_3 b_1 = \frac{1}{4} A^e$$

Then the penalty equilibrium determinant corresponding to Equation 5.47 is

$$\mathbf{H}_p = \frac{2\pi A^e}{r_0} \begin{bmatrix} 1 & 0 & 0 & -1 & 2r_0 & 0 & -r_0 & -z_0 \\ & 0 & 0 & 0 & 0 & 0 & 0 & 0 \\ & & 1 & 0 & 0 & 0 & 0 & 0 \\ & & & 1 & -2r_0 & 0 & r_0 & z_0 \\ & & & & 4r_0^2 & 0 & -2r_0^2 & -2r_0 z_0 \\ & \textit{Symmetric} & & & & 0 & 0 & 0 \\ & & & & & & r_0^2 & r_0 z_0 \\ & & & & & & & z_0^2 \end{bmatrix} \tag{5.49}$$

Correspondingly, thc stiffness matrix of the penalty equilibrium hybrid element Axi-8β[2] is

$$\mathbf{K}^e = \mathbf{G}^T \left(\mathbf{H} + \frac{\alpha}{E} \mathbf{H}_p \right)^{-1} \mathbf{G} \tag{5.50}$$

Because the penalty matrix (Equation 5.49) is singular, the penalty factor α may be taken from a very large range with no effect on the stability of the solution. In the following calculation we take $\alpha = 10^4$. For $\alpha = 0$, the Axi-8β(α) element is reduced to the ordinary axisymmetric hybrid element, Axi-8β.

An example of numerical analysis is given in the following. Figure 5.10 is the cross section of a circular tube of infinite length, under internal pressure. Five axisymmetric elements are used in the analysis. Table 5.2 lists the

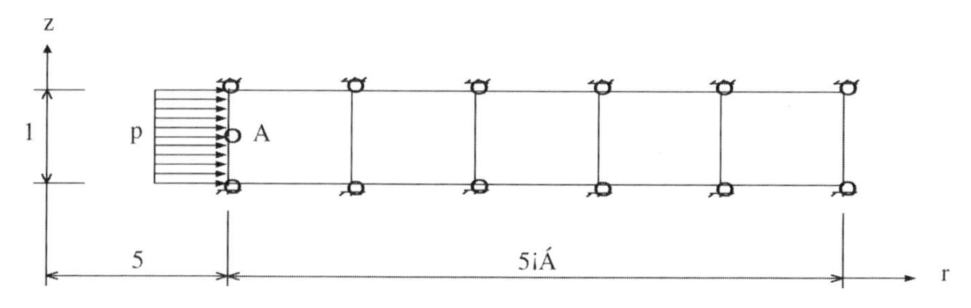

FIGURE 5.10
Thick wall tube under internal pressure and a finite element mesh pattern (E = 1, m = 0.3, p = 10/p).

solutions for displacements and stresses along the axial direction at point A of the inner surface. The solution by hybrid element Axi-8β is better than that by the four-node axisymmetric isoparametric element Axi-Q4, while the solution by penalty equilibrium hybrid element Axi-8β(α) is even better than that by Axi-8β. Figure 5.11 shows the distribution of the finite element solution for the axial direction stress σ_{rr} along the *r* direction. But the distribution of σ_{rr} solutions by the Axi-8β element and the Axi-Q4 element are quite different from the actual solution, while the solution by Axi-8β(α) falls right on the curve of the actual solution.

From this example we can see that the penalty equilibrium treatment is effective for increasing the accuracy of a hybrid element and for improving the stress solutions within the element.

TABLE 5.2

Axial Direction Displacement and Stress Solutions at Point A in the Inner Face of a Circular Tube

Element	u_A	σ_{rrA}	σ_{zzA}	τ_{rzA}	$\sigma_{\theta\theta A}$
Axi-Q4	30.15	–1.914	1.17	0.000	5.806
Axi-8β	30.185	–2.164	0.64	0.000	5.52
Axi-8β(α)	30.345	–3.119	0.637	0.000	5.241
Analytical solution	30.35	–3.183	0.637	0.000	5.305

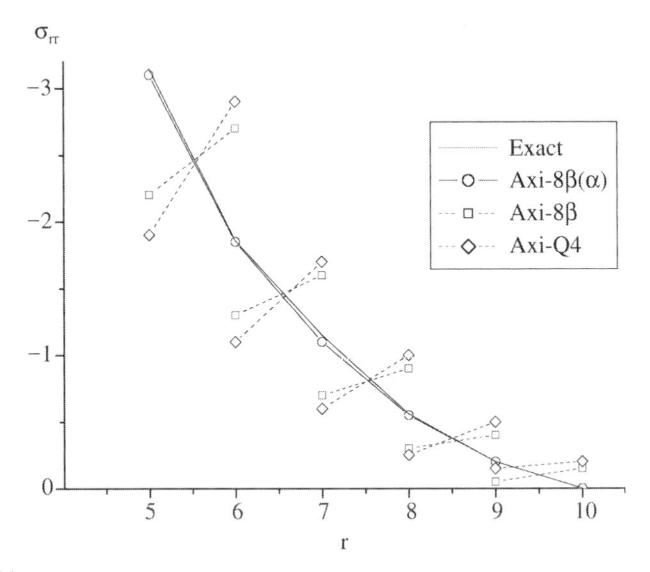

FIGURE 5.11
Distribution stress σ_{rr} along the axial direction.

5.5 Model Optimization of Hybrid Stress General Shell Element

5.5.1 Introduction

The finite element models for plates and shells based upon the Reissner–Mindlin plate bending theory have been increasingly developed in recent years. Theoretically, it describes the c^1 problem by c^0 trials. Unfortunately, thin plate or thin shell models based on the minimum potential energy principle result in shear locking effects. Some approaches have been developed [14, 15] to alleviate such difficulties. Many investigators have declared that the locking effects can be eliminated by using the hybrid models based upon various general complementary energy principles [16].

The present work is to extend the optimizing design approach of hybrid elements developed by the authors. A 20–degree of freedom (DOF) hybrid stress element based on Mindlin plate bending theory is constructed using the new approach for thin and moderately thick shells in orthogonal curvilinear coordinates. The element is a quadrilateral with only four corner nodes and can be used for the shallow or deep shell with improved performance. The shear locking behavior no longer exists.

5.5.2 The Optimizing Condition and the Variational Principle

To avoid the transformation between the local and global systems decrease or eliminate the discrete error of geometry and enhance the continuity of trials. It is appropriate to use the orthogonal curvilinear coordinates corresponding to the shell surface. All the formulations and discussions in this chapter are in the orthogonal curvilinear coordinates. For simplicity, it is assumed in the following discussion that the displacement boundary condition has been satisfied and the prescribed boundary traction $T^1 = 0$. The Hellinger–Reissner functional can be written in tensor notation for an assemblage of n finite elements as

$$\Pi_R = \sum_n \Pi_R^e = \sum_n \int_{V^e} \left[-\frac{1}{2}\sigma^{ij}\sigma^{kl}S_{ijkl} + \frac{1}{2}\sigma^{ij}(u_{i;j} + u_{j;i}) - f^i u_i \right] dV \quad (5.51)$$

where σ^{ij}, u_1, f^1, S_{ijkl}, and V^e are stresses, displacements, body forces, an elastic compliance tensor, and the volume of the element, respectively. The semicolons represent the covariant derivative.

When the incompatible displacements u_1^λ are introduced, the compatible displacements can be expressed as

$$u_1^q = u_1 - u_1^\lambda \quad (5.52)$$

In this case, the functional (Equation 5.51) is still applicable, provided the weak discontinuity condition [20] is satisfied. However, for the purpose of convergence, some limitation must be suppressed on the discontinuous function of Equation 5.51. Using the stationary condition, we can obtain

$$\delta\Pi_R(\sigma^{ij}, u_i^q, u_i^\lambda) = \sum_n \int_{V^e} \left[(\varepsilon_{ij} - S_{ijkl}\sigma^{kl})\delta\sigma^{ij} - (\sigma^{ij}{}_{;j} + f^i)\delta u_i \right] dV$$

$$+ \sum_{ab} \int_{S_{ab}} (T^i_{(a)} + T^i_{(b)})\delta u_i^q dS + \sum_n \int_{S_\sigma^e} T^i \delta u_i^q dS + \sum_n \int_{\partial V^e} T^i \delta u_i^\lambda dS = 0 \quad (5.53)$$

where $T^1 = \sigma^{ij} n_j$ is the boundary traction related to stresses, strains $\varepsilon_{ij} = ½(u_{i,j} + u_{j,i})$, S_{ab} is the boundary between elements a and b, S_σ^e is the boundary of element e with prescribed traction, and ∂V^e is the entire boundary of the element. With the variations $\delta\sigma^{ij}$, δu_i^q, and δu_i^λ being arbitrary, the Euler–Lagrange equation and natural boundary condition can be derived from Equation 5.53, but an unrational constraint can result from the last term in Equation 5.53:

$$T^i = \sigma^{ij} n_j = 0 \text{ on } \partial V^e \quad (5.54)$$

It is obviously not compatible for a nonzero variational solution to impose an improper condition on stresses. In other words, when the incompatible displacement u_1^λ is introduced, we cannot obtain the rational approximations of the true value from the stationary condition; therefore Equation 5.51 is untenable, unless

$$\sum_n \int_{\partial V^e} T^i \delta u_i^\lambda dS = 0 \quad (5.55)$$

Equation 5.51 can be used, and $\Pi_R = \sum \Pi_R^e$ only when Equation 5.55 is satisfied.

If we denote the element stresses as the sum of the constant part σ_c^{ij} and the high-order part σ_h^{ij}

$$\sigma^{ij} = \sigma_c^{ij} + \sigma_h^{ij}$$

then Equation 5.55 can be satisfied by the following two conditions:

$$\sum_n \int_{\partial V^e} \sigma_c^{ij} n_j u_i^\lambda dS = 0 \quad (5.56)$$

and

$$\sum_n \int_{\partial V^e} \sigma_h^{ij} n_j u_i^\lambda dS = 0 \tag{5.57}$$

Here the condition (Equation 5.56) is equivalent to Iron's patch test; it is the convergence condition for an incompatible finite element. When the size of the elements is reduced to zero, $\sigma_h^{ij} \to 0$, it is clear that Equation 5.57 is not the necessary condition for guaranteeing the convergence of solutions, but its imposition will improve element behavior effectively. If we drop the element summation symbol for the convenience of implementation, Equation 5.52 becomes

$$\int_{\partial V^e} \sigma_h^{ij} n_j u_i^\lambda dS = 0 \tag{5.58}$$

which can be called the optimizing condition of hybrid elements. By this condition, the optimizing formula of a hybrid element can be set up.

When the stresses are determined by using Equation 5.58 and designated as σ_h^{*ij}, the optimizing element stress pattern is

$$\sigma^{*ij} = \sigma_c^{*ij} + \sigma_h^{*ij}$$

If the incompatible displacements u_1^λ do not satisfy the patch test (Equation 5.56), the multiplier method must be used to guarantee the tenability of Equation 5.51. The energy function of the element is

$$\Pi_{mR}^e = \Pi_R^e - \int_{\partial V^e} \sigma_c^{ij} n_j u_i^\lambda dS$$

$$= \int_{\partial V^e} \left[-\frac{1}{2} \sigma^{*ij} \sigma^{*kl} S_{ijkl} + \sigma^{*ij} \varepsilon_{ij}^q + \sigma_h^{*ij} \varepsilon_{ij}^\lambda - \sigma_{c\ ;i}^{ij} u_j^\lambda \right] dV \tag{5.59}$$

in which ε_{ij}^q, ε_{ij}^λ are the strains corresponding to u_1^q and u_1^λ, respectively. According to the selecting principle of incompatible displacement [20], u_1^λ should be the higher-order part of the element displacement, so the third term in Equation 5.59 is smaller. If it is neglected in the finite element formulation, the patch test condition (Equation 5.56) can still be satisfied, because the higher-order part of stresses σ_h^{ij} will vanish for any constant stress state. If in addition we have

$$\int_{V^e} \sigma_{c;i}^{ij} u_j^\lambda dV = 0 \tag{5.60}$$

then Equation 5.59 can be further simplified as

$$\Pi^e_{mRs} = \int_{V^e} \left[-\frac{1}{2}\sigma^{*ij}\sigma^{*kl}S_{ijkl} + \sigma^{*ij}\varepsilon^q_{ij} \right] dV \tag{5.61}$$

Because the incompatible displacements u_1^λ are not included in Equation 5.61, the procedure for the formulation of element stiffness is simple, as will be illustrated by examples in Section 5.5.5.

5.5.3 The Discrete Formulation of the Energy Functional

In this section, we shall discuss the discrete expression of the energy functional in curvilinear coordinates. After denoting the Christoffel notation Γ^i_{jk}, the product of the stress tensor and displacement gradient tensor is

$$\sigma^{ij}u_{j;i} = \sigma^{ij}(u_{j,i} - u_k\Gamma^k_{ij}) = \bar{\sigma}^T(\bar{\mathbf{D}}_\varepsilon\bar{\mathbf{u}}) = \sigma^{ij}\varepsilon_{ij} \tag{5.62}$$

in which $\bar{D}_c$ is defined as the strain differential operator

$$\bar{\mathbf{D}}_\varepsilon = \bar{\mathbf{D}}_{\varepsilon 0} + \bar{\mathbf{D}}_{\varepsilon 1} \tag{5.63}$$

The zero-order and first-order operator matrices can be written as

$$\bar{\mathbf{D}}_{\varepsilon 0} = \begin{bmatrix} \Gamma_1 & \Gamma_2 & \Gamma_3 \end{bmatrix}$$

$$\Gamma_i = \begin{bmatrix} \Gamma^i_{11} & \Gamma^i_{22} & \Gamma^i_{33} & \Gamma^i_{12} & \Gamma^i_{21} & \Gamma^i_{23} & \Gamma^i_{32} & \Gamma^i_{31} & \Gamma^i_{13} \end{bmatrix}^T \tag{5.64}$$

$$\bar{\mathbf{D}}_{\varepsilon 1} = \begin{bmatrix} \frac{\partial}{\partial x_1} & 0 & 0 & \frac{\partial}{\partial x_2} & 0 & \frac{\partial}{\partial x_3} \\ 0 & \frac{\partial}{\partial x_2} & 0 & \frac{\partial}{\partial x_1} & \frac{\partial}{\partial x_3} & 0 \\ 0 & 0 & \frac{\partial}{\partial x_3} & 0 & \frac{\partial}{\partial x_2} & \frac{\partial}{\partial x_1} \end{bmatrix} \tag{5.65}$$

The product of the stress gradient tensor and displacement is

$$\sigma^{ij}{}_{;i}u_j = (\sigma^{ji}{}_{;i} + \sigma^{lj}\Gamma^i_{il} + \sigma^{il}\Gamma^j_{il})u_j = (\bar{\mathbf{D}}_\sigma\bar{\sigma})^T\bar{\mathbf{u}} \tag{5.66}$$

in which the operator matrix $\overline{D_\sigma}$ is also the summation of two parts

$$\bar{\mathbf{D}}_\sigma = \bar{\mathbf{D}}_{\sigma 0} + \bar{\mathbf{D}}_{\sigma 1} \tag{5.67}$$

where

$$\bar{\mathbf{D}}_{\sigma 1} = \bar{\mathbf{D}}_{\varepsilon 1}{}^T \tag{5.68}$$

$$\bar{\mathbf{D}}_{\sigma 0} = -\bar{\mathbf{D}}_{\varepsilon 0}{}^T + \bar{\mathbf{D}}'_{\sigma 0}{}^T \tag{5.69}$$

and

$$\bar{\mathbf{D}}'_{\sigma 0}{}^T = \begin{bmatrix} \Gamma_1 & \mathbf{0} & \mathbf{0} \\ \mathbf{0} & \Gamma_2 & \mathbf{0} \\ \mathbf{0} & \mathbf{0} & \Gamma_3 \\ \Gamma_2 & \Gamma_1 & \mathbf{0} \\ \mathbf{0} & \Gamma_3 & \Gamma_2 \\ \Gamma_3 & \mathbf{0} & \Gamma_1 \end{bmatrix}$$

$$\Gamma_1 = \Gamma^j_{ji} = \Gamma^1_{1i} + \Gamma^2_{2i} + \Gamma^3_{3i} \tag{5.69a}$$

In the case of Cartesian coordinates, all the Christoffel notations vanish, and then $\bar{D}_{co} = 0$ and $\bar{D}_{\sigma 0}$; the familiar relation can be obtained:

$$\bar{\mathbf{D}}_\sigma{}^T = \bar{\mathbf{D}}_\varepsilon \tag{5.70}$$

Using the relation (Equation 5.63 through Equation 5.69), the functional (Equation 5.59) can be rewritten as

$$\Pi^e_{mR} = \int_{V^e} \left[-\frac{1}{2} \bar{\sigma}^{*T} \bar{\mathbf{S}} \bar{\sigma}^* + \bar{\sigma}^{*T} (\bar{D}_\varepsilon \bar{u}_{\mathrm{q}}) + \bar{\sigma}^{*T}_{\mathrm{h}} (\bar{D}_\varepsilon \bar{u}_\lambda) - (\bar{D}_{\sigma 0} \bar{\sigma}_c)^T \bar{u}_\lambda \right] dV \tag{5.71}$$

This is the general discrete form of the energy functional (Equation 5.59) in the curvilinear coordinate system, where

$$\bar{\mathbf{S}} = \begin{bmatrix} \bar{\mathbf{S}}_1 & \mathbf{0} \\ \mathbf{0} & \bar{\mathbf{S}}_2 \end{bmatrix}$$

$$\bar{\mathbf{S}}_1 = \frac{1}{E} \begin{bmatrix} 1 & -\mu & 0 \\ -\mu & 1 & 0 \\ 0 & 0 & 2(1+\mu) \end{bmatrix}$$

$$\bar{\mathbf{S}}_2 = \frac{2(1+\mu)}{E}\begin{bmatrix} 1 & 0 \\ 0 & 1 \end{bmatrix} \tag{5.72}$$

and h is the thickness of the shell.

5.5.4 The Optimizing Formulation of Element Stresses

For a general shell, by defining the x_1 and x_2 coordinates along the lines of principal curvature in the midsurface of a shell, x_3, along the normal line of the midsurface, the orthogonal curvilinear coordinate system is set up (see Figure 5.12).

The principal curvature and *Lame'* coefficients of a point on the midsurface of a shell are denoted as K_1, K_2, and A, B. Along the normal line of the midsurface at this point, the *Lame'* coefficients are

$$H_1 = A(1+k_1x_3)$$

$$H_2 = B(1+k_2x_3)$$

$$H_3 = 1 \tag{5.73}$$

The coordinates, curvatures, thickness, and *Lame'* coefficients of any point on the midsurface of a shell can be evaluated by interpolating from their

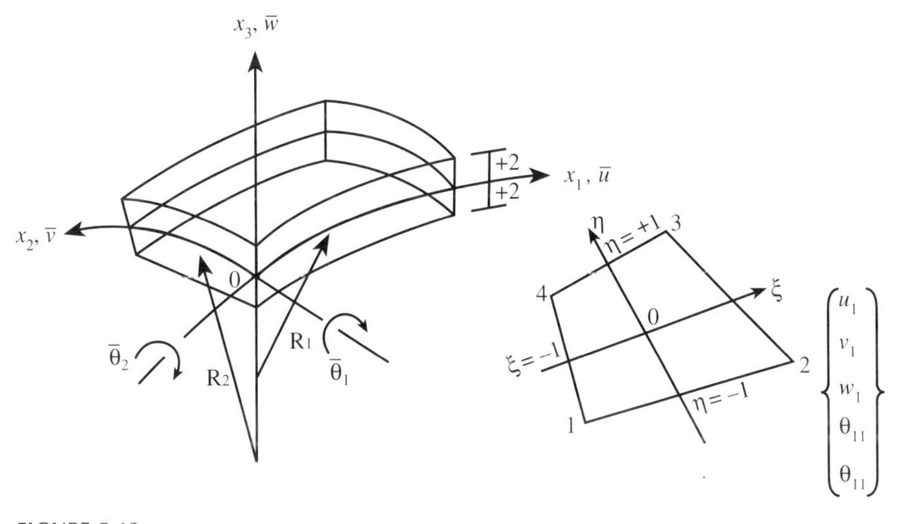

FIGURE 5.12
Curvilinear coordinates and element natural coordinates.

nodal value using the shape function. Most of the shells used in engineering can be exactly evaluated.

A quadrilateral element is shown in Figure 5.12, each node having five degrees of freedom. Using the independent displacement and rotation trial functions, the compatible displacement trials are represented in the form

$$\bar{u}_q = \begin{bmatrix} u \\ v \\ w \end{bmatrix}_q = \sum_{i=1}^{4} N_i(\xi,\eta) \begin{bmatrix} u_i - x_3\theta_{1i} \\ v_i - x_3\theta_{2i} \\ w_i \end{bmatrix} = \bar{\mathbf{N}}_q \bar{q} \tag{5.74}$$

where $\bar{q}$ is the nodal displacement vector of the midsurface

$$\bar{\mathbf{q}} = \begin{bmatrix} u_1 \; v_1 \; w_1 \; \theta_{11} \; \theta_{21} \; u_2 \; v_2 \; w_3 \; \theta_{12} \; \theta_{22} \; u_3 \; v_3 \; w_3 \; \theta_{13} \; \theta_{23} \; u_4 \; v_4 \; w_4 \; \theta_{14} \; \theta_{24} \end{bmatrix}^T$$

and the shape function matrix

$$\bar{\mathbf{N}}_\mathbf{q} = \begin{bmatrix} \bar{\mathbf{N}}_1 & \bar{\mathbf{N}}_2 & \bar{\mathbf{N}}_3 & \bar{\mathbf{N}}_4 \end{bmatrix}$$

$$\bar{\mathbf{N}}_i = \begin{bmatrix} N_i & 0 & 0 & -x_3N_i & 0 \\ 0 & N_i & 0 & 0 & -x_3N_i \\ 0 & 0 & N_i & 0 & 0 \end{bmatrix}$$

The bilinear interpolating function is $N_1(\xi,\eta) = (1+\xi_1\xi)(1+\eta_1\eta)/4$ used here.

The incompatible displacements are taken as

$$\bar{\mathbf{u}}_\lambda = \begin{bmatrix} u \\ v \\ w \end{bmatrix}_\lambda = \begin{bmatrix} \xi^2 & \eta^2 & 0 & 0 & 0 & 0 \\ 0 & 0 & \xi^2 & \eta^2 & 0 & 0 \\ 0 & 0 & 0 & 0 & \xi^2 & \eta^2 \end{bmatrix} \begin{bmatrix} \lambda_1 \\ \lambda_2 \\ \vdots \\ \lambda_6 \end{bmatrix} = \bar{\mathbf{N}}_\lambda \bar{\lambda} \tag{5.75}$$

The resultant element displacements of an element would be complete second-order trials.

When the functional (Equation 5.59) is used, the number of stress parameters should be

$$n_{\beta^*} \geq n_q + n_\lambda - r = 20 \tag{5.76}$$

where $n_q = \dim.(\bar{q})$, $n_\lambda = \dim.(\bar{\lambda})$, and r is the number of rigid motion. By considering the six constraint conditions, it can be determined that the parameter number of initial stresses $\bar{\sigma}$ should be 26 and the element stresses are defined by

$$
\bar{\sigma} = \begin{bmatrix} \sigma_{11} \\ \sigma_{22} \\ \sigma_{12} \\ \sigma_{23} \\ \sigma_{31} \end{bmatrix} = \begin{bmatrix} \beta_1 \\ \beta_2 \\ \beta_3 \\ \beta_4 \\ \beta_5 \end{bmatrix} +
$$

$$
\begin{bmatrix}
\xi & \eta & \zeta & \xi\eta & 0 & 0 & 0 & 0 & 0 & 0 & 0 & 0 & 0 & 0 & \xi\eta^2 & 0 & 0 & 0 & 0 & 0 \\
0 & 0 & 0 & 0 & \xi & \eta & \zeta & \xi\eta & 0 & 0 & 0 & 0 & 0 & 0 & 0 & \xi^2\eta & 0 & 0 & 0 & 0 \\
0 & 0 & 0 & 0 & 0 & 0 & 0 & 0 & \zeta & \xi\eta & \zeta(\xi+\eta) & 0 & 0 & 0 & 0 & 0 & \xi & \eta & 0 & 0 \\
0 & 0 & 0 & 0 & 0 & 0 & 0 & 0 & 0 & 0 & 0 & \zeta & \zeta^2 & 0 & 0 & 0 & 0 & 0 & \eta & 0 \\
0 & 0 & 0 & 0 & 0 & 0 & 0 & 0 & 0 & 0 & 0 & 0 & 0 & \zeta & 0 & 0 & 0 & 0 & 0 & \xi
\end{bmatrix}
\begin{bmatrix} \beta_6 \\ \beta_7 \\ \vdots \\ \beta_{20} \\ \cdots \\ \beta_{21} \\ \vdots \\ \beta_{26} \end{bmatrix}
$$

$$
= \beta_c + \begin{bmatrix} \bar{\phi}_{\mathrm{I}} & \bar{\phi}_{\mathrm{II}} \end{bmatrix} \begin{bmatrix} \bar{\beta}_{\mathrm{I}} \\ \bar{\beta}_{\mathrm{II}} \end{bmatrix} = \bar{\beta}_c + \bar{\phi}_h \bar{\beta}_h \tag{5.77}
$$

The stress components are uncoupled.

By using Gauss' theory of orthogonal curvilinear coordinates, the optimizing condition (Equation 5.58) becomes

$$
\begin{aligned}
\int_{\partial V^e} \sigma_h^{ij} n_i u_j^\lambda dS &= \int_{V^e} (\sigma_{h\ ;i}^{ij} u_j^\lambda + \sigma_h^{ij} u_{j;i}^\lambda) dV \\
&= \int_{-h/2}^{h/2} \int_{-1}^{1} \int_{-1}^{1} \left[(\bar{\boldsymbol{D}}_\sigma \bar{\sigma}_h)^T \bar{\boldsymbol{u}}_\lambda + \bar{\sigma}_h{}^T (\bar{\boldsymbol{D}}_\varepsilon \bar{\boldsymbol{u}}_\lambda) \right] H_1 H_2 |J| d\xi d\eta dx_3
\end{aligned} \tag{5.78}
$$

where $|J|$ is the determinant of the Jacobian matrix

$$
|J| = \det. \begin{bmatrix} x_{1,\xi} & x_{2,\xi} \\ x_{1,\eta} & x_{2,\eta} \end{bmatrix}
$$

With the natural coordinates $\zeta = \frac{2}{h}x_3$, and by using the trials (Equation 5.75 and Equation 5.77), the optimizing condition (Equation 5.58) becomes

$$\bar{\lambda}^T \begin{bmatrix} \bar{M}_{\mathrm{I}} & \bar{M}_{\mathrm{II}} \end{bmatrix} \begin{bmatrix} \bar{\beta}_{\mathrm{I}} \\ \bar{\beta}_{\mathrm{II}} \end{bmatrix} = 0 \tag{5.79}$$

where

$$\bar{M}_{\mathrm{I}} = \int_{-1}^{1}\int_{-1}^{1}\int_{-1}^{1} (\bar{N}_\lambda{}^T \bar{\Phi}_{\mathrm{I}}{}' + \bar{B}_\lambda{}^T \bar{\Phi}_{\mathrm{I}}) H_1 H_2 |J| \frac{h}{2} d\xi d\eta d\zeta$$

and

$$\bar{M}_{\mathrm{II}} = \int_{-1}^{1}\int_{-1}^{1}\int_{-1}^{1} (\bar{N}_\lambda{}^T \bar{\Phi}_{\mathrm{II}}{}' + \bar{B}_\lambda{}^T \bar{\Phi}_{\mathrm{II}}) H_1 H_2 |J| \frac{h}{2} d\xi d\eta d\zeta$$

in which $\overline{\Phi'_I} = \bar{D}_\sigma \bar{\Phi}_I, \overline{\Phi'_{IiI}} = \bar{D}_\sigma \bar{\Phi}_{II},$ whose explicit forms have been listed in the appendix.

Due to $\lambda \neq 0$, from Equation 5.79, we can obtain

$$\bar{M}_{\mathrm{I}}\bar{\beta}_{\mathrm{I}} + \bar{M}_{\mathrm{II}}\bar{\beta}_{\mathrm{II}} = 0 \tag{5.80}$$

Then we can eliminate the stress parameter $\bar{\beta}_{\mathrm{II}}$ and get the optimizing element stress pattern

$$\bar{\sigma}^* = \bar{\sigma}_c + \bar{\sigma}_h^* = \bar{\Phi}_c\bar{\beta}_c + (\bar{\Phi}_{\mathrm{I}} - \bar{\Phi}_{\mathrm{II}}\bar{M}_{\mathrm{II}}^{-1}\bar{M}_{\mathrm{I}})\bar{\beta}_{\mathrm{I}} = \begin{bmatrix} \bar{\Phi}_c & \bar{\Phi}_h^* \end{bmatrix} \begin{bmatrix} \bar{\beta}_c \\ \bar{\beta}_{\mathrm{II}} \end{bmatrix} = \bar{\Phi}^*\bar{\beta}^* \tag{5.81}$$

The inverse $\overline{M_{\mathrm{II}}^{-1}}$ always exists for an arbitrary configuration, so there is no difficulty in obtaining stress $\bar{\sigma}^*$ following the above procedure.

If the simplified function (Equation 5.61) is used, the parameter control condition becomes

$$n_{\beta^*} \geq n_q - r = 14 \tag{5.82}$$

The initial stresses of the element are chosen as

$$\bar{\sigma} = \begin{bmatrix} \sigma_{11} \\ \sigma_{22} \\ \sigma_{12} \\ \sigma_{23} \\ \sigma_{31} \end{bmatrix} = \begin{bmatrix} \beta_1 \\ \beta_2 \\ \beta_3 \\ \beta_4 \\ \beta_5 \end{bmatrix} + \begin{bmatrix} \eta & \zeta & \xi\eta & 0 & 0 & 0 & 0 & 0 & 0 \\ 0 & 0 & 0 & \xi & \zeta & \xi\eta & 0 & 0 & 0 \\ 0 & 0 & 0 & 0 & 0 & 0 & \zeta & \xi\eta & \zeta(\xi+\eta) \\ 0 & 0 & 0 & 0 & 0 & 0 & 0 & 0 & 0 \\ 0 & 0 & 0 & 0 & 0 & 0 & 0 & 0 & 0 \end{bmatrix}$$

$$\begin{matrix} \xi & 0 & 0 & 0 & 0 & 0 \\ 0 & \eta & 0 & 0 & 0 & 0 \\ 0 & 0 & \xi & \eta & 0 & 0 \\ 0 & 0 & 0 & 0 & \eta & 0 \\ 0 & 0 & 0 & 0 & 0 & \xi \end{matrix}\Bigg] \begin{bmatrix} \beta_1 \\ \vdots \\ \beta_{14} \\ \cdots \\ \beta_{15} \\ \vdots \\ \beta_{20} \end{bmatrix}$$

$$= \bar{\Phi}_c\bar{\beta}_c + \begin{bmatrix} \bar{\Phi}_{\mathrm{I}} & \bar{\Phi}_{\mathrm{II}} \end{bmatrix} \begin{bmatrix} \bar{\beta}_{\mathrm{I}} \\ \bar{\beta}_{\mathrm{II}} \end{bmatrix} = \bar{\Phi}_c\bar{\beta}_c + \bar{\Phi}_h\bar{\beta}_h \tag{5.83}$$

In this case, stress components are complete in linearity. Under a similar procedure, the $\bar{\sigma}^*$ can be obtained. However, in this case, the condition (Equation 5.60),

$$\int_{V^e} (\bar{\boldsymbol{D}}_{\sigma 0}\bar{\sigma}_c)^T \bar{\boldsymbol{u}}_\lambda dV = 0 \tag{5.60a}$$

must be satisfied. This is difficult in the general shell configuration. Fortunately, for most of the shells used in engineering, the curvatures and *Lame'* coefficients are constants, so Equation 5.60a can be satisfied by defining the incompatible displacement trial as

$$\bar{\boldsymbol{N}}_\lambda = \begin{bmatrix} \xi^2 - \frac{1}{3} & \eta^2 - \frac{1}{3} & 0 & 0 & 0 & 0 \\ 0 & 0 & \xi^2 - \frac{1}{3} & \eta^2 - \frac{1}{3} & 0 & 0 \\ 0 & 0 & 0 & 0 & \xi^2 - \frac{1}{3} & \eta^2 - \frac{1}{3} \end{bmatrix} \tag{5.75a}$$

Finally, using

$$\bar{\sigma}^* = \bar{\Phi}^*\bar{\beta}^*$$

and

$$\bar{\varepsilon}_q = \bar{\boldsymbol{B}}_q\bar{\boldsymbol{q}},\ \bar{\varepsilon}_\lambda = \bar{\boldsymbol{B}}_\lambda\bar{\lambda}$$

we have

$$\Pi^e_{mR} = -\frac{1}{2}\bar{\beta}^{*T}\bar{H}\bar{\beta}^* + \bar{\beta}^{*T}\bar{G}_q\bar{q} + \bar{\beta}_h^{*T}\bar{G}_{\lambda h}\bar{\lambda} - \bar{\beta}_c^T\bar{G}_c^T\bar{\lambda} \tag{5.84}$$

where

$$\bar{H} = \int_{V^e} \bar{\Phi}^{*T}\bar{S}\bar{\Phi}^* dV$$

$$\bar{G}_q = \int_{V^e} \bar{\Phi}^{*T}\bar{B}_q dV$$

$$\bar{G}_{\lambda h} = \int_{V^e} \bar{\Phi}^{*T}\bar{B}_{\lambda} dV$$

$$\bar{G}_{\lambda c} = \int_{V^e} -(\bar{D}_{\varepsilon 0}\bar{\Phi}_c)^T \bar{N}_{\lambda} dV$$

$$\int_{V^e}(\ldots)dV = \int_1^1\int_{-1}^1\int_{-1}^1 (\ldots)H_1H_2|J|\frac{h}{2}d\xi d\eta d\zeta \tag{5.85}$$

The Gauss quadrature is used and the explicit formulation of $\bar{B}_q$, $\bar{B}_\lambda$ is listed in the appendix.

The element formulation set up by the stationary condition is in the form

$$\begin{bmatrix} -\bar{H} & \bar{G}_\lambda & \bar{G}_q \\ \bar{G}_\lambda{}^T & 0 & 0 \\ \bar{G}_q{}^T & 0 & 0 \end{bmatrix}\begin{bmatrix} \bar{\beta}^* \\ \bar{\lambda} \\ \bar{q} \end{bmatrix} = \begin{bmatrix} 0 \\ 0 \\ \bar{r}_q \end{bmatrix} \tag{5.86}$$

where

$$\bar{G}_\lambda{}^T = \bar{G}_{\lambda c}{}^T + \bar{G}_{\lambda h}{}^T$$

$\bar{r}_q =$ the constant force vector

By eliminating the parameters $\bar{\beta}^*$ and $\bar{\lambda}$ from Equation 5.86, we have

$$\begin{bmatrix} \bar{U}_1 & * & * \\ 0 & \bar{U}_2 & * \\ 0 & 0 & \bar{K}_{qq} \end{bmatrix}\begin{bmatrix} \bar{\beta}^* \\ \bar{\lambda} \\ \bar{q} \end{bmatrix} = \begin{bmatrix} 0 \\ 0 \\ \bar{r}_q \end{bmatrix} \tag{5.87}$$

Here $\bar{U}_i$ is the upper-triangular matrix, and $\bar{K}_{qq}$ is the element stiffness matrix corresponding to the nodal degree of freedom $\bar{q}$. When $\bar{q}$ is obtained, we can calculate the parameter of incompatible displacement $\bar{\lambda}$ and stresses $\bar{\beta}^*$ backward from Equation 5.87, and then the stresses of element $\bar{\sigma}^*$ can be evaluated.

By using the elimination method mentioned above, the CPU time required for a computer to calculate the element stiffness matrix is only 60% of the standard matrix calculation [21].

5.5.5 Numerical Examples and Results

Five example solutions are presented in this section to test the numerical performance of the element presented in this chapter.

5.5.5.1 Thick and Thin Plates

A series of square plates with varied thicknesses, under a variety of boundary conditions and loading were tested to illustrate the uniformity in accuracy between thick and thin plate configurations in addition to the convergence of this element (see Table 5.4).

The presented element can be used to various plate thicknesses, and the convergence and accuracy are satisfactory, but the cost in terms of degree of freedom is less than that of the assumed displacement method listed in Table 5.3.

5.5.5.2 Locking Test

In order to show that the present formulation is free of shear locking, a simply supported square plate under loading was tested. The test covered the side length to the thickness ratios (L/h) from thick ($L/h = 10$) to very thin ($L/h = 10^6$). It should be noted that due to symmetry, only one quarter of the plate was treated.

The results of these tests are illustrated in Figure 5.13, where the percentage disagreement between the computed solution and the exact thin plate solution is plotted against the logarithm of the ratio (L/h). The large disagreement for ($L/h = 10^6$) is due to thin plate theory being inapplicable in this range of aspect ratios. From this figure it is evident that the element does not suffer from shear locking and possesses good convergent characteristics.

5.5.5.3 Thin Rhombic Plates

As a final example of the application of this element to plates, a strongly skewed plate subjected to a uniform normal load was selected. The chosen geometry is shown in Figure 5.14 with two types of boundary conditions. The quantities of comparison are the normal deflection and the moment results. It is important to understand the nature of this problem. There is singularity in the moments at the obtuse vertex of the plate, and for this particular geometry the singularity varies as $\gamma^{-4/5}$ [17]. The strength of the singularity depends on the magnitude of the enclosed angle and on the boundary conditions applied to the two edges that intersect to form the obtuse angle. For a given obtuse angle the singularity is most severe when the two edges are both clamped [17]. The case of one edge simply supported

TABLE 5.3

The Deflections at the Center of a Square Plate (simply supported under uniform loading q_0; $r_q = w_{max}Eh^3/q_0L^4$; $\mu = 0.3$)

Span-Thickness Ratio L/h	Thin Plate Solution	Reissner Plate Solution	Present Element 14β	Pryor et al. [23]	Rao et al. [24]
100	0.04437	0.04439	0.04448	0.04423	—
20	0.04437	0.04486	0.04496	0.04469	0.04483
10	0.04437	0.04632	0.04643	0.04612	0.04627
20/30	0.04437	0.05217	0.04889	0.04852	0.04866
5	0.04437	0.05217	0.05233	0.05186	0.05201
4	0.04437	0.05656	0.05675	0.05617	0.05631
Mesh of ¼ plate			4 × 4	6 × 6	2 × 2
DOF without Boundary Constraint			n = 75	n = 245	n = 108

TABLE 5.4

The Convergent Characteristics of the Deflection at the Center of a Square Plate ($L/h = 100$; $\mu = 0.3$)

	Simply Supported		Clamped	
Mesh of ¼ Plate	$w_{max}Eh^3/q_0L^4$	$w_{max}D/PL^2$	$w_{max}Eh^3/q_0L^4$	$w_{max}D/PL^2$
2 × 2	0.044827	0.012079	0.014104	0.005174
4 × 4	0.044483	0.011700	0.013885	0.005543
6 × 6	0.044452	0.011657	0.013859	0.005599
8 × 8	0.044405	0.011645	0.013851	0.005618
Exact	0.044437	0.01160	0.01376	0.0056

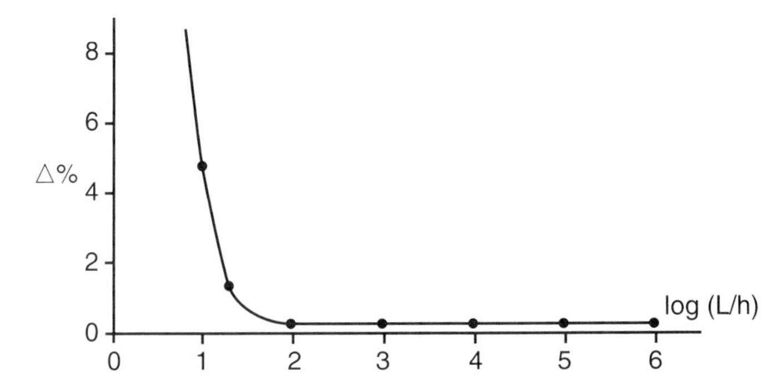

FIGURE 5.13
Locking test (uniform load on square plate, ¼ mesh 4 × 4).

and one edge clamped lies between these two limiting behaviors. In the current example, there are two obtuse corners, each of which will give rise to a singularity. These singularities make treating this problem by numerical techniques very difficult.

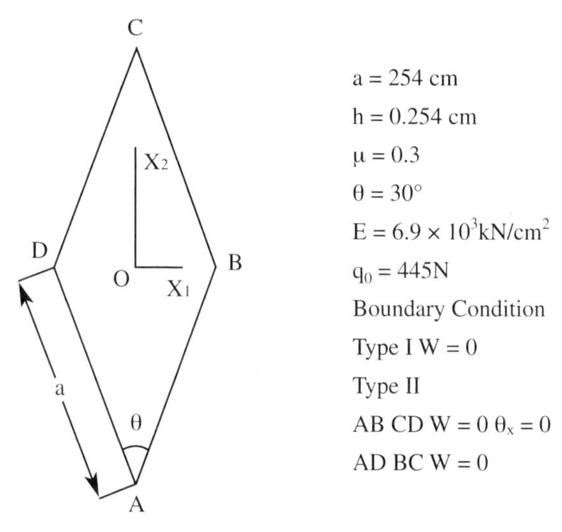

FIGURE 5.14
Rhombic plate test: geometry.

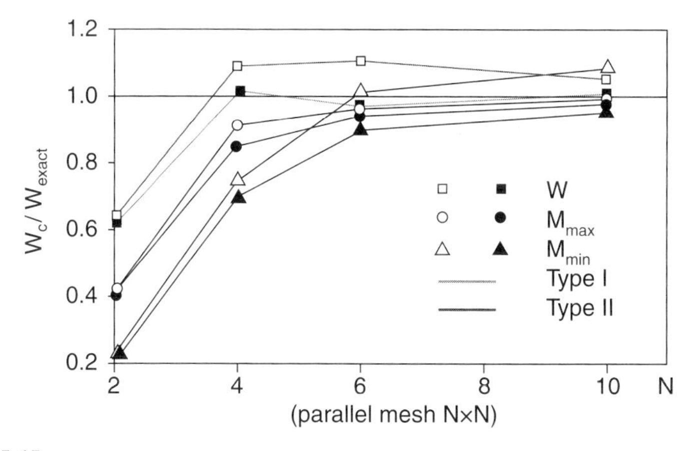

FIGURE 5.15
Rhombic plate test: convergent results.

The results of employing a regular mesh of a finite element whose boundaries are parallel to the edges of the plate are plotted in Figure 5.15. The element converges monotonically to the analytic results of Sander et al. 2[22].

5.5.5.4 *Pinched Cylinder Problem*

A benchmark problem widely used to study the convergent behavior of shell elements is a pinched cylinder with either open ends or rigid diaphragm ends shown in Figure 5.16. Due to symmetry, only one eighth of the cylinder is modeled.

Table 5.5 shows the results of the cylinder shell with free-free ends and different thicknesses: the thick shell with h = 0.239 cm and the thin shell with $h = 0.0349$ cm. The number of DOFs includes the constrained ones. The convergent characteristics of this element are illustrated for the vertical deflection under the load. For both the thick and thin shells, the results exhibit acceptable behavior.

Results are presented in Table 5.6 for the cylinder shell with shear diaphragm ends, and comparison is made between using the full functional (Equation 5.59), named the 20β element, and the simplified functional (Equation 5.61), the 14β element. This behavior shows that the higher-order terms in Equation 5.59 can be reasonably neglected.

5.5.5.5 *Scordelis Cylindrical Shell*

One final example, the so-called scordelis cylindrical shell, is the most typical test used to illustrate a new element performance in dealing with a thin,

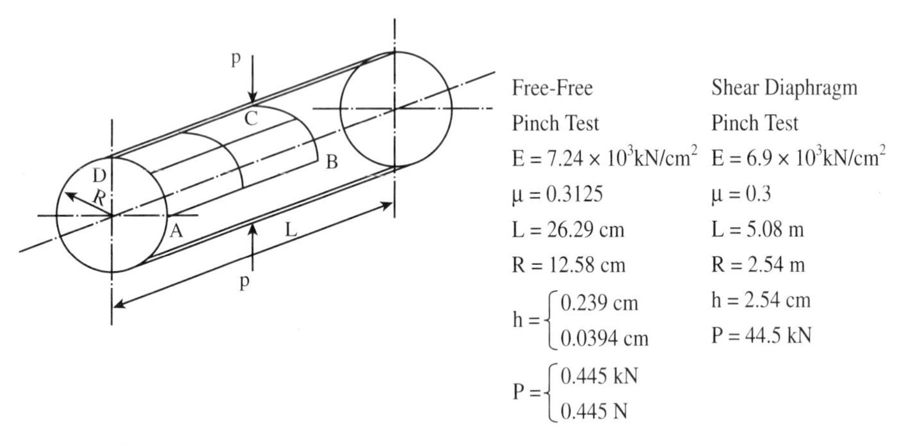

FIGURE 5.16
Pinch test geometry.

TABLE 5.5

Free-Free Pinch Test Results (Δ% is the Percentage of Disagreement between the Computed Solution and the Accepted Solution of W = –0.11390) [17]

		h = 0.0394 cm, P = 0.445 N		**h = 0.239 cm, P = 0.445 kN**	
Mesh	**DOF**	**Displacements under Load**	**Δ%**	**Displacements under Load**	**Δ%**
2 × 2	45	–0.01382	–43.3	–0.06289	–44.8
2 × 4	75	–0.02120	–13.1	–0.09420	–17.3
4 × 4	125	–0.02125	–12.9	–0.09853	–13.5
2 × 8	135	–0.02353	–3.5	–0.10943	–3.9
2 × 12	195	–0.02401	–1.6	–0.11258	–1.2
6 × 6	245	–0.02302	–5.6	–0.10691	–6.1
8 × 8	405	–0.02370	–2.8	–0.1100	–3.4
5 × 20	630	–0.02439	0.0	–0.1131	–0.7
Exact	—	0.02439[4]	—	0.1139 [4]	—

TABLE 5.6

Diaphragm Pinch Test Results, Displacement under Load (Exact Solution W = 164.24 P/Eh [17])

Mesh	DOF	14β Element		20β Element	
		$W_0 \times 10^{-1}$ P/Eh	Δ%	$W_0 \times 10^{-1}$ P/Eh	Δ%
2 × 2	45	−0.1755	−89.3	−0.1880	−88.6
4 × 4	125	−1.1677	−28.9	−1.1638	−28.3
6 × 6	245	−1.4577	−11.2	−1.4622	−11.0
8 × 8	405	−1.5640	−4.8	−1.5683	−4.5
10 × 10	605	−1.6083	−2.1	−1.6110	−1.9

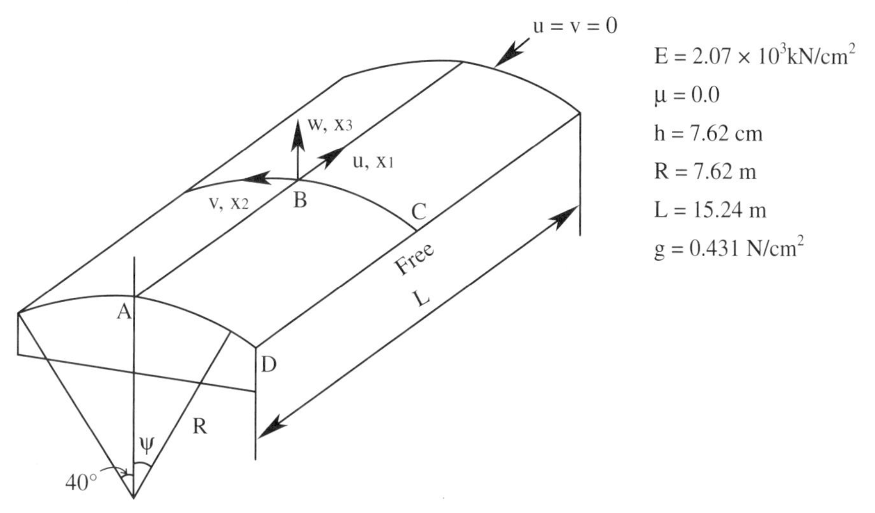

FIGURE 5.17
Cylindrical shell geometry.

deep, open shell section. The problem consists of an open circular cylindrical segment that is supported at its two ends by rigid shear diaphragms and has its longitudinal edges free. The shell is subjected to gravitational self-weight loading, shown in Figure 5.17. Taking advantage of the double symmetry, only a quarter of the shell is discretized.

The comparison figure of convergence results of a cylindrical shell, drawn from Reference 23, presents the deflection of the middle point of the free edges of the shell for various finite elements extracted from the literature and for the present element. It can be seen that the present 20-DOF 14β element converges extremely rapidly outwards from the analytical solution. The results for the vertical displacement around the midspan meridian and the axial displacement are shown in Figure 5.18 and Figure 5.19. The distribution of stresses and moments at the midspan are presented in Figure 5.20 and Figure 5.21. The agreement with the exact solution is very good, especially for the stresses. Comparison between the 20β

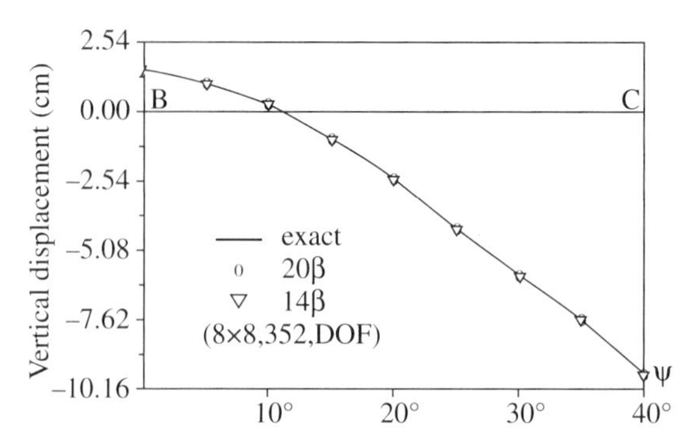

FIGURE 5.18
Cylindrical shell: displacement at the midspan.

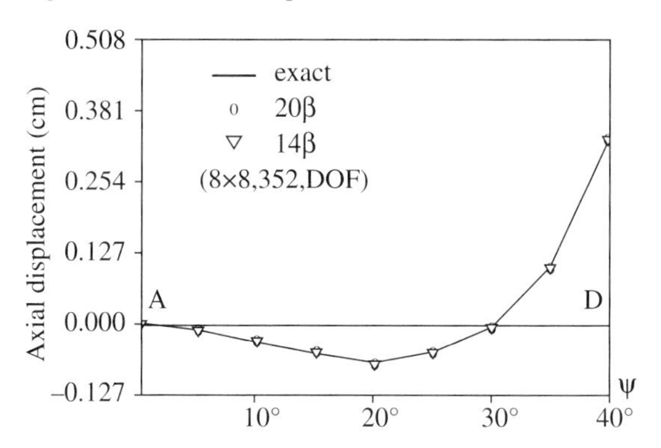

FIGURE 5.19
Cylindrical shell: displacement at a diaphragm.

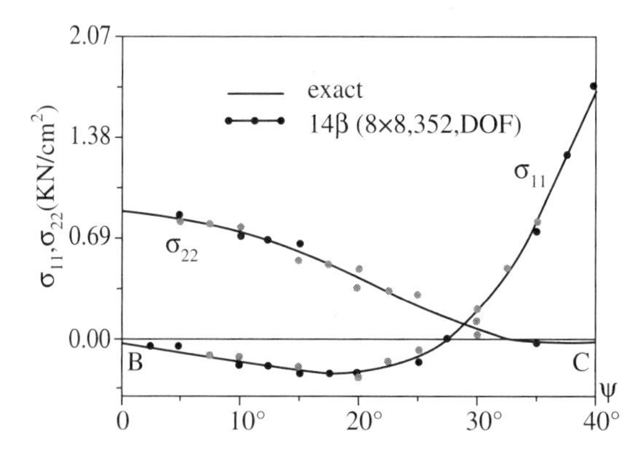

FIGURE 5.20
Cylindrical shell: surface stresses at the midspan.

and 14β elements using the functionals in Equation 5.59 and Equation 5.61, respectively, again indicates that the disagreement is very small.

In this section, the three samples of Gauss quadrature are used for each of the three natural coordinates ξ, η, and ζ. The results are generally given by the 14β element unless otherwise specified.

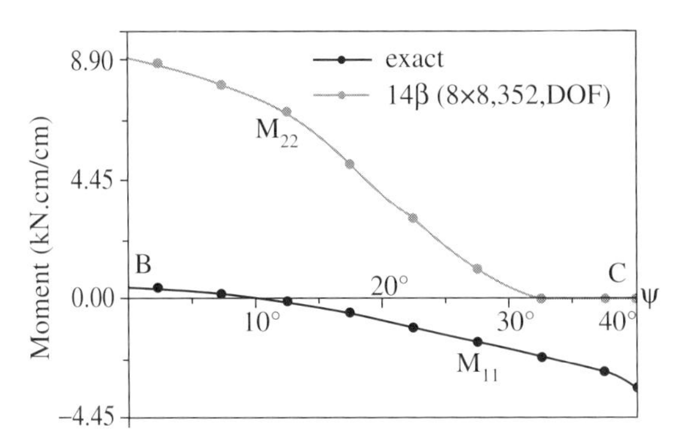

FIGURE 5.21
Cylindrical shell: moment results at the midspan.

5.6 Appendix

5.6.1. $\overline{\Phi'_I}$ and $\overline{\Phi'_{II}}$

Assume the general form of the uncoupled stress trial to be

$$\bar{\sigma} = \bar{\Phi}\bar{\beta} = \begin{bmatrix} \bar{\Phi}_1 & 0 & 0 & 0 & 0 \\ 0 & \bar{\Phi}_2 & 0 & 0 & 0 \\ 0 & 0 & \bar{\Phi}_3 & 0 & 0 \\ 0 & 0 & 0 & \bar{\Phi}_4 & 0 \\ 0 & 0 & 0 & 0 & \bar{\Phi}_5 \end{bmatrix} \bar{\beta} \tag{5.88}$$

Then we have

$$\bar{\Phi}' = \begin{bmatrix} \frac{1}{H_1}\bar{\Phi}_{1,1} + \frac{B_{,1}}{H_2 A}\bar{\Phi}_1 & -\frac{B_{,1}}{H_2 A}\bar{\Phi}_2 & \frac{1}{H_2}\bar{\Phi}_{3,2} + \frac{2A_{,2}}{H_1 B}\bar{\Phi}_3 & 0 & \left(\frac{Bk_2}{H_2} + \frac{2Ak_1}{H_1}\right)\bar{\Phi}_5 \\ -\frac{A_{,2}}{H_1 B}\bar{\Phi}_1 & \frac{1}{H_2}\bar{\Phi}_{2,2} + \frac{A_{,2}}{H_1 B}\bar{\Phi}_2 & \frac{1}{H_1}\bar{\Phi}_{3,1} + \frac{2B_{,1}}{H_2 A}\bar{\Phi}_3 & \left(\frac{Ak_1}{H_1} + \frac{2Bk_2}{H_2}\right)\bar{\Phi}_4 & 0 \\ -\frac{Ak_1}{H_1}\bar{\Phi}_1 & -\frac{Bk_2}{H_2}\bar{\Phi}_2 & 0 & \frac{1}{H_1}\bar{\Phi}_{4,2} + \frac{A_{,2}}{H_1 B}\bar{\Phi}_4 & \frac{1}{H_1}\bar{\Phi}_{5,1} + \frac{B_{,1}}{H_2 A}\bar{\Phi}_5 \end{bmatrix} \tag{5.89}$$

Corresponding to $\overline{\Phi_I'}$ and $\overline{\Phi_{II}'}$, the matrices can be split from Equation 5.89.

5.6.2. The Geometry Matrices $\overline{B_q}$ and $\overline{B_\lambda}$

Corresponding to the compatible displacement, the geometry matrices are in the form

$$\bar{B}_q = \begin{bmatrix} \bar{B}_{q1} & \bar{B}_{q2} & \bar{B}_{q3} & \bar{B}_{q4} \end{bmatrix} \tag{5.90}$$

where

$$\bar{B}_{qi} = \left[\begin{matrix} \frac{1}{H_1}N_{i,1} & \frac{A_{,2}}{H_1B}N_i & \frac{Ak_1}{H_1}N_i \\ \frac{B_{,1}}{H_2A}N_i & \frac{1}{H_2}N_{i,2} & \frac{Bk_2}{H_2}N_i \\ \frac{1}{H_2}N_{i,2}-\frac{A_{,2}}{H_1B}N_i & \frac{1}{H_2}N_{i,1}-\frac{B_{,1}}{H_2A}N_i & 0 \\ 0 & -\frac{Bk_2}{H_2}N_i & \frac{1}{H_2}N_{i,2} \\ -\frac{Ak_1}{H_1}N_i & 0 & \frac{1}{H_1}N_{i,1} \end{matrix}\right.$$

$$\left.\begin{matrix} -\frac{1}{H_1}x_3N_{i,1} & -\frac{A_{,2}}{H_1B}x_3N_i \\ -\frac{B_{,1}}{H_2A}x_3N_i & -\frac{1}{H_2}x_3N_i \\ \left(-\frac{1}{H_2}N_{i,2}+\frac{A_{,2}}{H_1B}N_i\right)x_3 & \left(-\frac{1}{H_1}N_{i,1}+\frac{B_{,1}}{H_2A}N_i\right)x_3 \\ 0 & \left(\frac{Bk_2}{H_2}x_3-1\right)N_i \\ \left(\frac{Ak_1}{H_1}x_3-1\right)N_i & 0 \end{matrix}\right] \tag{5.91}$$

The incompatible displacements can be generally expressed as

$$\bar{u}_\lambda = \begin{bmatrix} \bar{N}_\lambda & 0 & 0 \\ 0 & \bar{N}_\lambda & 0 \\ 0 & 0 & \bar{N}_\lambda \end{bmatrix} \tag{5.92}$$

In this chapter, only the translational displacements are introduced in the incompatible displacements, so the geometry matrix $\overline{B_\lambda}$ can be obtained simply by taking the last two columns off Equation 5.92 and then replacing N_1 for N_λ in the remaining one.

References

1. Pian, T.H.H. and Sumihara, K., Rational approach for assumed stress finite elements, *Int. J. Numer. Methods Eng.*, 20, 1685, 1984.
2. Wilson, E.L. et al., Incompatible displacement models, in *Numerical Computer Methods in Structural Mechanics*, Fenves, S.J. et al., Eds., Academic Press, New York, 1973.
3. Pian, T.H.H., Derivation of element stiffness matrices by assumed stress distributions, *AIAA J.*, 2, 1333, 1964.
4. Jiao, Z.P., An approach for eliminating H matrix of hybrid stress element, *Comput. Struct. Mech. Appl.*, 8, 214, 1991.
5. Wu, C.C. and Cheung, Y.K., On optimization approaches of hybrid stress elements, *Finite Elements in Analysis & Design* 21, 111–128, 1995.
6. Simo, J.C. and Rifai, M.S., A class of mixed assumed strain methods and the method of incompatible modes, *Int. J. Numer. Methods Eng.*, 29, 1559, 1990.
7. Weissman, S.L. and Taylor, R.L., A unified approach to mixed finite element methods: application to in-plane problems, *Comp. Methods Appl. Mech. Eng.*, 98, 127, 1992.
8. Yuan, K.Y., Huang, Y.S., and Pian, T.H.H., New strategy for assumed stresses for 4-node hybrid stress membrane element, *Int. J. Numer. Methods Eng.*, 36, 1747, 1993.
9. Bouzeghoub, M.C. and Gunn, M.J., On stress interpolation for hybrid models, *Int. J. Numer. Methods Eng.*, 37, 895, 1994.
10. MacNeal, R.H. and Harder, R.H., A proposed standard set of problems to test finite element accuracy, *Finite Elements Anal. Des.*, 1, 3, 1985.
11. Pian, T.H.H. and Chen, D.P., On the suppression of zero energy deformation modes, *Int. J. Numer. Methods Eng.*, 19, 1741, 1983.
12. Loikkanen, M.J. and Irons, B.M., A 8-node brick finite element, *Int. J. Numer. Methods Eng.*, 20, 523, 1984.
13. Punch, E.F. and Atluri, S.N., Application of isoparametric three dimensional hybrid stress finite elements with least-order stress fields, *Comput. Struct.*, 19, 409, 1984.
14. Zinkiwicz, O.C., Tayor, R.L., and Too, J.M., Reduced integration technique in general analysis of plates and shells, *Int. J. Numer. Methods Eng.*, 3, 275, 1971.
15. Wemper, G.A., Oden, J.T., and Kross, D.K., Finite element analysis of thin shells, *J. Eng. Mech. Div., Proc. ASCE*, 94, 1273, 1968.
16. Robison, J., LORA: an accurate four node stress plate bending element, *Int. J. Numer. Methods Eng.*, 14, 296, 1979.
17. Heppler, G.R. and Hansen, J.S., A Mindlin element analysis of thin shells, *J. Eng. Mech. Div., Proc. ASCE*, 54, 21, 1986.
18. Wu, C.C., Di, S.L., and Huang, M.G., The optimizing design of hybrid element, *KEXUE TONGBAO*, 15, 1986.

19. Wu, C.C., Di, S.L., and Pian, T.H.H., Optimizing formulation of axisymmetry hybrid stress element, *ACTA Aeronaut. Astronaut. Sin.*, 8, 9, 1987.
20. Wu, C.C., The Incompatible Principle of Discrete System and Multivariable Method: The Optimizing Theory and Practice of Hybrid Element, Ph.D. thesis, University of Science and Technology of China.
21. Irons, B., A letter to the editor, *Int. J. Numer. Methods Eng.*, 20, 780, 1984.
22. Sander, G. and Idelsohn, S., A family of conforming finite elements for deep shell analysis, *Int. J. Numer. Methods Eng.*, 18, 1982.
23. C.W. Pryor et al., Finite element bending analysis of Reissner plates, *J. Eng. Mech., ASCE*, 96, 967–983, 1970.
24. G.V. Rao et al., A high precision triangular plate bending element for the analysis of thick plates, *Nucl. Eng. Design*, 30, 408–412, 1974.

6

Numerical Stability: Zero Energy Mode Analysis

6.1 Introduction

A challenging problem in the hybrid and mixed element methods is numerical stability. An exact mathematical condition for the existence and uniqueness of the saddle point solution, that is, the Babuška-Brezzi condition, was presented many years ago [1, 2], but when the condition with an inf-sup version of the energy inequality is used for a discrete system, the analysis becomes very complex [3]. The condition does not provide engineers with an actual way or procedure [4] with which to inspect and remedy some instability problems that appear in the applications of finite elements.

However, the stability problem is the question about zero energy modes (ZEMs): how to find the modes and how to control them from the mechanical viewpoint [5, 6]. The method for analyzing ZEMs possesses a vivid physical concept and a mechanical background, and so it would be more convenient for it to be used directly.

We carry out our discussion on the basis of variational principles with two kinds of field variables. However, it is easy to extend to the three-field finite elements based on the Hu–Washizu principle [15]. Some mathematical analysis for the stability has been presented in [13–17].

6.2 Definition of ZEM

In order to distinguish the zero energy displacement mode from the rigid body modes that correspond to the zero strains of elements, the element displacement u, in which the rigid body degree of freedom (DOF) of the element has been eliminated in advance, will be introduced and marked by u_*. The sets of element trial function u_* and σ are denoted by $U_* = \{u_*\}$ and $\Sigma = \{\sigma\}$, respectively [3].

The so-called ZEM is a special nonzero mode included in element trial functions in such a manner that it has no capacity to do work with respect to its dual variable, and it does not contribute anything to the energy functional $\Pi(\sigma, u_*)$. Thus the related energy principle, or the virtual work equation, cannot provide ZEM with any constraints, so the amplitudes of ZEM must be indefinite. This is why the solutions of mixed-hybrid elements with ZEM are always unstable. Conversely, the numerical solutions are stable if the related discrete system has no ZEM.

Because of this, we have the following general definitions for ZEM.

Definition 1: A nonzero displacement mode $\overset{\circ}{\mathbf{u}} \in \mathbf{U}_*$ is said to be ZEM(u) if the functional increment

$$\Delta\Pi(\overset{\circ}{\mathbf{u}}) \equiv \Pi(\sigma,\ \mathbf{u}_* + \overset{\circ}{\mathbf{u}}) - \Pi(\sigma, \mathbf{u}_*) = 0 \quad \forall \sigma \in \Sigma \tag{6.1}$$

Definition 2: A nonzero stress mode $\overset{\circ}{\sigma} \in \Sigma$ is said to be ZEM(σ) if the functional increment

$$\Delta\Pi(\overset{\circ}{\sigma}) \equiv \Pi(\sigma + \overset{\circ}{\sigma},\ u_*) - \Pi(\sigma, u_*) = 0 \quad \forall\, u_* \in U_* \tag{6.2}$$

In other words, $\overset{\circ}{u}$ is the nonzero solution of $\Delta\Pi(u) = 0\ \forall\sigma \in \Sigma$, and $\overset{\circ}{\sigma}$ is the nonzero solution of $\Delta\Pi(\sigma) = 0\ \forall u_* \in \boldsymbol{U}_*$. For a typical mixed-hybrid element, if

$$\Delta\Pi(\mathbf{u}_*) = 0 \quad \forall u_* \in \Sigma \ \Rightarrow \mathbf{u}_* = 0 \tag{6.3a}$$

then the element has no ZEM(u), and if

$$\Delta\Pi(\sigma) = 0 \quad \forall u_* \in U_* \ \Rightarrow \sigma = 0 \tag{6.3b}$$

then the element has no ZEM(σ). Equation 6.3a and Equation 6.3b, which guarantee the absence of all ZEM, should be our element stability conditions.

6.3 Rank Conditions for Two-Field Mixed-Hybrid Elements

For a mixed-hybrid element with two kinds of variables, the element stress σ and the displacement u, the energy functional $\Pi = \Pi(\sigma, u)$ consists of two basic parts besides the terms related to applied loads and prescribed displacements. They are:

Part I: The positive definite quadratic form, that is, the element strain energy

$$\frac{1}{2}\langle \boldsymbol{Lu},\boldsymbol{u}\rangle = \frac{1}{2}\int_{V^e} (\boldsymbol{Du})^T \boldsymbol{C}(\boldsymbol{Du})dV \tag{6.4}$$

or the element complementary

$$\frac{1}{2}\langle \boldsymbol{S}\sigma,\sigma\rangle = \frac{1}{2}\int_{V^e} \sigma^T \boldsymbol{S}\sigma dV \tag{6.5}$$

where Du is the element strain, $S = C^{-1}$, the elastic compliance matrix, and $L = D^T CD$, the elliptic operator of elasticity.

Part II: The bilinear integral term $I(\sigma,u)$, which is one of the following energy integrals in Equation 6.6, or a certain combination of them:

$$\boldsymbol{I}(\sigma,\mathbf{u}) = \left\{\int_{V^e} \sigma^T(\mathbf{Du})dV, \int_{V^e} (\partial\sigma)^T \mathbf{u} dV, \int_{S^e} \mathbf{T}^T \tilde{\mathbf{u}} dS\right\} \tag{6.6}$$

$\boldsymbol{T} = \boldsymbol{T}(\sigma)$ is the surface traction on the element surface, and $\tilde{\boldsymbol{u}}$ the displacements on the element surface S^e.

The mixed-hybrid elements may be classified into two essential kinds in stability analysis, with the help of Equation 6.4 to Equation 6.6. First let us consider the mixed-hybrid element I, which is based on various modified complementary energy principles and Hellinger–Reissner principles, such as Fraeijs de Veubeke's equilibrium element [5], Herrmann's mixed element [8], and Pian's various hybrid stress elements [9, 10]. For this kind of element, without rigid body DOF, the element energy functional can be expressed as

$$\Pi_{\text{I}}(\sigma,\boldsymbol{u}_*) = \boldsymbol{I}(\sigma,\boldsymbol{u}_*) - \frac{1}{2}\langle \boldsymbol{S}\sigma,\sigma\rangle \tag{6.7}$$

Accordingly, we have a functional increment corresponding to a stress increment $\sigma' \in \Sigma$:

$$\begin{aligned}\Delta\Pi_{\text{I}}\left[\sigma'\right] &\equiv \Pi_{\text{I}}(\sigma+\sigma',\boldsymbol{u}_*) - \Pi(\sigma,\boldsymbol{u}_*)\\ &= \boldsymbol{I}(\sigma',\boldsymbol{u}_*) - \frac{1}{2} < \boldsymbol{S}(\sigma+\sigma'),(\sigma+\sigma') > + \frac{1}{2} < \boldsymbol{S}\sigma,\sigma >\end{aligned}$$

The last two terms are positive definite quadratic forms of the stresses, and thus

$$\Delta\Pi_{\text{I}}(\sigma') = 0 \quad \forall \boldsymbol{u}_* \in \mathrm{U}_* \quad \Rightarrow \sigma' = 0$$

that is, Equation 6.3b holds.

This means that the mixed-hybrid element I has no ZEM (σ) However,

$$\Delta\Pi_{\text{I}}(\boldsymbol{u}'_*) = \Pi_{\text{I}}(\sigma,\boldsymbol{u}_* + \boldsymbol{u}'_*) - \Pi_{\text{I}}(\sigma,\boldsymbol{u}_*) = I(\sigma,\boldsymbol{u}'_*)$$

or generally we can rewrite the expression as $\Delta\Pi_I[u_*] = I(\sigma, u_*)$. The element stability condition in Equation 6.3a is now reduced to

$$I(\sigma, u_*) = 0 \quad \forall \sigma \in \Sigma \quad \Rightarrow u_* = 0 \tag{6.8}$$

Let the element stress trial be $\sigma = \sigma(\beta), \beta$, which is the element node stress values or internal stress parameters, and the element displacement trial be $u = u(q), q$, which is the element node displacement or internal displacement parameters. In order to prevent the element rigid body motion, at least γ nodal displacements must be restricted in q, and the rest of the node displacements in q are denoted by q_*. Substitution of $\sigma = \sigma(\beta)$ and $u_* = u_*(q_*)$ into Equation 6.7 results in a discrete form of $\Pi_I(\sigma, u_*)$:

$$\Pi_I(\beta, q_*) = \beta^T G_* q_* - \frac{1}{2}\beta^T H \beta \tag{6.9}$$

Instead of the stability condition (Equation 6.8), one has

$$\beta^T G_* q_* = 0 \quad \forall \beta \in \{\beta\} \quad \Rightarrow \quad q_* = 0$$

Namely,

$$G_* q_* = \mathbf{0} \Rightarrow q_* = \mathbf{0} \tag{6.10}$$

This is a necessary and sufficient condition for the absence of ZEM(u) in the element level. In view of its linear algebraic essence, we call Equation 6.10 the rank condition I. The basic way to suppress ZEM(u) has been presented by Pian and Chen [6].

Another kind of element, mixed-hybrid element II, is based on the modified potential energy principles, for example, some mixed displacement elements. See for instance Reference 11, where the related energy formulation of an element without the rigid body DOF takes the form

$$\Pi_{II}(\sigma, u_*) = \frac{1}{2}(Lu_*, u_*) - W(\sigma, u_*) \tag{6.11}$$

Note that

$$\Delta\Pi_{II}(u'_*) = \Pi_{II}(\sigma, u_* + u'_*) - \Pi_{II}(\sigma, u_*) =$$

$$\frac{1}{2}\langle L(u_* + u'_*), (u_* + u'_*)\rangle + \frac{1}{2}\langle Lu_*, u_*\rangle - I(\sigma, u'_*)$$

$$\Pi_{II}(u'_*) = 0 \quad \forall \sigma \in \Sigma \Rightarrow u' = 0 \tag{6.12}$$

This means that the mixed-hybrid element II has no (u). It can also be shown that

$$\Delta\Pi_{II}(\sigma) = -I(\sigma, u_*)$$

Then condition (Equation 6.3b) is reduced to a dual form of Equation 6.8, that is,

$$W(\sigma, u_*) = 0 \quad \forall u_* \in U_* \quad \Rightarrow \sigma = 0 \tag{6.13}$$

If the discrete type of Equation 6.11 is expressed as

$$\Pi_{II}(\beta, q_*) = \frac{1}{2} q_*^T A_* q_* - q_*^T F_* \beta \tag{6.14}$$

In accordance with Equation 6.13, one has the rank condition II

$$F_* \beta = 0 \Rightarrow \beta = 0 \tag{6.15}$$

6.4 Determination of the Zero Energy Modes

Denote the number of element parameters as

$$n_\beta = \dim(\beta),$$

$$n_q = \dim(q)$$

$$n_{q^*} = \dim(q_*) = n_q - r \quad (r \text{ is the element rigid body DOF}).$$

Consider the mixed-hybrid element I if the keeping rank condition (Equation 6.10) is not passed; ZEM(u) will appear, and the number should be

$$n_0 = n_{q^*} - rank(G_*)$$

According to the usual elimination by substitution used to solve $G_* q_* = 0$ in Equation 6.10, it is not difficult to employ $\overset{o}{q} = \{\overset{o}{q}_i\}$, which consists of some linear independent elements in $q_* = \{q_i\}_*$, to express all other elements in q_*. In this way, the general solution of the homogeneous equation can be expressed by $\overset{o}{q}$ that is,

$$q_* = T_q \overset{o}{q} \tag{6.16}$$

where $\overset{o}{q}$ can be decomposed as follows:

$$\overset{o}{q} = \begin{Bmatrix} q_1 \\ 0 \\ \vdots \\ 0 \end{Bmatrix} + \begin{Bmatrix} 0 \\ q_2 \\ 0 \\ \vdots \\ 0 \end{Bmatrix} + \cdots + \begin{Bmatrix} 0 \\ \vdots \\ 0 \\ q_{\overset{o}{n}} \end{Bmatrix} = \overset{o}{q}(1) + \overset{o}{q}(2) + \cdots + \overset{o}{q}(n_0) \tag{6.17}$$

Substituting Equation 6.16 into the element displacement trial, which is defined as $\boldsymbol{u}_* = \boldsymbol{N}_* \boldsymbol{q}_*$, the explicit expression of each component of ZEM(u) can be determined. Corresponding to Equation 6.17, they are

$$\overset{o}{u}(i) = \boldsymbol{N}_* \boldsymbol{T}_q \overset{o}{q}(i), \quad i = 1, 2, \ldots, n_0 \tag{6.18}$$

and their sum is

$$\overset{o}{u} = \sum_i \overset{o}{u}(i) = \underset{\sim}{N}_* \underset{\sim}{T}_q \underset{\sim}{\overset{o}{q}} \tag{6.19}$$

In the case of a mixed-hybrid element II, if the rank condition (Equation 6.15) is not satisfied the n_0 ZEM(σ) will appear and

$$n_0 = n_\beta - rank(\boldsymbol{F}_*)$$

Similarly the general solution of the homogeneous equation $\boldsymbol{F}_*\beta = 0$ in Equation 6.15 can be expressed as

$$\beta = \mathbf{T}_\beta \overset{o}{\beta} = \mathbf{T}_\beta(\overset{o}{\beta}(1) + \overset{o}{\beta}(2) + \cdots + \overset{o}{\beta}(\overset{o}{n})) \tag{6.20}$$

Here $\overset{o}{\beta}$ consists of the linear independent elements in β. If the element stress trial $\sigma = \varphi\beta$, we can obtain the expressions of ZEM(σ) corresponding to Equation 6.20. They are

$$\overset{o}{\sigma}(i) = \varphi \boldsymbol{T}_\beta \overset{o}{\beta}(i), \quad i = 1, 2, \ldots, n_0 \tag{6.21}$$

and their sum is

$$\overset{o}{\sigma} = \sum_i \overset{o}{\sigma}(i) = \varphi \mathbf{T}_\beta \overset{o}{\beta} \tag{6.22}$$

6.5 Control of the Zero Energy Displacement Modes

The ZEMs can be formulated by Equation 6.19 and Equation 6.22. Now let us consider how to control them. It has been shown that the ZEM(u) may appear in the mixed-hybrid element I. To suppress it, a control stress $\sigma_\Delta = \varphi_\Delta\beta_\Delta$ is employed and added to the primitive element stress $\sigma = \varphi\beta$. We have a modified one:

$$\sigma_m = \sigma + \sigma_\Delta = [\varphi \; \varphi_\Delta] \begin{bmatrix} \beta \\ \beta_\Delta \end{bmatrix} \tag{6.23}$$

where the basis function φ_Δ is linearly independent of φ.

THEOREM 6.1

Suppose that $\overset{\circ}{\boldsymbol{u}}$ is the ZEM(u) in the mixed-hybrid element I based on $(\sigma, \boldsymbol{u}_*)$, if σ_Δ provides with an energy control,

$$\boldsymbol{I}(\sigma_\Delta, \overset{\circ}{\boldsymbol{u}}) = 0 \quad \forall \sigma_\Delta \in \Sigma_m \quad \Rightarrow \overset{\circ}{\boldsymbol{u}} = 0 \tag{6.24}$$

then the modified element based on $(\sigma_m, \boldsymbol{u}_*)$ has no ZEM(u).

Proof

Since $\sigma_\Delta = \varphi_\Delta\beta_\Delta$ and $\sigma = \varphi\beta$ are linearly independent of each other, the following zero energy constraint

$$\boldsymbol{I}(\sigma_m, \boldsymbol{u}_*) = 0 \quad \forall \sigma_m \in \Sigma_m \tag{6.25}$$

is equivalent to

$$\begin{cases} \boldsymbol{I}(\sigma, \boldsymbol{u}_*) = 0 \quad \forall \sigma \in \Sigma_m & (6.26a) \\ \boldsymbol{I}(\sigma_\Delta, \boldsymbol{u}_*) = 0 \quad \forall \sigma_\Delta \in \Sigma_m & (6.26b) \end{cases}$$

From Equation 6.26a we have the nonzero solution $\boldsymbol{u}_* = \overset{\circ}{\boldsymbol{u}}$, which can be expressed as Equation 6.19. Equation 6.26b is then

$$\boldsymbol{I}(\sigma_\Delta, \overset{\circ}{\boldsymbol{u}}) \forall \sigma_\Delta \in \Sigma$$

By virtue of the energy control (Equation 6.24), we obtain

$$\overset{\circ}{\boldsymbol{u}} = \boldsymbol{u}_* = 0$$

The result is

$$I(\sigma_m, u_*) = 0 \quad \forall \sigma_m \in \Sigma_m \quad \Rightarrow u_* = 0 \tag{6.27}$$

The modified element based on (σ_m, u_*) has passed the element stability condition (Equation 6.8), and no ZEM(u) appears.

It is easy to choose the control stress σ_Δ for a determined ZEM as shown in Equation 6.19. Let

$$I(\sigma_\Delta, \overset{o}{u}) = \beta_\Delta^t G_c \overset{o}{q} \tag{6.28}$$

where the control matrix G_c is a square matrix of order n_0 when $\dim(\beta_\Delta) = \dim(\overset{o}{q})$. By regulating the basic function φ_Δ so that $|G_c| \neq 0$, the energy control (Equation 6.24) will be achieved.

As shown in Reference 7, there is a necessary condition for the parameters of the mixed-hybrid element I. That is

$$n_\beta \geq n_{q^*} \tag{6.29}$$

which has been involved in the keeping rank condition (Equation 6.10). When Equation 6.29 is satisfied, the number of ZEM(u), generally speaking, is very small, so that the choice of σ_Δ is convenient.

6.6 Control of the Zero Energy Stress Modes

The ZEM(σ) may appear in the mixed-hybrid element II. By introducing a control displacement $u_\Delta = N_\Delta q_\Delta$ into the primitive trial $u_* = N_* q_*$, the combination trial becomes

$$u_m = u_* + u_\Delta = \begin{bmatrix} N_* & N_\Delta \end{bmatrix} \begin{Bmatrix} q_* \\ q_\Delta \end{Bmatrix} \square$$

where N_Δ must be independent of N_*.

THEOREM 6.2

Suppose that $\overset{o}{\sigma}$ is the ZEM(σ) appeared in the mixed-hybrid element II based on (u_*, σ). If u_Δ provides $\overset{o}{\sigma}$ an energy control:

$$I(u_\Delta, \overset{o}{\sigma}) = 0 \quad \forall u_\Delta \in U_{*m} \Rightarrow \overset{o}{\sigma} = 0 \tag{6.30}$$

then the modified element based on (u_m, σ) has no ZEM(σ).

Proof

Because $u_{\Delta}(q_{\Delta})$ and $u_{*}(q_{*})$ are linearly independent of each other, the zero energy constraint, $I(u_m, \sigma) = 0 \forall u_m \in U_*$, should be equivalent to

$$I(u_*, \sigma) = 0 \quad \forall u_* \in U_* \tag{6.31a}$$

$$I(u_{\Delta}, \sigma) = 0 \quad \forall u_{\Delta} \in U_* \tag{6.31b}$$

From Equation 31a, $\sigma = \overset{o}{\sigma}$ as shown in Equation 6.22, and Equation 6.31b is then

$$I(u_{\Delta}, \overset{o}{\sigma}) = 0 \quad \forall u_{\Delta} \in U_* \tag{6.31c}$$

Finally, only a trivial solution $\overset{o}{\sigma} = 0$ is obtained from the given condition (Equation 6.30). So far we have demonstrated that the following element stability condition is satisfied:

$$I(u_m, \sigma) = 0 \quad \forall u_m \in U_* \Rightarrow \sigma = 0 \tag{6.31d}$$

Therefore the modified element based on (u_m, σ) has no ZEM(σ).

For the mixed-hybrid element II there is another necessary condition for the element parameters opposed to the condition (Equation 6.29), that is,

$$n_{\beta} \leq n_{q^*} \tag{6.32}$$

which has been involved in Equation 6.15. Under the condition (Equation 6.32) the number of ZEM(σ) is small, and it is easy to choose the control displacement $u_{\Delta} = N_{\Delta} q_{\Delta}$. Now we can discretize the bilinear energy integral in Equation 6.30 by means of $u_{\Delta}(q_{\Delta})$ and $\overset{o}{\sigma}(\overset{o}{\beta})$ given by Equation 6.22 as follows:

$$I(u_{\Delta}, \overset{o}{\sigma}) = q_{\Delta}^t F_c \overset{o}{\beta} \tag{6.33}$$

Let $\dim(q_{\Delta}) = \dim(\overset{o}{\beta})$ and regulate the basic function N_{Δ} so that $|F_c| \neq 0$. Thus the energy control (Equation 6.30) will be succeeded.

Now we can sum up the main results obtained in the above sections in Table 6.1.

A study to the zero energy modes of 3-field finite element has been presented in [15].

TABLE 6.1

Dual Analysis of the Zero Energy Modes in Mixed/Hybrid Elements

Classification	Mixed/Hybrid element	Mixed/Hybrid element
Variational principle	Modified complementary energy or Hellinger–Reissner principle $\Pi_I = W(\sigma, u_*) - \frac{1}{2}\langle S\sigma, \sigma\rangle$	Modified potential energy principle $\Pi_{II} = \frac{1}{2}\langle Lu_*, u_*\rangle - W(\sigma, u_*)$
Element stability condition	$W(\sigma, u_*) = 0, \forall \sigma \in \Sigma \Rightarrow u_* = 0$	$W(\sigma, u_*) = 0,$ $\forall u_* \in u_* \to \sigma = 0$
Rank condition	$G_* q_* = 0 \Rightarrow q_* = 0$	$F_* \beta = 0 \Rightarrow \beta = 0$
Parameter matching condition	$n_\beta \geq n_{q^*}$	$n_\beta \leq n_{q^*}$
ZEM (u) formula	$\overset{o}{u} = N_* T_q \overset{o}{q}$	(no ZEM(u))
ZEM (σ) formula	(no ZEM(σ))	$\overset{o}{\sigma} = \varphi T_\beta \overset{o}{\beta}$
Control of ZEM	**(Theorem 6.1)** Take a control stress σ_Δ, so that $W\left(\sigma_\Delta, \overset{o}{u}\right) = 0 \Rightarrow \overset{o}{u} = 0$	**(Theorem 6.2)** Take a control displacement u_Δ, so that $W\left(u_\Delta, \overset{o}{\sigma}\right) = 0 \Rightarrow \overset{o}{\sigma} = 0$

6.7 Patch Stability Test

So far all of our discussions have been based on a single element. Now let us consider the effect of the element assembly of the mixed-hybrid element I. Two factors should be considered simultaneously. The first factor is the displacement continuity at the common nodes of neighboring elements, with which it is possible that some ZEM (u) appearing in the individual element are suppressed. The other factor is the stress continuity at the common nodes, that is, the equilibrium relationship between the elements, with which it is possible that some element displacement restrictions are relaxed such that new ZEM (u) can be induced.

In the hybrid stress element method there exists only the first factor, and the element assembly will not bring out any extra ZEM(u). Thus we have the following conclusion: The stability of a hybrid element discrete system can be guaranteed by the element rank condition, provided that the parameter-necessary condition (Equation 6.29) always holds for the individual element and for the various possible combinations of elements.

In the hybrid stress element case, it will no longer be necessary to consider the affection of element assembly.

For mixed finite elements, since the above two factors of element assembly exist simultaneously, the system stability problem will be very complicated.

The node equilibrium relationships may bring us some patch ZEM(u) that have not yet appeared in the individual element. In the mixed element methods with the node stress continuity, the element rank condition does not necessarily guarantee the numerical stability of a whole discrete system, and a patch stability test, in which all effects of the element assembly have been included, must be considered and passed.

Several combinations of plane elements are shown in Figure 6.1. They can all be used as a basic patch in stability tests. In order to represent the element assembly effects, the connections between individual elements and the patch have to be set up. In the patch analysis we denote q and $\boldsymbol{\beta}$ as the patch displacement parameters and the stress parameters, respectively. There are two kinds of correspondent relationships:

$$\begin{bmatrix} q_1 \\ \vdots \\ q_n \end{bmatrix} = \boldsymbol{M}\boldsymbol{q} \quad \text{and} \quad \begin{bmatrix} \beta_1 \\ \vdots \\ \beta_n \end{bmatrix} = \boldsymbol{L}\beta \tag{6.34}$$

where q_e and β_e ($e = 1,\ldots,n$) are the element displacement and the element stress parameters, respectively. Different from the mark q_* used as above, here q_e includes the rigid body DOF of the element e.

For the mixed-hybrid element I, if the bilinear energy integral of an individual element is discretized as

$$\boldsymbol{I}_e(\sigma,\boldsymbol{u}) = \beta_e^t \boldsymbol{G}_e \boldsymbol{q}_e \quad (e = 1,\ldots,n)$$

the patch's bilinear energy integral should be of the form

$$\boldsymbol{I}(\sigma,\boldsymbol{u}) = \sum_e \boldsymbol{I}_e(\sigma,\boldsymbol{u}) = \begin{bmatrix} \beta_1 \\ \vdots \\ \beta_n \end{bmatrix}^t \begin{bmatrix} \boldsymbol{G}_1 & & \\ & \ddots & \\ & & \boldsymbol{G}_n \end{bmatrix} \begin{bmatrix} q_1 \\ \vdots \\ q_n \end{bmatrix} = \beta^t \boldsymbol{G}_{patch}\boldsymbol{q}$$

where

$$\boldsymbol{G}_{patch} = \boldsymbol{L}^t \begin{bmatrix} \boldsymbol{G}_1 & & \\ & \ddots & \\ & & \boldsymbol{G}_n \end{bmatrix} \boldsymbol{M} \tag{6.35}$$

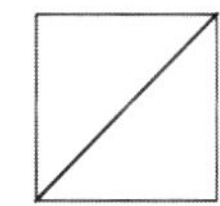
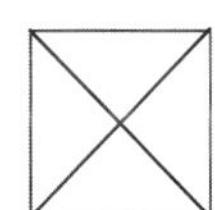
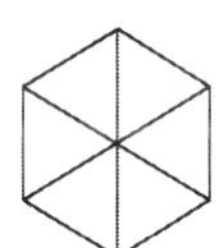

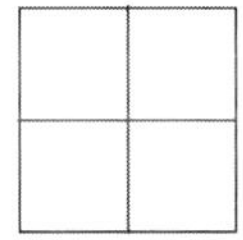

FIGURE 6.1
Some typical patches in the two-dimensional case.

In order to extend the element-keeping rank condition (Equation 6.10) to the patch, it is necessary to prevent the patch rigid body motion. Assume, at least, that r node displacements must be restricted in q and the rests in q are denoted by q_*. Thus, instead of Equation 6.10, one has the patch-keeping rank condition

$$G^*_{patch} q_* = 0 \quad \Rightarrow \quad q_* = 0 \tag{6.36}$$

Here the matrix G^*_{patch} can be yielded directly by the cancellation of r columns, which correspond to the r node displacements restricted in q, from the matrix G_{patch} defined in Equation 6.35. For the hybrid stress elements, since the matrix L in Equation 6.34 is a unit matrix, the patch condition (Equation 6.36) should be weaker than the individual element condition (Equation 6.10), so that possibly some hybrid elements with ZEM(u) may still pass the patch condition and can be used.

Similarly, one can establish the patch-keeping rank condition for the mixed-hybrid element II. According to the element condition (Equation 6.15), the patch should be of the form

$$F^*_{patch} \beta = 0 \quad \Rightarrow \quad \beta = 0 \tag{6.37}$$

where F^*_{patch} is formed by the cancellation of r rows, which correspond to the patch's rigid body motion, from the following matrix:

$$F_{patch} = M^t \begin{bmatrix} F_1 & & \\ & \ddots & \\ & & F_n \end{bmatrix} L \tag{6.38}$$

where M and L are defined by Equation 6.34, and $F_e\, (e = 1,\ldots,n)$ are defined by the energy integral of the mixed-hybrid element II as follows:

$$I_e(\sigma, u) = q_e^t F_e \beta_e \quad (e = 1,\ldots,n) \tag{6.39}$$

6.8 Examples

Examples of the stability analysis of the mixed-hybrid element I are given here to illustrate the application of keeping-rank conditions and show how the ZEM can be determined and suppressed.

The appearance of the ZEM is most likely for the elements that have regular geometrical forms, so one always takes elements with the simplest shapes as the analysis objects [6]. Consider a four-node rectangular membrane element

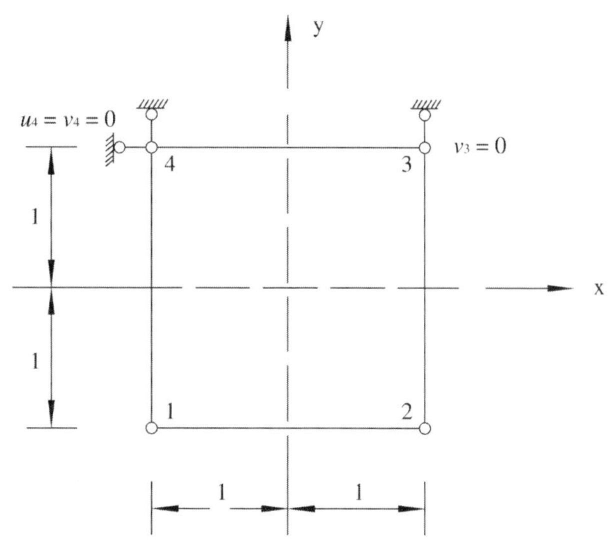

FIGURE 6.2
Restricted plane hybrid element.

of 2 × 2 dimension with the reference axis X-Y passing the centroid, and parallel to the element sides (Figure 6.2). We use an element with insufficient rank to begin our illustration. The assumed hybrid element has a constant stress trial:

$$\sigma = \begin{bmatrix} \sigma_x \\ \sigma_y \\ \sigma_{xy} \end{bmatrix} = \begin{bmatrix} 1 & & \\ & 1 & \\ & & 1 \end{bmatrix} \begin{bmatrix} \beta_1 \\ \beta_2 \\ \beta_3 \end{bmatrix} = \varphi\, \beta \tag{6.40}$$

and a bilinear displacement trial $u = Nq$. The minimum restrictions that prevent the element rigid body motion are shown in Figure 6.2. Accordingly,

$$\boldsymbol{u}_* = \mathbf{N}_* \boldsymbol{q}_*, \quad \boldsymbol{q}_* = \{u_1 \quad v_1 \quad u_2 \quad v_2 \quad u_3\}$$

Because $n_\beta < n_{q^*}$, the element must have ZEM $(\boldsymbol{u})$. Element formulations can be established by the aid of the Reissner principle, and we have the matrix

$$\boldsymbol{G}_* = \int_{-1}^{1}\int_{-1}^{1} \varphi^t (\boldsymbol{D}\boldsymbol{N}_*)\, dx dy = \begin{bmatrix} -1 & 0 & 1 & 0 & 1 \\ 0 & -1 & 0 & -1 & 0 \\ -1 & -1 & -1 & 1 & 1 \end{bmatrix}$$

The homogeneous equation $\boldsymbol{G}_* \boldsymbol{q}_* = \boldsymbol{0}$ in Equation 6.10 has nontrivial solutions. They are

$$\boldsymbol{q}_* = \boldsymbol{T}_q \overset{\circ}{\boldsymbol{q}} = \begin{bmatrix} 1 & 1 \\ -1 & 0 \\ 1 & 0 \\ 1 & 0 \\ 0 & 1 \end{bmatrix} \begin{bmatrix} v_2 \\ u_3 \end{bmatrix}$$

From Equation 6.19 the ZEM(u) can be determined as follows:

$$\overset{\circ}{\boldsymbol{u}} = \begin{bmatrix} \overset{\circ}{u} \\ \overset{\circ}{v} \end{bmatrix} = \boldsymbol{N}_* \boldsymbol{T}_q \overset{\circ}{\boldsymbol{q}} = \begin{bmatrix} \frac{1}{2}(1-y) & \frac{1}{2}(1+xy) \\ \frac{1}{2}x(1-y) & 0 \end{bmatrix} \overset{\circ}{\boldsymbol{q}}$$

Two ZEM(u) corresponding to $\overset{\circ}{\boldsymbol{q}}(1) = \begin{bmatrix} v_2 = 1 \\ 0 \end{bmatrix}$ and $\overset{\circ}{\boldsymbol{q}}(2) = \begin{bmatrix} 0 \\ u_3 = 1 \end{bmatrix}$ are shown in Figure 6.3a and Figure 6.3b, respectively. For the sake of controlling the above two ZEM(u), a control stress with two parameters is employed:

$$\sigma_\Delta = \begin{bmatrix} \sigma_x^\Delta \\ \sigma_y^\Delta \\ \sigma_{xy}^\Delta \end{bmatrix} = \varphi_\Delta \beta_\Delta, \quad \beta_\Delta = \begin{bmatrix} \beta_4 \\ \beta_5 \end{bmatrix}$$

Taking into account the symmetry of the trial function σ_Δ, the following two possible schemes are considered:

$$\varphi_\Delta = \begin{bmatrix} x & 0 \\ 0 & y \\ 0 & 0 \end{bmatrix} \tag{6.41}$$

$$\varphi_\Delta = \begin{bmatrix} y & 0 \\ 0 & x \\ 0 & 0 \end{bmatrix} \tag{6.42}$$

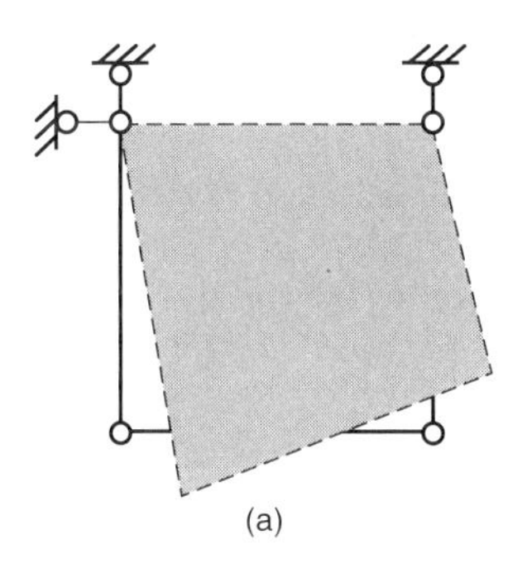

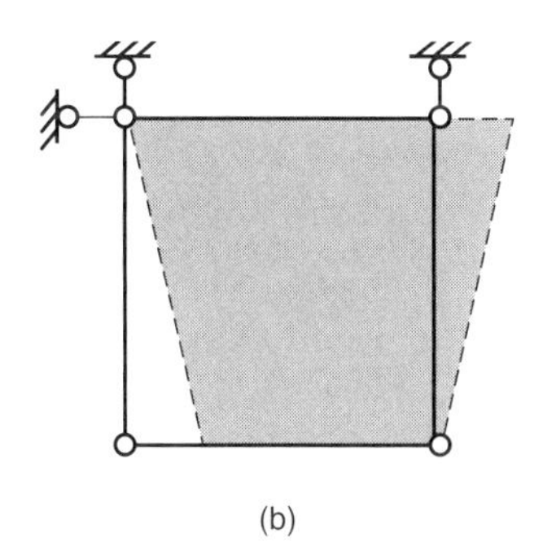

FIGURE 6.3
ZEM(u) of the constant stress hybrid element.

By the definition in Equation 6.28, the control matrix should be

$$\boldsymbol{G}_c = \int_{-1}^{1}\int_{-1}^{1} \varphi_\Delta^T (\boldsymbol{DN} \cdot \boldsymbol{T}_q) dx dy$$

In accordance with the scheme (Equation 6.41), $\boldsymbol{G}_c = \begin{bmatrix} 0 & 0 \\ 0 & 0 \end{bmatrix}$. It cannot be used to control ZEM(u).

In accordance with the scheme (Equation 6.42), $\boldsymbol{G}_c = \begin{bmatrix} 0 & \frac{2}{3} \\ -\frac{2}{3} & 0 \end{bmatrix}$, $|\boldsymbol{G}_c| \neq 0$, and then the ZEM($u$) will be controlled.

The modified stress trial is now a sum of σ in Equation 6.40 and the control stress σ_Δ:

$$\sigma_m = \sigma + \sigma_\Delta = \begin{bmatrix} 1 & 0 & 0 & y & 0 \\ 0 & 1 & 0 & 0 & x \\ 0 & 0 & 1 & 0 & 0 \end{bmatrix} \begin{bmatrix} \beta_1 \\ \vdots \\ \beta_5 \end{bmatrix}$$

This is Pian's 5β hybrid stress pattern [14].

References

1. Babuška, The finite element methods with Lagrange multipliers, *Numer. Math.*, 20, 179–192, 1973.
2. Brezzi, F., On the existence, uniqueness and approximation of saddle-point problems arising from Lagrange multipliers, *RAIRO Ser. Rouge*, 8, 129, 1974.

3. Wu, C.C., Dual zero energy modes in mixed/hybrid elements—definition, analysis and control, *Comp. Meth. Appl. Mech. Eng.*, 81, 39–56, 1990.
4. Rubinstein, R., Punch, E.F., and Atluri, S.N., An analysis of, and remedies for, kinematic modes in hybrid-stress finite elements: selection of stable, invariant stress fields, *Comput. Methods Appl. Mech. Eng.,* 38, 63, 1983.
5. Fraeijs de Veubeke, B., Displacement and equilibrium models in the finite element method, in *Stress Analysis,* Zienkiewicz, O.C. and Holister, G.C., Eds., Wiley, London, 1965, pp. 145–197.
6. Pian, T.H.H. and Chen, D.P., On the suppression of zero-energy deformation modes, *Int. J. Numer. Methods Eng.,* 19, 1741, 1983.
7. Tong, P. and Pian, T.H.H., A variational principle and the convergence of a finite element method based on assumed stress distribution, *Int. J. Solids Struct.,* 5, 463, 1969.
8. Herrmann, L.R., A bending analysis for plates, *Proc. Conf. on Matrix Methods in Structural Mechanics,* AFFDL-TR-66-80, 1965, pp. 577–604.
9. Pian, T.H.H. and Chen, D.P., Alternative ways for formulation of hybrid stress elements, *Int. J. Numer. Methods Eng.,* 18, 1679, 1982.
10. Pian, T.H.H. and Wu, C.C., A rational approach for choosing stress terms for hybrid finite element formulations, *Int. J. Numer. Methods. Eng.,* 26, 2331, 1988.
11. Jones, R.E., A generalization of the direct-stiffness method of structural analysis, *AIAA J.,* 2, 821, 1964.
12. Harvey, J.W. and Kelsey, S., Triangular plate bending element with enforced compatibility, *AIAA J.,* 9, 1023, 1971.
13. Oden, J.T. and Reddy, J.N., *Variational Methods in Theoretical Mechanics,* 2nd ed., Springer, Berlin, 1983.
14. Pian, T.H.H., Derivation of element stiffness matrices by assumed stress distribution, *AIAA J.,* 2, 1333, 1964.
15. Cheung, Y.K. and Wu, C.C., A study on the stability of 3-field finite elements by the theory of zero energy modes, *Int. J. Solids Struct.,* 29, 215, 1992.
16. Zhongci Shi, Convergence properties of two nonconforming finite elements, *Comp. Meth. Appl. Mech. Eng.*, 48, 123–137, 1985.
17. Ciarlet, P.G., Numerical Analysis of the finite element method, 1975.

7

Plastic Analysis of Structures

7.1 Introduction

For the plastic analysis of structures, the ordinary finite element displacement methods often encounter numerical difficulties. In three-dimensional (3-D) problems, plane stress problems, and axisymmetric problems, the displacement isoparametric elements always increase the actual stiffness of the elastic-plastic structure such that the limit load of the plastic system can never be determined [1, 2]. This chapter introduces the incompatible and hybrid element methods for the solution of incompressible problems and introduces the solution of visco-plastic material by hybrid elements.

7.2 Form of Incompressible Elements and Analysis of Plane Stress Plastic Analysis

For 3-D problems, plane stress problems, and axisymmetric problems, the stiffness equations for the finite element analysis can all be written in the following form:

$$(\mathbf{K}_S + \alpha \mathbf{K}_V)\mathbf{q} = \bar{\mathbf{Q}} \tag{7.1}$$

$\mathbf{K}_S$ and $\mathbf{K}_V$ are, respectively, distortion stiffness and bulk stiffness, and the factor $\alpha = (k - G)/2$, $k = (1 + \mu)G/(1 - 2\mu)$ is the material bulk modulus. When μ is close to 0.5, $k \to \infty$, and $\alpha \to \infty$, Equation 7.1 can be

$$\mathbf{K}_V \mathbf{q} = 0$$

By eliminating the rigid body degree of freedom (DOF) of the element, the above equation is changed to

$$\boldsymbol{K}_V^* \boldsymbol{q}_* = 0 \tag{7.2}$$

This is the geometric constraint of the element caused by material incompressibility. The deforming DOFs of an incompressible element or plastic element are

$$n_d = \dim(\mathbf{q}_*) - \operatorname{rank}(\mathbf{K}_V^*) \tag{7.3}$$

We now use a four-node isoparametric Q4 element as an example to determine the element deformation under incompressible conditions. For a rectangular Q4 element the displacement can be expressed in terms of Cartesian coordinates as follows:

$$\mathbf{u} = \begin{bmatrix} u & v \end{bmatrix} = \begin{bmatrix} x & xy & y & 1 \end{bmatrix} \begin{bmatrix} A_1 & B_1 \\ A_2 & B_2 \\ A_3 & B_3 \\ A_4 & B_4 \end{bmatrix} \tag{7.4}$$

The factors A_i and B_i are defined as follows:

$$\begin{bmatrix} A_1 & B_1 \\ A_2 & B_2 \\ A_3 & B_3 \\ A_4 & B_4 \end{bmatrix} = \frac{1}{4} \begin{bmatrix} -1 & 1 & 1 & -1 \\ 1 & -1 & 1 & -1 \\ 1 & -1 & 1 & 1 \\ 1 & 1 & 1 & 1 \end{bmatrix} \begin{bmatrix} u_1 & v_1 \\ u_2 & v_2 \\ u_3 & v_3 \\ u_4 & v_4 \end{bmatrix} \tag{7.5}$$

By eliminating the rigid body DOFs in $\boldsymbol{u}$ and introducing the following incompressibility condition,

$$e = \frac{\partial u}{\partial x} + \frac{\partial v}{\partial y} = 0 \tag{7.6}$$

the form of the incompressible displacement corresponding to Q4 can be determined:

$$\mathbf{u}_d = \begin{Bmatrix} u_d \\ v_d \end{Bmatrix} = \begin{bmatrix} x & y \\ -y & x \end{bmatrix} \begin{Bmatrix} c_1 \\ c_2 \end{Bmatrix} \tag{7.7}$$

This form is the nonzero solution of the equation of constraint (Equation 7.2). The two basic deformation patterns for the independent factors c_1 and c_2 are shown in Figure 7.1.

From the above result we can conclude that under plane stress and incompressible conditions, the Q4 element can be used to model stretching deformation and shearing deformation, but it cannot be used to model bending deformation. Thus, in analyzing plane-strain incompressible (or nearly incompressible) bending problems, the self-locking phenomena for the Q4 element cannot be avoided.

A simple and effective method for the Q4 element improving the numerical solutions is to add an internal displacement defined by $\mathbf{u}_\lambda = \mathbf{N}_\lambda\lambda$. This can effectively increase the DOFs of incompressible deformation. For this approach, the four-node plane-incompatible element NQ6, which was introduced in Section 3.1, is a successful example. Figure 7.2 presents six incompressible deformation patterns. The Q4 element has stretching and shearing patterns, a pair of linear bending patterns, and a pair of second-order bending patterns. This indicates that the NQ6 element can be used to model stretching, compressing, and bending of structures without any appearance of the self-locking phenomena.

Figure 7.3 is an example of plane stress and pure bending. The numerical solution shows that when μ is close to 0.5, the solutions for displacements (u_A and v_A) all approach zero. But the solutions for stresses σ_{xB} (exact solution = 3000) are expanded two times. However, by using the NQ6 element, reasonable solutions for incompressible bending displacement can be obtained.

We now consider the plastic analysis of structures. The assumption of incompressibility in plasticity requires that the finite element formulation have enough incompressible DOFs. We note that in the process of gradually increasing loading, many elements enter the plastic incompressible states gradually. This means that the elements should be applicable to compressible and incompressible materials. The accuracy of the stress solution and its

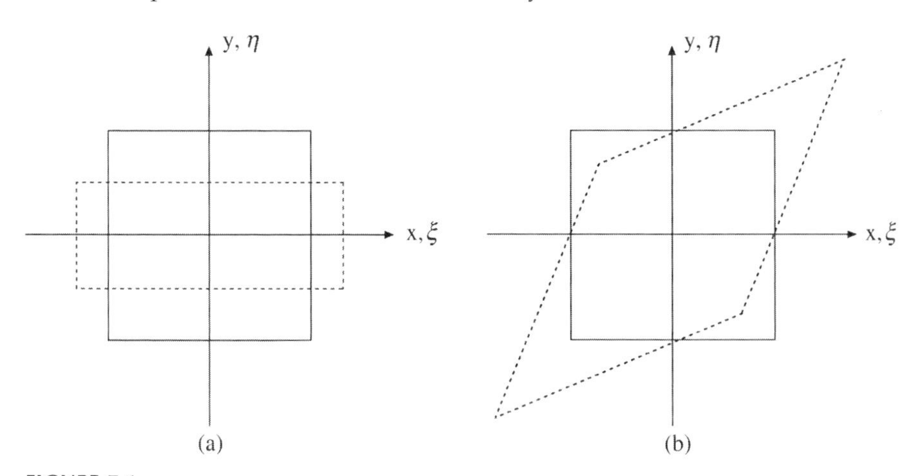

FIGURE 7.1
Incompressible displacement patterns of Q4 element. (a) Stretching pattern ($c_1 \neq 0$, $c_2 = 0$); (b) shearing pattern ($c_1 = 0$, $c_2 \neq 0$).

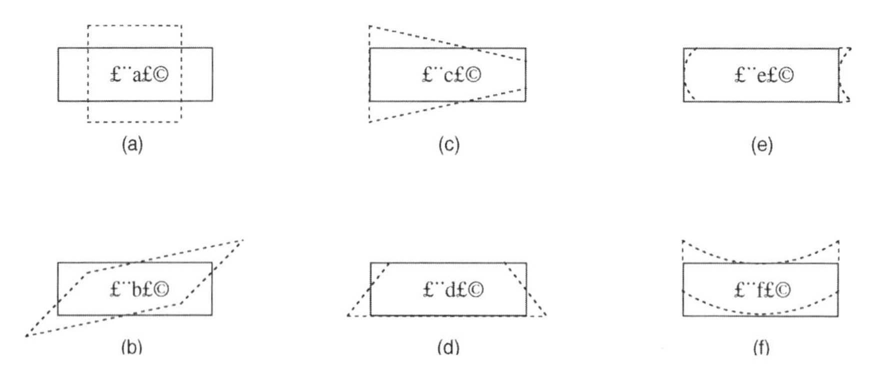

FIGURE 7.2
Incompressible displacement patterns of NQ6 element. (a) Stretching; (b) shearing; (c, d) linear bending; (e, f) second-order bending.

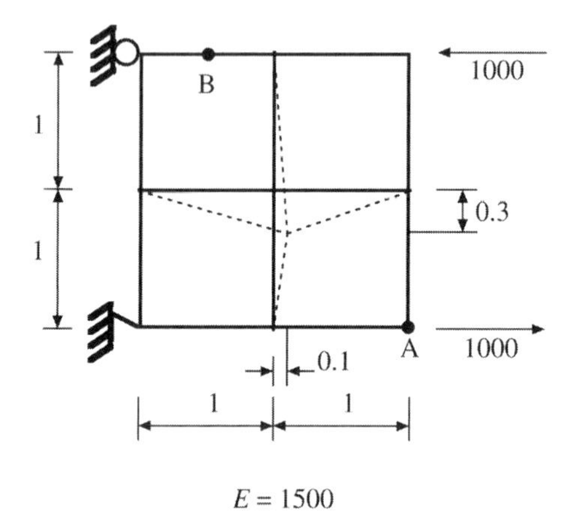

FIGURE 7.3
Numerical experiment on plane stress pure bending (numbers in brackets are obtained by using irregular meshes).

distribution within each element are important in elastic-plastic analysis. The commonly used procedure is to find out whether the stress values at the Gauss integration points within each element have reached the plastic condition.

7.3 Incompatible Elements in Plasticity Analysis

7.3.1 Introduction

In structural plastic analysis it is often difficult to find a reliable numerical result by using the conventional finite element methods. In plane strain, axisymmetric, and 3-D problems, the incremental iterative process is not always suitable for perfectly plastic materials [3]. The value of the limit load (i.e., collapse load) of a plastic structure may be amplified even when it does not exist in an iterative process using isoparametric elements. The reason is that the assumption of plastic incompressibility in the flow theory of plasticity, according to which the volumetric strain $e = u_{i,i}$ equals zero, provides the discrete system with restrictions that are so strong that the structural stiffness is amplified essentially. This problem may be solved by using some special approaches, for instance, the penalty-selective reduced-integration method suggested by Zienkiewicz [4], the direct elimination technique presented by Needleman and Shih [5], and various mixed element methods [6–8]. But these approaches make the finite element algorithm special and complicated, and it may be difficult to implement them in a current nonlinear program.

In this section we introduce the incompatible displacement functions that satisfy certain requirements for an isoparametric element such that the element behavior improves. The new model is universal for both compressible and incompressible materials and allows a normal algebraic formulation, like that obtained by the direct stiffness method. In this way no special treatment for the algorithm is needed, and the element can be implemented directly into a current program.

7.3.2 Incompressibility of the Isoparametric Element (Elastic Case)

In the incompressible elastic problem, based on the potential energy principle, the element equation can be expressed as

$$\mathbf{Kq} = (\mathbf{K}_s + k\mathbf{K}_v)\mathbf{q} = \mathbf{0} \tag{7.8}$$

Here the element stiffness matrix $\mathbf{K}$ is separated into two parts: the distortion part $\mathbf{K}_s$ and the dilatation part $\mathbf{K}_v$. The problem is to solve Equation 7.8 when the material bulk modulus $k \to \infty$ or, equivalently, when the Poisson's ratio $\upsilon \to 0.5$. In the limit case of $k = \infty$, Equation 7.8 will degenerate into a homogeneous equation:

$$\mathbf{K}_v\mathbf{q} = \mathbf{0} \tag{7.9}$$

In order to analyze the element deformations, at least r nodal displacements must be restricted in q such that the element rigid-body motion is prevented. The remaining displacement components in $\mathbf{q}$ are denoted by $\mathbf{q}^*$, so instead of Equation 7.9 we have

$$\mathbf{K}^*_v\mathbf{q}^* = \mathbf{0} \tag{7.10}$$

Equation 7.10 is the element deformation restriction caused by the incompressibility assumption; therefore the deformation DOFs of an incompressible element should be

$$n_d = \dim(\mathbf{q}^*) - \text{rank}(\mathbf{K}^*_v) \tag{7.11}$$

If $|\tilde{\mathbf{K}}^*_v| \neq 0$, Equation 7.10 has only a trivial solution, $\mathbf{q}^* = 0$. So an element cannot be used, and a kind of element with a singular dilatation matrix is expected. But it must be emphasized that the singularity of $\mathbf{K}^*_v$ is not sufficient for solving incompressibility problems. Indeed, there are many elements that cannot be applied to plastic calculations even though $|\mathbf{K}^*_v| = 0$ for them.

Let us take the well-known bilinear isoparametric element Q_4 as an example. For simplicity, a 2×2 square element is considered in Figure 7.4. Denote the element displacement

$$u = \begin{Bmatrix} u \\ v \end{Bmatrix} = \mathbf{N}\mathbf{q}, \quad q = \{u_1 \quad v_1 \quad \cdots \quad v_4\}^T \tag{7.12}$$

Element	ν	u_A	v_A	σ_{xB}
Q4	0.25	3.33 (3.33)	3.33 (3.21)	–2833 (–2733)
	0.49	0.75 (1.00)	0.75 (0.90)	–5168 (–4092)
	0.49999	0.00 (0.00)	0.00 (0.00)	–5999 (–4780)
NQ6	0.25	3.75 (3.68)	3.75 (3.68)	–3000 (–3046)
	0.49	3.04 (2.96)	3.04 (2.98)	–3000 (–2960)
	0.4999	3.01 (2.91)	3.01 (2.92)	–3002 (–2942)

(Numbers in the parentheses are obtained by using irregular meshes)

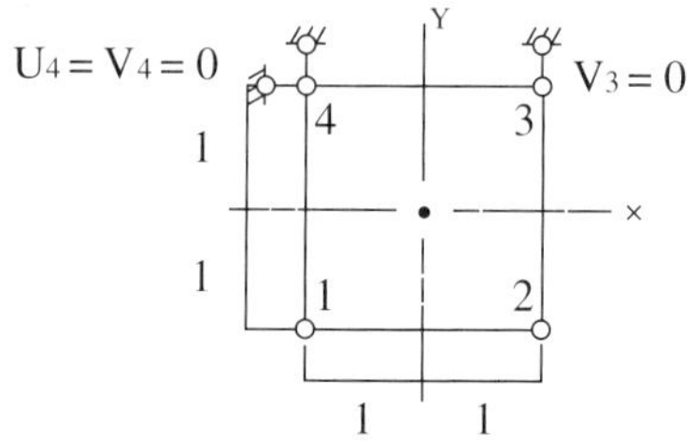

FIGURE 7.4
Constrained Q_4 square element 2×2.

where **N** is the 2-D bilinear interpolation function. The minimum restrictions to prevent the element rigid body motion are shown in Figure 7.4. The remaining element displacement is now

$$\mathbf{u}^* = \begin{Bmatrix} u^* \\ v^* \end{Bmatrix} = \mathbf{N}^*\mathbf{q}^*, \; \mathbf{q}^* = \{u_1 \quad v_1 \quad u_2 \quad v_2 \quad u_3\}^T \tag{7.13}$$

The volumetric strain is

$$e = \frac{\partial u^*}{\partial x} + \frac{\partial v^*}{\partial y} = \begin{bmatrix} \dfrac{\partial}{\partial x} & \dfrac{\partial}{\partial y} \end{bmatrix} \mathbf{N}^*\mathbf{q}^* = \mathbf{B}^*\mathbf{q}^* \tag{7.14}$$

and the element dilatation stiffness is

$$\mathbf{K}^*_v = \int_{-1}^{1}\int_{-1}^{1} \mathbf{B}^{*T}\mathbf{B}^* dxdy = \frac{1}{12}\begin{bmatrix} 4 & 3 & -4 & 3 & -2 \\ 3 & 4 & -3 & 2 & -3 \\ -4 & -3 & 4 & -3 & 2 \\ 3 & 2 & -3 & 4 & -3 \\ -2 & -3 & 2 & -3 & 4 \end{bmatrix} \tag{7.15}$$

Since rank $\left(\underset{\sim}{\mathbf{K}}^*_v\right) = 3$, the deformation DOFs are $n_d = 5 - 3 = 2$. Solving the homogeneous Equation 7.10, we obtain the following nonzero solution:

$$\mathbf{q}^* = \begin{Bmatrix} u_1 \\ v_1 \\ u_2 \\ v_2 \\ u_3 \end{Bmatrix} = \begin{bmatrix} 1 & -1 \\ 0 & 1 \\ 1 & 0 \\ 0 & 1 \\ 0 & 1 \end{bmatrix} \begin{Bmatrix} u_2 \\ v_2 \end{Bmatrix} \tag{7.16}$$

Substituting it into Equation 7.13, we find a possible deformation solution under the incompressibility assumption:

$$\mathbf{u}^* = \begin{bmatrix} (1-y)/2 & (x+y)/2 \\ 0 & (1-y)/2 \end{bmatrix} \begin{Bmatrix} u_2 \\ v_2 \end{Bmatrix} \tag{7.17}$$

The element deformation modes corresponding to $\begin{Bmatrix} u_2 = 1 \\ 0 \end{Bmatrix}$ and $\begin{Bmatrix} 0 \\ v_2 = 1 \end{Bmatrix}$ are shown in Figure 7.5a and Figure 7.5b, respectively.

FIGURE 7.5
Deformation modes of the Q_4 element in the incompressible case.

Besides the three rigid body modes, the Q_4 element should render the following five deformation modes [9]: the compression mode, stretch mode, shear mode, and two independent flexure modes. But under the incompressibility assumption, only the shear and stretch modes are possible. Thus the element has no capacity to adequately approximate the incompressible plastic deformations.

7.3.3 Incompatible Discrete Model

In order to relax the restrictions caused by the incompressibility we introduce a set of incompatible trials, called the element inner displacements, into the Q_4 isoparametric element. For example, we choose an incompatible function

$$u_\lambda = \begin{bmatrix} \xi^2 & \eta^2 \end{bmatrix} \begin{Bmatrix} \lambda_1 \\ \lambda_2 \end{Bmatrix} \tag{7.18}$$

where λ_1 and λ_2 are the element inner parameters and ξ and η are the element isoparametric coordinates (see Figure 7.6). The incompatible function (Equation 7.18) does not pass the patch test unless the element mesh is regular. Let us consider the "strong form of the patch test" shown by Wu et al. [10]:

$$\oint_{\partial A} \begin{Bmatrix} l \\ m \end{Bmatrix} u_\lambda dS = 0 \tag{7.19}$$

where $\{l \quad m\}$ are the direction cosines of the outward normal on the element sides. In order to absorb the condition (Equation 7.19) a modifying term with the virtual parameters α and β is added to Equation 7.18:

$$u_\lambda = \begin{bmatrix} \xi^2 & \eta^2 \end{bmatrix} \begin{Bmatrix} \lambda_1 \\ \lambda_2 \end{Bmatrix} + \begin{bmatrix} \xi & \eta \end{bmatrix} \begin{Bmatrix} \alpha \\ \beta \end{Bmatrix} \tag{7.20}$$

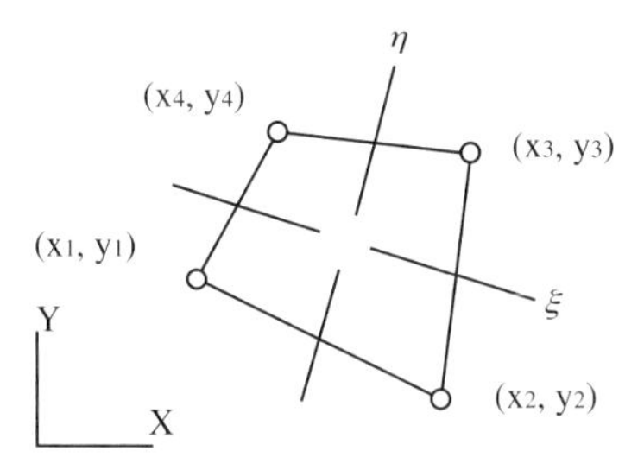

FIGURE 7.6
Quadrilateral element.

By substituting Equation 7.20 into Equation 7.19 and then eliminating [α β], the desirable incompatible function is obtained as follows:

$$u_\lambda = \begin{bmatrix} \xi^2 + \Delta & \eta^2 - \Delta \end{bmatrix} \begin{Bmatrix} \lambda_1 \\ \lambda_2 \end{Bmatrix} \tag{7.21}$$

where

$$\Delta = \frac{2}{3}\left[\left(a_1 b_2 - b_1 a_2\right)\xi - \left(a_2 b_3 - b_2 a_3\right)\eta\right] / \left(a_3 b_1 - b_3 a_1\right)$$

$$a_1 = \left(-x_1 + x_2 + x_3 - x_4\right)/4$$

$$a_2 = \left(x_1 - x_2 + x_3 - x_4\right)/4$$

$$a_3 = \left(-x_1 - x_2 + x_3 + x_4\right)/4$$

$$b_1 = \left(-y_1 + y_2 + y_3 - y_4\right)/4$$

$$b_2 = \left(y_1 - y_2 + y_3 - y_4\right)/4$$

$$b_3 = \left(-y_1 - y_2 + y_3 + y_4\right)/4$$

The incompatible basis function in Equation 7.21 is used also for the other displacement component, so we have

$$\mathbf{u}_\lambda = \begin{Bmatrix} u_\lambda \\ v_\lambda \end{Bmatrix} = \begin{bmatrix} \xi^2 + \Delta & \eta^2 - \Delta & 0 & 0 \\ 0 & 0 & \xi^2 + \Delta & \eta^2 - \Delta \end{bmatrix} \begin{Bmatrix} \lambda_1 \\ \vdots \\ \lambda_4 \end{Bmatrix} = \mathbf{N}_\lambda \lambda \tag{7.22}$$

In Figure 7.7 a patch test is presented for an elastic cylinder under a constant stress state. The numerical results show that the incompatible element passed

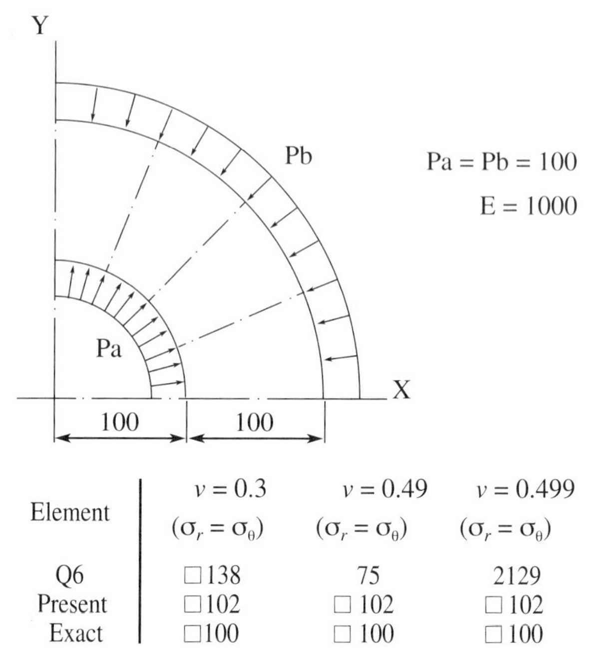

Element	$\nu = 0.3$ $(\sigma_r = \sigma_\theta)$	$\nu = 0.49$ $(\sigma_r = \sigma_\theta)$	$\nu = 0.499$ $(\sigma_r = \sigma_\theta)$
Q6	□138	75	2129
Present	□102	□ 102	□ 102
Exact	□100	□ 100	□ 100

FIGURE 7.7
Elastic cylinder under opposite pressures (dimensionless quantities) (P = 300, E = 1000).

the test (the 2% stress error comes from using four straight-sided elements to approximate a curvilinear body). But the Q_6 incompatible element [11] cannot pass the test.

Considering the incompressibility of the incompatible element in accordance with Equation 7.22, we again take the constrained element in Figure 7.4 as an analysis example. Instead of Equation 7.13, the element trial function can be formulated as

$$\mathbf{u}^* = \begin{Bmatrix} u_\lambda^* \\ v_\lambda^* \end{Bmatrix} = \underset{\sim}{\mathbf{N}}^* \mathbf{q}^* + \mathbf{N}_\lambda \lambda = \begin{bmatrix} \underset{\sim}{\mathbf{N}}^* & \underset{\sim}{\mathbf{N}}_\lambda \end{bmatrix} \begin{Bmatrix} \mathbf{q}^* \\ \lambda \end{Bmatrix} \tag{7.23}$$

the volumetric strain,

$$e = \begin{bmatrix} \dfrac{\partial}{\partial x} & \dfrac{\partial}{\partial y} \end{bmatrix} \begin{bmatrix} \mathbf{N}^* & \mathbf{N}_\lambda \end{bmatrix} \begin{Bmatrix} \mathbf{q}^* \\ \lambda \end{Bmatrix} = \begin{bmatrix} \mathbf{B}^* & \mathbf{B}_\lambda \end{bmatrix} \begin{Bmatrix} \mathbf{q}^* \\ \lambda \end{Bmatrix} \tag{7.24}$$

and the dilatation stiffness matrix,

$$
\mathbf{K}_v^* = \int_{-1}^{1}\int_{-1}^{1} \begin{bmatrix} \mathbf{B}^* & \mathbf{B}_\lambda \end{bmatrix}^T \begin{bmatrix} \mathbf{B} & \mathbf{B}_\lambda \end{bmatrix} dxdy
$$

$$
= \frac{1}{12}\begin{bmatrix}
4 & 3 & -4 & 3 & -2 & 0 & 0 & 0 & 8 \\
3 & 4 & -3 & 2 & -3 & 8 & 0 & 0 & 0 \\
-4 & -3 & 4 & -3 & 2 & 0 & 0 & 0 & -8 \\
3 & 2 & -3 & 4 & -3 & -8 & 0 & 0 & 0 \\
-2 & -3 & 2 & -3 & 4 & 0 & 0 & 0 & 8 \\
0 & 8 & 0 & -8 & 0 & 64 & 0 & 0 & 0 \\
0 & 0 & 0 & 0 & 0 & 0 & 0 & 0 & 0 \\
0 & 0 & 0 & 0 & 0 & 0 & 0 & 0 & 0 \\
8 & 0 & -8 & 0 & 8 & 0 & 0 & 0 & 64
\end{bmatrix} \tag{7.25}
$$

We find that rank $(\underset{\sim}{\mathbf{K}}_v^*) = 3$ and the deformation DOF are now $n = 9 - 3 = 6$. Various linear or square stretch, shear, and flexure modes can be composed in terms of the six-element deformation DOF, such that the incompatible discrete model can approximate the incompressible materials and plastic deformations. A simple test for the incompressibility calculations is given in Figure 7.8. This is an elastic plane-strain cantilever with four rectangular elements. When $\upsilon \to 0.5$, the Q_4 element locks while the incompatible element does not lock.

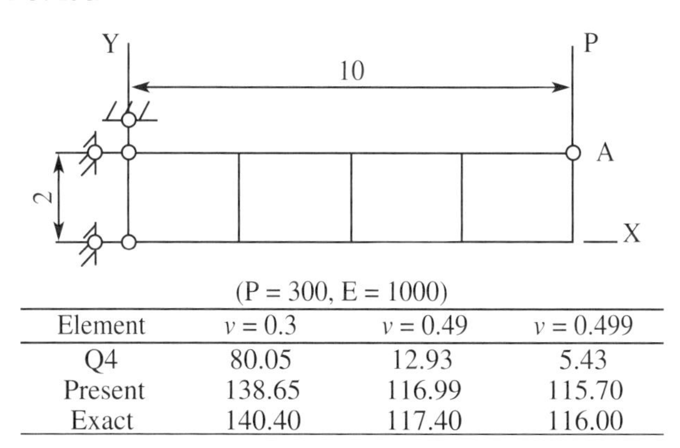

Element	$v = 0.3$	$v = 0.49$	$v = 0.499$
Q4	80.05	12.93	5.43
Present	138.65	116.99	115.70
Exact	140.40	117.40	116.00

FIGURE 7.8
Deflection solution u_A of the elastic plane strain cantilever dimensionless quantities.

7.3.4 Application to the Limit Analysis

According to the flow theory of plasticity with the von Mises yield condition, a relation between the increments of stress and strain can be established. For ideal plasticity it takes the simple form,

$$
\Delta\sigma = \mathbf{C}\Delta\varepsilon \tag{7.26}
$$

with

$$\mathbf{C} = \begin{cases} \mathbf{C}_e & \text{for } \bar{\sigma} < \sigma_s \\ \mathbf{C}_{ep} = \mathbf{C}_e - \mathbf{C}_p & \text{for } \bar{\sigma} = \sigma_s \end{cases}$$

Here $\mathbf{C}_e$ and $\mathbf{C}_p$ are the elastic and plastic tangent stiffness matrices, respectively, $\mathbf{C}_p = \mathbf{C}_p(\sigma')$ is a function of the deviator stress, σ', $\bar{\sigma}$ is the element effective stress, and σ_s is the material yield stress.

For our incompatible element the strain is now

$$\underset{\sim}{\varepsilon} = \underset{\sim}{\mathbf{D}}(\underset{\sim}{\mathbf{u}} + \underset{\sim}{\mathbf{u}}_\lambda) = \mathbf{D}\begin{bmatrix} \underset{\sim}{\mathbf{N}} & \underset{\sim}{\mathbf{N}}_\lambda \end{bmatrix} \begin{Bmatrix} \underset{\sim}{\mathbf{q}} \\ \underset{\sim}{\lambda} \end{Bmatrix} \tag{7.27}$$

where u is the Q_4-compatible displacement field and q is the corresponding node displacement. The relevant element stiffness matrix is given by

$$\mathbf{K} = \int_A (\mathbf{D}\begin{bmatrix} \mathbf{N} & \mathbf{N}_\lambda \end{bmatrix})^T \mathbf{C}(\mathbf{D}\begin{bmatrix} \mathbf{N} & \mathbf{N}_\lambda \end{bmatrix})\,dA \tag{7.28}$$

By making a static condensation on the element level, the reduced system stiffness matrix $\bar{K}$ can be constructed. Accordingly, the system incremental equation reads

$$\underset{\sim}{\bar{\mathbf{K}}}\Delta\underset{\sim}{\bar{\mathbf{q}}} = \Delta\underset{\sim}{\bar{\mathbf{Q}}} \tag{7.29}$$

where $\Delta\bar{\mathbf{Q}}$ and $\Delta\bar{\mathbf{q}}$ are increments of the system load $\underset{\sim}{\bar{\mathbf{Q}}}$ and the system displacement $\bar{\mathbf{q}}$, respectively. The stiffness matrix $\bar{\mathbf{K}} = \bar{\mathbf{K}}(\bar{\mathbf{q}})$ is dependent on the node displacements unless the whole system is in an elastic state.

In order to solve Equation 7.29 the usual incremental iterative procedures can be employed, and many current nonlinear programs may be applied directly. In the following we apply the incompatible discrete model to the limit analysis for plane strain problems.

7.3.4.1 Example I: Thick Elastic-Plastic Cylinder under Internal Pressure

The element meshes of a quarter of a cylinder section are shown in Figure 7.9, where the developing process of the plane plastic range is also indicated by the numbers inside the meshes. The numerical results in Table 7.1 show that for the compatible element Q_4, after the yielding of all elements, the internal pressure p can still increase without destroying the cylinder. In other words, there is no collapse load in the compatible isoparametric element solutions.

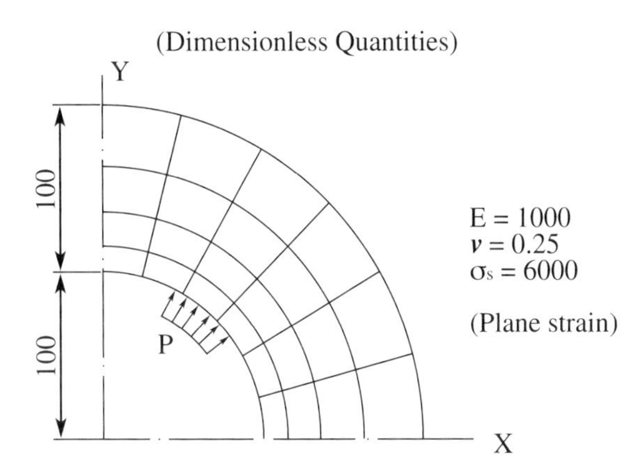

FIGURE 7.9
Thick elastic-plastic cylinder under internal pressure p.

TABLE 7.1
Dimensionless Quantities

Element	Pe (Elastic limit load)	Ps (Collapse load)
Q_4	2979	(Does not exist)
NQ6	2975	4778
Prager [44]	2964	4802

On the contrary, for the incompatible element, when the value of p increases to $p_s = 4778$, the radial displacement of the cylinder tends to be infinite and the system cannot bear additional loads. So p_s is just the collapse load of the cylinder. The elastic limit load p_e, the whole structure to be a plastic one, are also listed in Table 7.1.

7.3.4.2 *Example II: Elastic-Plastic Square Tube with a Circular Hole under Internal Pressure*

The structural size and the finite element meshes are shown in Figure 7.10, and the plastic incompatible element is indicated in the meshes. After the fourth set of elements is in the plastic range, the displacement u_A suddenly becomes infinite. When the collapse of the structure takes place, some elements are still in an elastic state. In Table 7.2 we see that for the Q_4 element there is no collapse pressure, and p can further increase even after all elements are in the plastic region. Figure 7.11 shows that the real structural stiffness is overestimated by a compatible isoparametric element. The results given by Casciaro and Cascini [7] and Gao [8] are included in Table 7.2 for the sake of comparison. The numbers in parentheses are the numbers of the mesh nodes used in the cited papers.

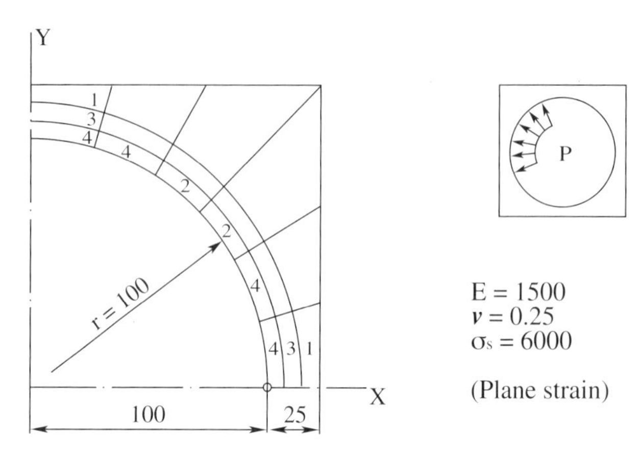

FIGURE 7.10
Elastic-plastic square tube with a circular hole.

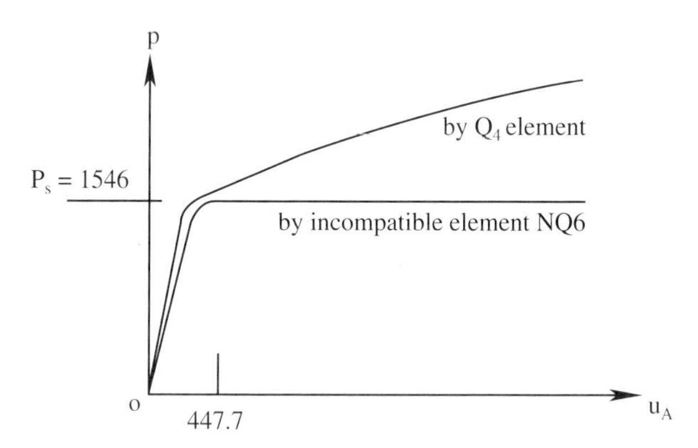

FIGURE 7.11
Pressure-displacement curve for the example in Figure 7.10.

TABLE 7.2

Dimensionless Quantities

Element	Pe (Elastic limit load)	Ps (Plastic limit load)
Q_4	1503	(Does not exist)
NQ6 [10]	1382	1572(28)
Casciaro [7]	—	1612(?)
Gao [8]	—	1572(36)
Prager [45]	1338	1546

7.3.4.3 Example III: Elastic-Plastic Tension Specimen with Semicircular Edge Notches

The data and meshes are given in Figure 7.12. The numbers marked in the individual elements indicate the yielding order of the incompatible elements.

Table 7.3 lists the results obtained for this tension specimen. The incompatible solutions are compared with those based on the Q_4 element and some other results, where P_s^+ and P_s^- are the upper and lower bounds of the limit load, respectively. From the numerical results and from Figure 7.13 we see that with a few nodes, a rational solution of the collapse load is obtained once again by the incompatible element NQ6.

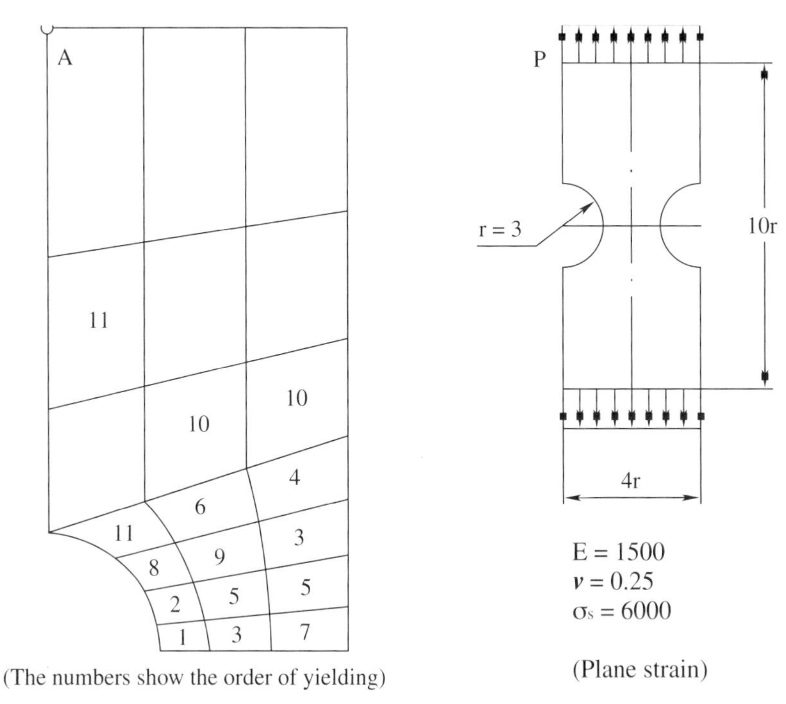

FIGURE 7.12
Elastic-plastic tension specimen with semicircular edge notches.

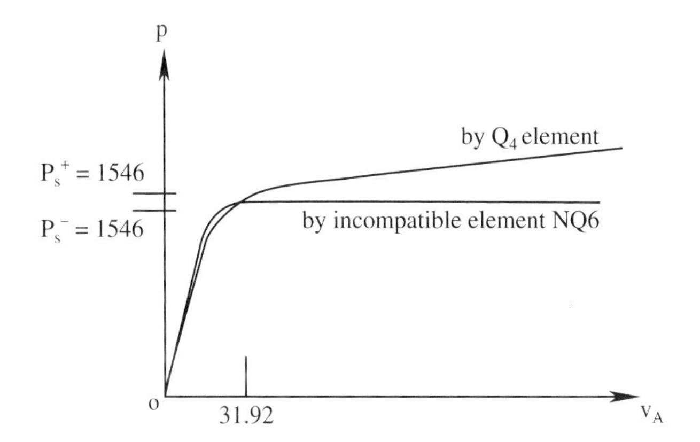

FIGURE 7.13
Tension load-displacement curve for the example in Figure 7.12.

TABLE 7.3

Dimensionless Quantities

Element	Pe	Ps	Pss
Q_4	2748	(Does not exist)	5875(32)
Present	2747	4792(32)	*
Casciaro	—	4822(113)	—
Gao	—	4704(56)	—
Prager		$P_s^+ = 4365$	
		$P_s^{\pounds} = 4815$	

7.4 Deviatoric Hybrid Model for the Incompressible Medium

7.4.1 Introduction

A significant amount of research has recently been devoted to finite element methods in incompressible problems (including solids and fluids) for both theoretical implications and the underlying engineering background. Since Herrmann's initial work was published [12], various of variational principles [13–15] and discrete models have been proposed for incompressible calculations as surveyed in Reference 16.

Displacement-based finite elements, especially isoparametric elements, have been used in incompressible analysis for their simple formulations and standard solution procedures. Displacement locking and instability of the stress solution may, however, arise in nearly incompressible situations. Approaches to overcome these problems are the penalty method with selective integration [17, 18] and the incompatible-element method [19, 20], both of which can be applied in nearly incompressible analysis with satisfactory results, although not effective for fully incompressible analysis. The fully incompressible medium was first dealt with by Needleman and Shih [21], wherein the dependent displacement parameters stemming from the zero volumetric constraint ($e = 0$) were eliminated from the system equations. In this approach, the band character of the global stiffness matrix is destroyed; therefore, an algorithm elimination was proposed by Zhang [22] and Li [23] to incorporate the incompressible condition beforehand. Both elimination approaches nevertheless run into trouble in the estimation of stress.

Meanwhile, the mixed element approximations with pressure and displacement ($p - u$) as independent fields [see References 12, 14, 18, 24, 25] are suited to both the compressible and incompressible analyses. The approaches, however, always lead to a nonpositive-definite element matrix with a zero diagonal block that affects their efficiency for obtaining a solution. Furthermore, in the context of mixed element formulations based on the modified potential energy principle, zero energy modes for stress (ZEM[σ]) can be induced by the displacement continuity between elements [see Reference 26 for details], and the stability difficulty mentioned above attracted much attention in the past few years [27]. An effective three-field ($u - p - e$) mixed model was thus given by Simo et al. [28] and has been extended to the finite deformation plasticity flow with near incompressibility.

In line with these investigations, it has been found that assumed-stress hybrid elements do not usually lock and can offer reasonable stress solutions at a nearly incompressible limit. In these models, stress parameters are eliminated at the element level such that the system of displacement stiffness equations, which can be solved more easily, can be obtained. In spite of these fully incompressible materials with $\mu = 0.5$, the flexible matrix $\boldsymbol{H}$ becomes singular, which results in failure for hybrid element models [29, 30]. For the numerical simulation of incompressible Stokes flow, Tong [31] and Bratianu and Atluri [32] proposed an alternative approach in which the stress field is split into the deviatoric stress field and the hydrostatic pressure field, they are assumed independently in element approximation. Although this gives rise to the avoidance of the singularity of the flexible matrix, pressure variables cannot be eliminated at the element level, and thus the resultant equation system remains mixed.

This chapter formulates a three-field hybrid model, in which the incompatible displacement is introduced reasonably such that the pressure variable can be eliminated at the element level. This model is based on an unconditional variational principle for both the compressible and the incompressible medium, in which the complementary energy of the element is split into deviatoric and volumetric parts to be approximated. The formulation therefore unifies the compressible and incompressible (including Stokes flow) calculations, and the band and sparse characters remain in the system equations.

The outline of the section is as follows. In Section 7.4.2, a deviatoric variational principle is deduced, based on which a hybrid element is proposed in Section 7.4.3. Section 7.4.4 discusses the internal variable elimination at the element level. As a concrete example, a plane hybrid element model is constructed in Section 7.4.5 and a technique for suppressing zero energy modes is introduced in Section 7.4.6. Numerical examples are presented in Section 7.4.7.

7.4.2 Deviatoric Variational Principle

The Hellinger–Reissner functional, in the case of small deformations and linear elastic materials, is given by

$$\pi_R(\sigma_{ij}, u_i) = \int_{V_e} \left[\sigma_{ij}\varepsilon_{ij} - B(\sigma_{ij})\right] dv - l.t. \tag{7.30}$$

where $\varepsilon_{ij} = ½(u_{i,j} + u_{j,i})$, $B(\sigma_{i,j})$, and *l.t.* are the strain, the complementary energy density, and the loading term, respectively.

Meanwhile, the deviatoric stress and the deviatoric strain can be expressed as

$$\sigma'_{ij} = \sigma_{ij} - p\delta_{ij} \tag{7.31}$$

$$\varepsilon'_{ij} = \varepsilon_{ij} - \frac{1}{3}e\delta_{ij} \tag{7.32}$$

where $p = \sigma_{ii}$ is the hydrostatic pressure, and $e = u_{i,i}$ is the volume strain. Substitution of Equation 7.31 and Equation 7.32 into Equation 7.30 yields a three-field functional:

$$\pi'(\sigma'_{ij}, p, u_i) = \int_{V_e} \left[\sigma'_{ij}\varepsilon'_{ij} + pe - B_s(\sigma'_{ij}) - B_v(p)\right] dv - l.t. \tag{7.33}$$

where $B_s(\sigma'_{ij})$ and $B_v(p)$ are the deviatoric complementary energy and the volumetric energy functions, respectively, in the form of

$$B_s\left(\sigma'_{ij}\right) = \frac{1}{2}\frac{1}{2G}\sigma'_{ij}\sigma'_{ije}, \text{ G = shear modulus} \tag{7.34}$$

$$B_v(p) = \frac{1}{2}\frac{1}{k}p^2, \; k = \frac{E}{3(1-2\mu)} = \text{bulk modulus} \tag{7.35}$$

It is easily certified that if the prescribed displacement boundary condition is satisfied, the stationary condition $\delta\pi'(\sigma'_{ij}, p, u_i) = 0$ will lead to the equilibrium equation

$$\sigma'_{ij,j} + p_{,i} = 0 \text{ in } V \tag{7.36}$$

and the constitutive relationships

$$e = \frac{p}{k} \text{ in } V \tag{7.37}$$

$$\varepsilon'_{ij} = \frac{1}{2G}\sigma'_{ij} \text{ in } V \tag{7.38}$$

Besides a traction boundary condition can also be obtained.

As for general compressible materials, the Hellinger–Reissner principle in relation to the functional (Equation 7.30) is an unconditional variational problem [33]. The principle for incompressible materials is, however, converted into a conditional variational problem with the zero volumetric constraint

$$e = u_{i,j} = 0 \tag{7.39}$$

which causes an enormous amount of difficulty in unifying the compressible and the incompressible calculations. However, as a Euler equation, the $e - p$ relation (Equation 7.37) derived from the functional $\pi'(\sigma'_{ij}, p, u_i)$ (Equation 7.33) may naturally degenerate into zero volume constraint (Equation 7.39) when the medium becomes incompressible. Being independent of compressibility, the functional (Equation 7.33) is, hence, always an unconditional variational, which makes it possible to develop a universal numerical approach for both the compressible and incompressible analyses.

The above deviatoric variational principle can also be used to solve the incompressible Stokes flow, wherein u_i, σ'_{ij}, and p are the velocity field, the deviatoric stress, and the hydrostatic pressure, respectively, while constant G in Equation 7.34 represents the fluid viscosity. Stokes flow allows the volume complementary energy to vanish, i.e. $B_v(p) = 0$ and the energy functional (Equation 7.33) is therefore simplified as

$$\pi'(\sigma'_{ij}, p, u_i) = \int_{V_e} \left[\sigma'_{ij}\varepsilon'_{ij} + pe - B_s(\sigma'_{ij}) \right] dv - l.t. \tag{7.40}$$

where $B_s(\sigma'_{ij})$ is the deviatoric complementary energy function:

$$B_s(\sigma'_{ij}) = \frac{1}{2}\frac{1}{2G}\sigma'_{ij}\sigma'_{ij}, G = \text{shear modulus} \tag{7.41}$$

$$B_v(p) = \frac{1}{2}\frac{1}{K}P^2, \quad K = \frac{E}{3(1-2\mu)} = \text{Bulk modulus} \tag{7.42}$$

7.4.3 Three-Field Hybrid Element Formulation

Similar to conventional hybrid element formulations, the element stress fields for σ'_{ij} and p are assumed to be

$$\sigma' = \{\sigma'_{ij}\} = \phi_\beta \beta \tag{7.43}$$

$$p = \phi_\alpha \alpha \tag{7.44}$$

where α and β are the parameters for the pressure and the deviatoric stress, respectively. The displacement field is assumed to be

$$\mathbf{u} = \{u_i\} = \mathbf{u}_q + \mathbf{u}_\lambda = \left[\mathbf{N}_q \ \mathbf{N}_\lambda \right] \begin{Bmatrix} \mathbf{q} \\ \lambda \end{Bmatrix} \tag{7.45}$$

in which $\mathbf{u}_q$ is the conventional compatible displacement interpolation in terms of element nodal displacement $\mathbf{q}$

$$\mathbf{u}_q = \mathbf{N}_q \mathbf{q} \tag{7.46}$$

However, the incompatible displacements are defined by the element parameter λ within the element

$$\mathbf{u}_\lambda = \mathbf{N}_\lambda \lambda \tag{7.47}$$

the introduction of $\mathbf{u}_\lambda$ may improve the performance of the element [34, 35]. Usually the employed incompatible displacement should satisfy the constant stress patch test condition (PTC) [36–39].

In accordance with Equation 7.47, the deviatoric and the volumetric strains can be, respectively, written as

$$\varepsilon' = \begin{bmatrix} \mathbf{B}_q & \mathbf{B}_\lambda \end{bmatrix} \begin{Bmatrix} \mathbf{q} \\ \lambda \end{Bmatrix} \tag{7.48}$$

$$\mathbf{e} = \begin{bmatrix} \psi_q & \psi_\lambda \end{bmatrix} \begin{Bmatrix} \mathbf{q} \\ \lambda \end{Bmatrix} = \mathbf{e}_q + \mathbf{e}_\lambda \tag{7.49}$$

Substituting these assumed fields into the functional (Equation 7.33) gives

$$\begin{aligned} \pi'(\alpha,\beta,\mathbf{q},\lambda) = {} & \beta^T \begin{bmatrix} \mathbf{G}_{\beta q} & \mathbf{G}_{\beta\lambda} \end{bmatrix} \begin{Bmatrix} \mathbf{q} \\ \lambda \end{Bmatrix} + \mathbf{a}^T \begin{bmatrix} \mathbf{G}_{\alpha q} & \mathbf{G}_{\alpha\lambda} \end{bmatrix} \begin{Bmatrix} \mathbf{q} \\ \lambda \end{Bmatrix} \\ & - \frac{1}{2}\beta^T \mathbf{H}_{\beta\beta}\beta - \frac{1}{2}\alpha^T \mathbf{H}_{\alpha\varepsilon}\alpha - \mathbf{q}^T \mathbf{Q} \end{aligned} \tag{7.50}$$

where

$$\begin{bmatrix} \mathbf{G}_{\beta q} & \mathbf{G}_{\beta\lambda} \end{bmatrix} = \int_{V_e} \phi_\beta^T \begin{bmatrix} \mathbf{B}_q & \mathbf{B}_\lambda \end{bmatrix} dv$$

$$\begin{bmatrix} \mathbf{G}_{\alpha q} & \mathbf{G}_{\alpha\lambda} \end{bmatrix} = \int_{V_e} \phi_\alpha^T \begin{bmatrix} \psi_q & \psi_\lambda \end{bmatrix} dv$$

$$\mathbf{H}_{\beta\beta} = \frac{1}{2G} \int_{V_e} \phi_\beta^T \phi_\beta dv$$

$$\mathbf{H}_{\alpha\alpha} = \frac{1}{k} \int_{V_e} \phi_\beta^T \phi_\beta dv$$

In Equation 7.47, Q is the equivalent nodal loads. The stationary condition $\partial\pi' / \partial\beta = 0$ gives rise to

$$\beta = \mathbf{H}_{\beta\beta}^{-1} \begin{bmatrix} \mathbf{G}_{\beta q} & \mathbf{G}_{\beta\lambda} \end{bmatrix} \begin{Bmatrix} \mathbf{q} \\ \lambda \end{Bmatrix} \tag{7.51}$$

The substitution of Equation 7.48 into Equation 7.47 yields

$$\pi'(\alpha, \mathbf{q}, \lambda) = \frac{1}{2} \begin{Bmatrix} \mathbf{q} \\ \lambda \end{Bmatrix}^T \begin{bmatrix} \mathbf{K}_{qq} & \mathbf{K}_{q\lambda}^T \\ \mathbf{K}_{q\lambda} & \mathbf{K}_{\lambda\lambda} \end{bmatrix} \begin{Bmatrix} \mathbf{q} \\ \lambda \end{Bmatrix} + \alpha^T \begin{bmatrix} \mathbf{G}_{\alpha q} & \mathbf{G}_{\alpha\lambda} \end{bmatrix} \begin{Bmatrix} \mathbf{q} \\ \lambda \end{Bmatrix} - \frac{1}{2} \alpha^T \mathbf{H}_{\alpha\alpha} \alpha \tag{7.52}$$

where the distortion stiffness matrix is given by

$$\begin{bmatrix} \mathbf{K}_{qq} & \mathbf{K}_{q\lambda}^T \\ \mathbf{K}_{q\lambda} & \mathbf{K}_{\lambda\lambda} \end{bmatrix} = \begin{bmatrix} \mathbf{G}_{\beta qe} & \mathbf{G}_{\beta\lambda} \end{bmatrix}^T \mathbf{H}_{\beta\beta}^{-1} \begin{bmatrix} \mathbf{G}_{\beta q} & \mathbf{G}_{\beta\lambda} \end{bmatrix} \tag{7.53}$$

By using the stationary condition for π' in Equation 7.49 with respect to $\{\mathbf{q} \quad \lambda\}$, we have

$$\begin{bmatrix} \mathbf{K}_{qq} & \mathbf{K}_{q\lambda}^T & \mathbf{G}_{\alpha q}^T \\ \mathbf{K}_{q\lambda} & \mathbf{K}_{\lambda\lambda} & \mathbf{G}_{\alpha\lambda}^T \\ \mathbf{G}_{\alpha q} & \mathbf{G}_{\alpha\lambda} & -\mathbf{H}_{\alpha\alpha} \end{bmatrix} \begin{Bmatrix} \mathbf{q} \\ \lambda \\ \alpha \end{Bmatrix} = \begin{Bmatrix} \mathbf{Q} \\ 0 \\ 0 \end{Bmatrix} \tag{7.54}$$

For the sake of convenience, let

$$\mathbf{M} = \begin{bmatrix} \mathbf{K}_{\lambda\lambda} & \mathbf{G}_{\alpha\lambda}^T \\ \mathbf{G}_{\alpha\lambda} & -\mathbf{H}_{\alpha\alpha} \end{bmatrix} \tag{7.55}$$

For a given compressible or nearly incompressible medium, $\mathbf{H}_{\alpha\alpha}$ is positive definite, so that $|\mathbf{M}| \neq 0$. Parameters λ and α henceforth turn out as

$$\begin{Bmatrix} \lambda \\ \alpha \end{Bmatrix} = -\mathbf{M}^{-1} \begin{bmatrix} \mathbf{K}_{q\lambda} \\ \mathbf{G}_{\alpha q} \end{bmatrix} \mathbf{q} \tag{7.56}$$

and can be eliminated from Equation 7.51. Conclusively, Equation 7.51 is transformed into the conventional stiffness equation only with respect to $\mathbf{q}$, that is,

$$\mathbf{K}_{qq}^{*}\mathbf{q} = \mathbf{Q} \tag{7.57}$$

where

$$\mathbf{K}_{qq}^{*} = \mathbf{K}_{qq} - \begin{bmatrix} \mathbf{K}_{q\lambda} \\ \mathbf{G}_{\alpha\lambda} \end{bmatrix}^{T} \mathbf{M}^{-1} \begin{bmatrix} \mathbf{K}_{q\lambda} \\ \mathbf{G}_{\alpha\lambda} \end{bmatrix} \tag{7.58}$$

7.4.4 Incompressible Elimination at the Element Level

Although the hybrid element formulation in the previous section can be implemented to calculate compressible and nearly incompressible materials, it is still hard to use the formulations for solving fully incompressible problems, in which $k = \infty$ and $\mathbf{H}_{\alpha\alpha} = 0$, the matrix in Equation 7.52 being of the form

$$\mathbf{M} = \begin{bmatrix} \mathbf{K}_{\lambda\lambda} & \mathbf{G}_{\alpha\lambda}^{T} \\ \mathbf{G}_{\alpha\lambda} & \mathbf{0} \end{bmatrix} \tag{7.59}$$

As will be shown later, the matrix (Equation 7.56) is usually rank-deficient, and one has

THEOREM 7.1

If incompatible displacement $\mathbf{u}_{\lambda}$ passes the PTC, then it must satisfy the element incompressible condition in the sense of integration, specified as

$$\int_{V_e} \mathbf{e}_{\lambda} dv = 0 \tag{7.60}$$

where $\mathbf{e}_{\lambda} = div\ \mathbf{u}_{\lambda}$ is the incompatible volume strain.

Proof

Observing that

$$\mathbf{D}\mathbf{u}_{\lambda} = 0 \Rightarrow \text{div}\ \mathbf{u}_{\lambda} = 0 \tag{7.61}$$

accordingly gives us

$$\int \mathbf{D}\mathbf{u}_{\lambda} dv = 0 \Rightarrow \int \text{div } \mathbf{u}_{\lambda} dv = 0 \tag{7.62}$$

or, equivalently,

$$\int \varepsilon_{\lambda} dv = 0 \text{ (i.e. } \mathbf{u}_{\lambda} \text{ passes PTC)} \Rightarrow \int \mathbf{e}_{\lambda} dv = 0 \tag{7.63}$$

and the theorem is proved. In addition, the incompressible condition (Equation 7.57), in view of $\mathbf{e}_{\lambda} = \psi_{\lambda}\lambda$ in Equation 7.46, can also be written as

$$\int_{V_e} \psi_{\lambda} dv = 0 \tag{7.64}$$

Without a loss of generality, the assumed pressure trial function can be expressed in terms of the element's natural coordinates as

$$p = \phi_{\alpha}\alpha = \begin{bmatrix} 1 & \xi & \eta & \zeta & \cdots \end{bmatrix} \begin{Bmatrix} \alpha_1 \\ \alpha_2 \\ \vdots \end{Bmatrix} \tag{7.65}$$

In accordance with Equation 7.47 and Equation 7.60, we have

$$\mathbf{G}_{\alpha\lambda} = \int_{V_e} \phi_{\alpha}^{T} \psi_{\lambda} dv = \int_{V} \begin{Bmatrix} 1 \\ \cdots \\ \xi \\ \eta \\ \zeta \\ \vdots \end{Bmatrix} \psi_{\lambda} dv \xrightarrow{(7.60)} \begin{bmatrix} 0 \\ \cdots\cdots\cdots\cdots\cdots \\ \displaystyle\int_{V} \begin{Bmatrix} \xi \\ \eta \\ \zeta \\ \vdots \end{Bmatrix} \psi_{\lambda} dv \end{bmatrix} \tag{7.66}$$

which shows that the row related to the constant pressure vanishes, the matrix $\boldsymbol{M}$ in Equation 7.56 thereby becoming singular. Consequently, Equation 7.53 will no longer hold, and the elimination of variables $\{\lambda \ \alpha\}$ cannot be carried out.

An effective strategy to avoid this problem is to introduce a small disturbance to the incompatible mode $\mathbf{u}_{\lambda}$, as shown in the next section, such that the condition (Equation 7.57) is no longer met, and the parameters $\{\lambda \ \alpha\}$ can be eliminated within an element.

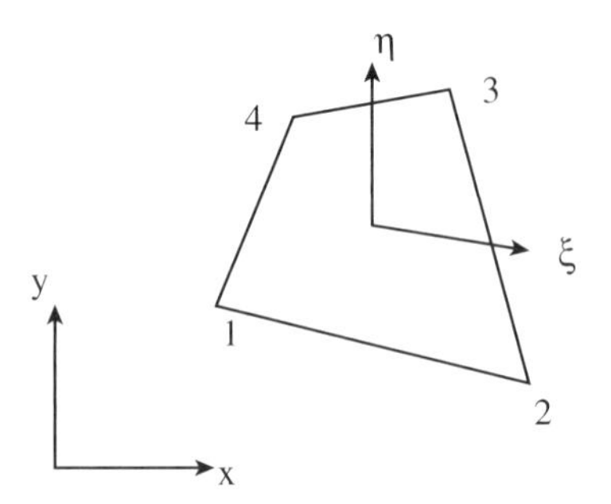

FIGURE 7.14
Four-node plane hybrid element.

7.4.5 Hybrid Model

Let us consider a four-node plane hybrid model [21] (shown in Figure 7.14), the element geometry of which is the same as that of the well-known bilinear isoparametric element Q_4. The compatible displacement is naturally assumed as a standard bilinear interpolation. In order to pass the PTC in this case, incompatible displacement $\mathbf{u}_\lambda$ can be generated by using the general incompatible formulation [40], by which an alternative incompatible model is initially assumed to be

$$\begin{bmatrix} u_\lambda & v_\lambda \end{bmatrix} = \begin{bmatrix} \xi^2 - \Delta & \left|\eta^2 + \Delta\right| & \xi + \eta - \left(\xi^3 + \eta^3\right) \end{bmatrix} \begin{bmatrix} \lambda_1 & \lambda_4 \\ \lambda_2 & \lambda_5 \\ \lambda_3 & \lambda_6 \end{bmatrix} \tag{7.67}$$

where

$$\Delta = \frac{2}{3}\left(\frac{J_1}{J_0}\xi - \frac{J_2}{J_0}\eta\right)$$

and the coefficients J_0, J_1, and J_2 are defined by the Jacobian determinant as follows:

$$|\mathbf{J}| = J_0 + J_1\xi + J_2\eta \tag{7.68}$$

It can be verified that the incompatible displacement $\mathbf{u}_\lambda$ (Equation 7.63) satisfies the incompressible condition (Equation 7.57), and there must be deficiency in the rank of matrix M in Equation 7.56. To solve the singular problem, we introduce a small parameter δ into the high-order component of Equation 7.63, which gives

$$\begin{bmatrix} u_\lambda & v_\lambda \end{bmatrix} = \begin{bmatrix} \xi^2 - \Delta & |\eta^2 + \Delta| & \xi + \eta - (\xi^3 + \eta^3)(1+\delta) \end{bmatrix} \begin{bmatrix} \lambda_1 & \lambda_4 \\ \lambda_2 & \lambda_5 \\ \lambda_3 & \lambda_6 \end{bmatrix} \tag{7.69}$$

The incompatible function (Equation 7.65) does not meet Equation 7.57 any longer and keeps the rank of matrix $\mathbf{G}_{\alpha\lambda}$ sufficient. Moreover, in order to keep $|\mathbf{M}| \neq 0$, the following condition for matching parameters should also be satisfied:

$$\dim(\lambda) \geq \dim(\alpha) \tag{7.70}$$

In accordance with the inequality (Equation 7.66), a bilinear pressure trial function is adopted here:

$$p = \begin{bmatrix} 1 & \xi & \eta & \xi\eta \end{bmatrix} \begin{Bmatrix} \alpha_1 \\ \vdots \\ \alpha_4 \end{Bmatrix} \tag{7.71}$$

For the three-field hybrid model based on the functional (Equation 7.33), besides the condition (Equation 7.66), there exists another parameter matching condition [29, 41–42] that should be satisfied to avoid the appearance of the zero energy mode (ZEM), that is,

$$\dim(\beta) + \dim(\alpha) \geq \dim(\mathbf{q}^*) + \dim(\lambda) \tag{7.72}$$

in which $\mathbf{q}^*$ is the vector of element nodal displacements excluding DOFs of the rigid body. This condition must be obeyed by the rationally assumed deviatoric stress field. However, many stress trial functions, such as the bilinear one, which satisfy the condition (Equation 7.68), may still induce some zero energy displacement mode ZEM($\mathbf{u}$). Indeed, Equation 7.68 is just a necessary condition for avoiding ZEM($\mathbf{u}$), and this results in the problem of how to select a rational deviatoric stress mode.

7.4.6 Suppression of the ZEM

General discussion of the ZEM was given by Pian et al. [20, 26, 29, 39, 41], some of which are still applicable for present hybrid models. We suggest a more effective method to prevent all the possible ZEM(**u**) stems from the bilinear term $\int_{ve} \sigma'_{ije}\varepsilon'_{ij}dv$ in the functional (Equation 7.33).

The exclusion of the element rigid body DOF from nodal displacement **q** yields **q***, and the deviatoric strain (Equation 7.45) then becomes

$$\varepsilon^* = \begin{bmatrix} \mathbf{B}_q^* & \mathbf{B}_\lambda \end{bmatrix} \begin{Bmatrix} \mathbf{q}^* \\ \lambda \end{Bmatrix} \tag{7.73}$$

For the purpose of preventing ZEM(u), we define the element deviatoric stresses in the manner of

$$\sigma' = \phi_\beta \beta = \begin{bmatrix} \mathbf{B}_q^* & \mathbf{B}_\lambda \end{bmatrix} \beta \tag{7.74}$$

and this automatically satisfies dim(β) = dim($\mathbf{q}^*$) + dim(λ) and constrains Equation 7.66 and Equation 7.68. The rigid body displacement does not contribute to the energy functional (Equation 7.33), and in terms of Equation 7.69 and Equation 7.70, the term $\int_{ve} \sigma'_{ije}\varepsilon'_{ij}dv$ in the functional can be formulated as

$$\int_{V^\varepsilon} \sigma'^T \varepsilon' dv = \int_{V^\varepsilon} \sigma'^T \varepsilon^* dv = \beta^T \mathbf{G}^* \begin{Bmatrix} \mathbf{q}^* \\ \lambda \end{Bmatrix} \tag{7.75}$$

where

$$\mathbf{G}^* = \int_{V_e} \begin{bmatrix} \mathbf{B}_q^* & \mathbf{B}_\lambda \end{bmatrix}^T \begin{bmatrix} \mathbf{B}_q^* & \mathbf{B}_\lambda \end{bmatrix} dv \tag{7.76}$$

Observing the positive-definite attribute of strain energy, we have

$$\int_{V_e} \varepsilon^{*T} \varepsilon^* dv = \begin{Bmatrix} \mathbf{q}^* \\ \lambda \end{Bmatrix}^T \mathbf{G}^* \begin{Bmatrix} \mathbf{q}^* \\ \lambda \end{Bmatrix} \geq 0 \tag{7.77}$$

where matrix $\mathbf{G}^*$ is positive definite, and the resulting hybrid element must be free from ZEM(**u**). ε' may be replaced with ε due to the identity $\sigma'^T \varepsilon' \equiv \sigma'^T \varepsilon$ in Equation 7.71.

With the understanding of the ZEM, take a rectangular element as an example, since ZEM(**u**) is more easily induced by a regular element, for which $\Delta = 0$ in Equation 7.65. Additionally, the small parameter δ in Equation 7.65 can be ignored since it is independent of our element rank analysis. The element displacements become

$$\mathbf{u} = \{u \quad v\} = \mathbf{u}_q + \mathbf{u}_\lambda = \begin{bmatrix} 1 & \xi & \eta & \xi\eta & \xi^2 & \eta^2 & \xi+\eta & -(\xi^3+\eta^3) \end{bmatrix} \begin{bmatrix} a_1 & b_1 \\ \vdots & \vdots \\ a_7 & b_7 \end{bmatrix} \tag{7.78}$$

where a_i and b_i are general parameters. For convenience, let the element natural coordinates coincide with the physical coordinates, such that the element strain can be formulated as

$$\varepsilon = \begin{Bmatrix} \dfrac{\partial u}{\partial \xi} \\ \dfrac{\partial v}{\partial \eta} \\ \dfrac{\partial u}{\partial \eta} + \dfrac{\partial v}{\partial \xi} \end{Bmatrix} = \begin{bmatrix} 0 & 1 & 0 & \eta & 2\xi & 0 & 1-3\xi^2 & 0 & 0 & 0 & 0 & 0 & 0 & 0 \\ 0 & 0 & 0 & 0 & 0 & 0 & 0 & 0 & 0 & 1 & \xi & 0 & 2\eta & 1-3\eta^2 \\ 0 & 0 & 1 & \xi & 0 & 2\eta & 1-3\eta^2 & 0 & 1 & 0 & \eta & 2\xi & 0 & 1-3\xi^2 \end{bmatrix} \begin{Bmatrix} a_1 \\ \vdots \\ a_7 \\ b_1 \\ \vdots \\ b_7 \end{Bmatrix} \tag{7.79}$$

Subtracting two zero columns and one of the duplicate columns from the matrix of order (3 × 14), and imposing a linear transform to the matrix, the desired deviatoric stress mode can finally be obtained:

$$\sigma' = \phi_\beta \beta = \begin{bmatrix} 1 & 0 & 0 & \eta & \xi & 0 & \xi^2 & 0 & 0 & 0 & 0 \\ 0 & 1 & 0 & 0 & 0 & 0 & 0 & \xi & 0 & \eta & \eta^2 \\ 0 & 0 & 1 & \xi & 0 & \eta & \eta^2 & \eta & \xi & 0 & \xi^2 \end{bmatrix} \begin{Bmatrix} \beta_1 \\ \vdots \\ \beta_{11} \end{Bmatrix} \tag{7.80}$$

So far, all the element trial functions are determined as specified by Equation 7.42, Equation 7.65, Equation 7.76, and Equation 7.67, respectively. The resultant deviatoric element hybrid model is able to unify the compressible and the fully incompressible calculations.

7.4.7 Numerical Examples

In numerical calculations, the following four-node plane elements are employed: (a) the bilinear isoparametric Q4 element with 2 × 2 Gauss quadrature, denoted as Q4(2 × 2); (b) modified bilinear element Q4(2 × 2 and 1 × 1), in which the deviatoric strain energy and the volumetric strain energy take 2 × 2 and 1 × 1 Gauss quadrature, respectively [18]; (c) the enhanced strain-incompatible element QUAD suggested by Simo and Rifai [43]; and (d) the four-node deviatoric hybrid element with 3 × 3 Gauss quadrature.

7.4.7.1 Constant Stress Patch Test

Since disturbing parameter δ was introduced in the incompatible formulation (Equation 7.65), so that the elimination at the element level can be carried out, let us inspect the effect of the parameter on the constant stress patch test. A plane stress/strain panel with size 10 × 10 and unit thickness is shown in Figure 7.15. The panel was modeled by four irregular elements, for which the pure stretch patch test and the pure shear test were considered.

Some of the resultant stress solutions are listed in Table 7.4. With the range of $10^{-5} \leq \delta \leq 10^{-3}$, regardless of the compressibility of the material, the

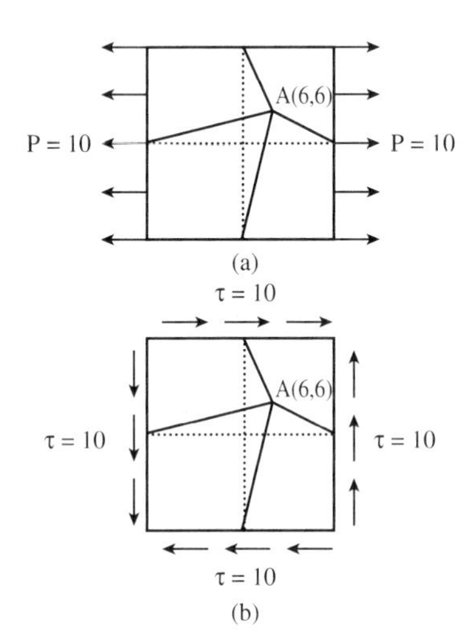

FIGURE 7.15
Constant stress patch test for plane stress/strain square panel (10 × 10). (a) Pure stretch patch test; (b) pure shear patch test.

TABLE 7.4

Patch Test for the Plane Strain Panel (Figure 7.15) by the Hybrid Model [21]

	Stretch stress σ_{xAe}		Shear stress τ_{xyAe}	
δ	$\mu = 0.3$	$\mu = 0.5$	$\mu = 0.3$	$\mu = 0.5$
10^{-3}	9.9958	9.9958	9.9915	9.9914
10^{-4}	9.9996	9.9996	9.9992	9.9992
10^{-5}	9.9999	9.9999	9.9999	9.9998
Exact	10	10	10	10

present hybrid element is able to reproduce the constant stress states for both σ_x and τ_{xy} with high precision. In the following computations, δ was taken to be 10^{-5}.

7.4.7.2 Numerical Stability Tests

Figure 7.16 shows a plane strain circular with a pair of radial concentrated loads and its finite element meshes. Three sets of stress solutions, given by Q4(2 × 2), and the proposed deviatoric hybrid element, are all listed in Table 7.5 for comparison. The isoparametric element, Q4(2 × 2), provides us

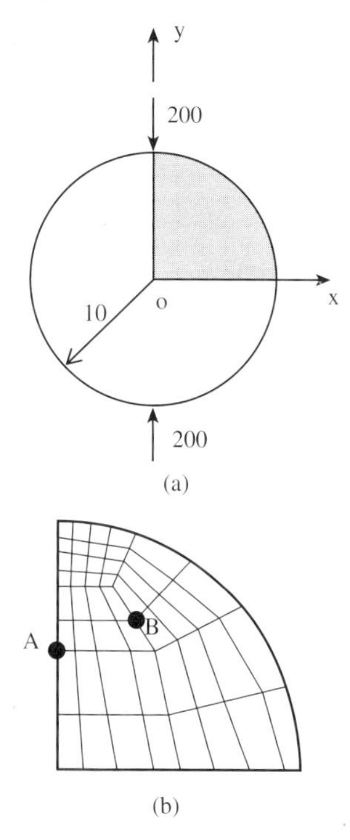

FIGURE 7.16
A plane strain circular disc with radial concentrated loads.

TABLE 7.5
Stress Solutions of a Circular Disc (Figure 7.16)

	Q4 (2 × 2)				**Q4 (2 × 2 and 1 × 1)**				**Hybrid Model**			
Stress	σ_{xAe}	σ_{yAe}	σ_{xBe}	σ_{yBe}	σ_{xAe}	σ_{yAe}	σ_{xBe}	σ_{yBe}	σ_{xAe}	σ_{yAe}	σ_{xBe}	σ_{yBe}
u = 0.3	5.41	–26.1	–0.88	–12.6	6.10	–25.3	–1.89	–14.0	5.98	–26.1	1.58	–13.0
u = 0.49	–28.7	–59.2	8.51	–4.04	6.10	–25.3	–1.89	–13.5	5.78	–26.3	1.88	–13.3
u = 0.5									5.76	26.3	1.99	13.4
Exact	6.37	–26.1	–1.28	–12.9	6.37	–26.1	–1.28	–12.9	6.37	–26.1	1.28	–12.9

with bad stress solutions for a nearly incompressible material $\mu = 0.49$, but no incompressible solutions. From Table 7.5, we find that only the proposed element can offer rational stress solutions for compressible and nearly and fully incompressible materials and exhibit a fine numerical stability.

Another example is the pure bending of a plane strain cantilever modeled by five elements as shown in Figure 7.17. The results in Table 7.6 indicate that some numerical problems, such as the displacement locking phenomenon, the poor stress solutions and incompressible solutions, etc., cannot be avoided by Q4(2 × 2). The element Q4(2 × 2), Q4(2 × 2 and 1 × 1) provides very poor stress solutions too. The proposed hybrid element, however, uni-

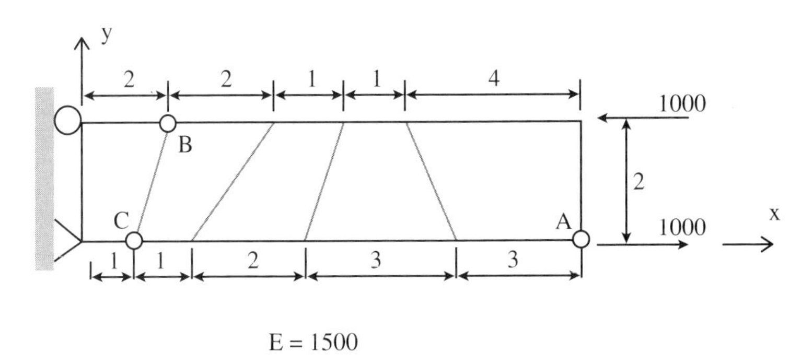

FIGURE 7.17
Pure bending plane strain beam of five elements.

TABLE 7.6
Pure Bending Solutions of the Plane Strain Beam (Figure 7.17)

Element	u = 0.25		u = 0.499		u = 0.5	
	v_A	$\sigma_{xB} / \sigma_{xCe}$	v_{Ae}	$\sigma_{xB} / \sigma_{xCe}$	v_{Ae}	$\sigma_{xB} / \sigma_{xCe}$
Q4(2 × 2)	42.05	1721/1710	3.77	2327/4228	No solutions	
Q4(2 × 2 and 1 × 1)	73.15	509/281	87.18	526/268	No solutions	
Present	90.54	2987/2989	73.60	2993/2996	73.54	2994/2997
Exact	93.75	3000/3000	75.10	3000/3000	75.00	3000/3000

TABLE 7.7
Elastic and Plastic Limit Loads of a Thick-Walled Cylinder (Figure 7.18)

Element	Elastic Limit Load	Plastic Limit Load
Q4(2 × 2)	3000	No solutions
QUAD[43]	2900	4750
Present	2900	4900
Exact	2964	4802

fied the compressible and incompressible calculations and produced excellent bending solutions for both the displacements and the bending stresses.

7.4.7.3 Plastic Analysis

For plane strain problems, it is hard for conventional isoparametric elements to find a rational plastic limit solution due to the assumption of plastic incompressibility. Let us consider an elastic-plastic thick cylinder under inner pressure p. The analytical solution for a plastic limit load is 4802 N/cm^2. The discrete meshes for a quarter of the cross-sectional part of the cylinder are shown in Figure 7.18a, and the usual incremental iterative formulae are adopted. For the plastic solutions, the loading-displacement curves by the Q4(2×2) element are drawn in Figure 7.18b. The figure shows that there is no plastic limit load, and the real structural stiffness is overestimated by the isoparametric element. However, as shown in Figure 7.18c, the present hybrid element and the enhanced-strain incompatible element QUAD offer rational plastic solutions, and the obtained plastic limit loads are consistent with the analytical one. The results of each element are shown in Table 7.7.

7.4.7.4 Stokes Flow Analysis

Except for the deviatoric hybrid model, the other elements mentioned above, that is, Q4(2×2), Q4(2×2 and 1×1), and QUAD, cannot be directly used in Stokes flow calculations due to the strict incompressible constraint. For Stokes flow analysis, two examples are offered here to verify the efficiency of the proposed element. First, we consider the stationary Couette flow between infinite plates, which is approximated by a finite flow field (length 10 and height 1) with 16×4 finite element meshes. As shown in Figure 7.19a, the upper surface pulls the fluid flow with velocity $U = 1$, and the change of the pressure radiant P on the left-hand side will result in the variation of velocity.

Figure 7.19b shows five sets of velocity distributions along the middle section $A - A'$, each of which corresponds to a given pressure P. All of the numerical solutions are consistent with the analytical ones.

Another Stokes flow problem is the lubrication in a bearing. The essential features of this type of motion can be understood with the example of a slide block moving on a plane guide surface (Figure 7.20a). In order to obtain a steady-state problem, let us assume that the slide block is at rest and that the plane guide is forced to move with a constant velocity U with respect to it. Pressure P_0 is then enforced at two sides of the bearing. A set of 10×4 discrete meshes were employed to simulate the flow in the wedge between the slide block and the guide surface.

The obtained pressure distribution over the slide block is shown in Figure 7.20b, and the related velocity profiles are shown in Figure 7.20c. All these numerical results demonstrate the efficiency of the suggested deviatoric hybrid element approach for incompressible fluid analysis.

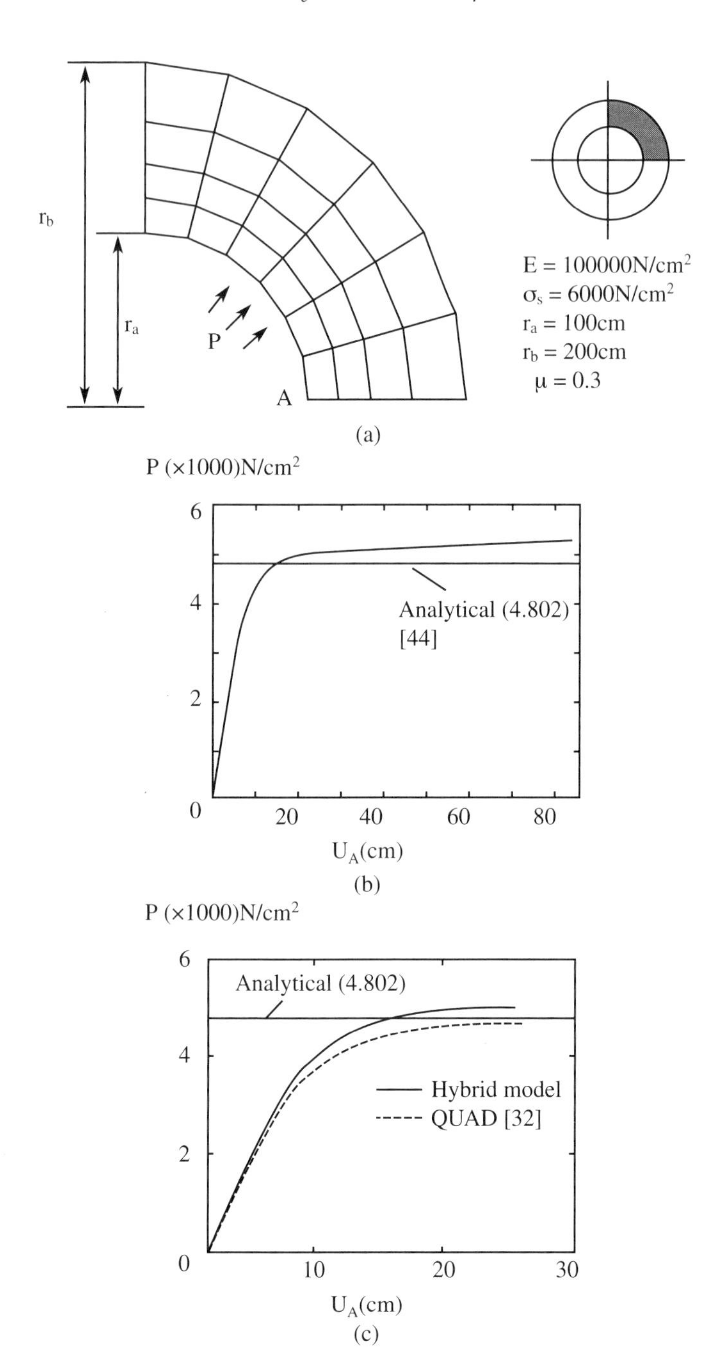

FIGURE 7.18
Elastic-plastic thick cylinder under inner pressure *P*. (a) Finite element meshes; (b) loading-displacement curve by the Q4 element; (c) loading-displacement curves by the present hybrid element and QUAD [43].

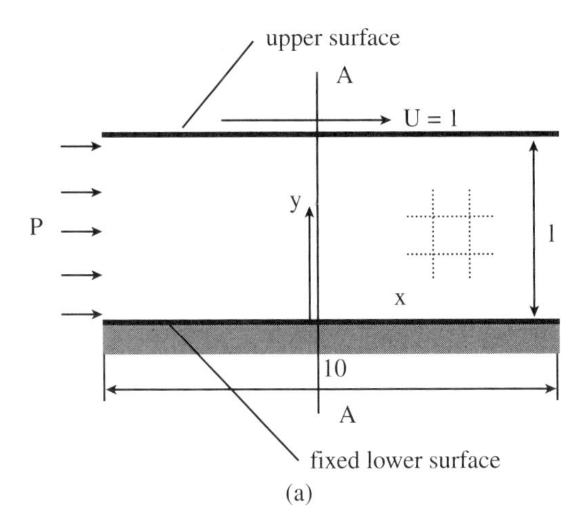

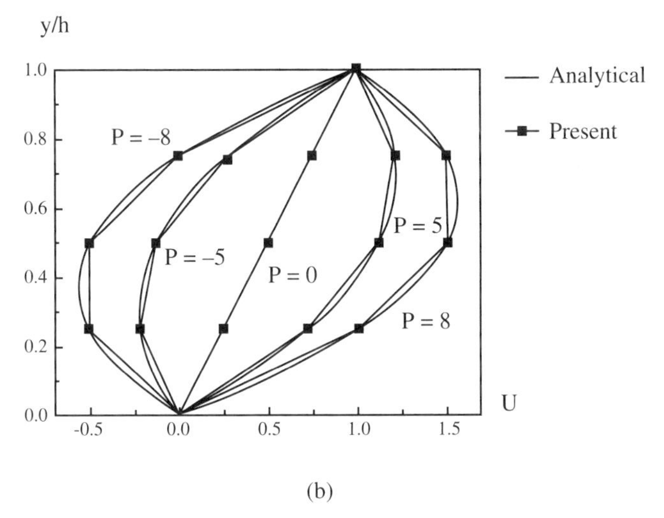

FIGURE 7.19
Incompressible Couette flow between infinite plates. (a) Computational for Couette flow (mesh 16 × 4); (b) velocity distributions on the middle section A–A′ (for the pressure radiant $P = -8/-5/0/5/8$).

7.4.8 Conclusion

This section has proposed the hybrid model based on the deviatoric variational formulation $\pi'(\sigma'_{ij}, p, u_i)$, which is universal for compressible and incompressible media. The greatest advantage of the model is that it is able to provide quite reliable fully incompressible solutions for both the stresses and the displacements.

A dramatic result of this section would be that the zero-volumetric strain constraints are enforced at the element level by introduction of the incompatible displacement function. It follows that the pressure variables will not

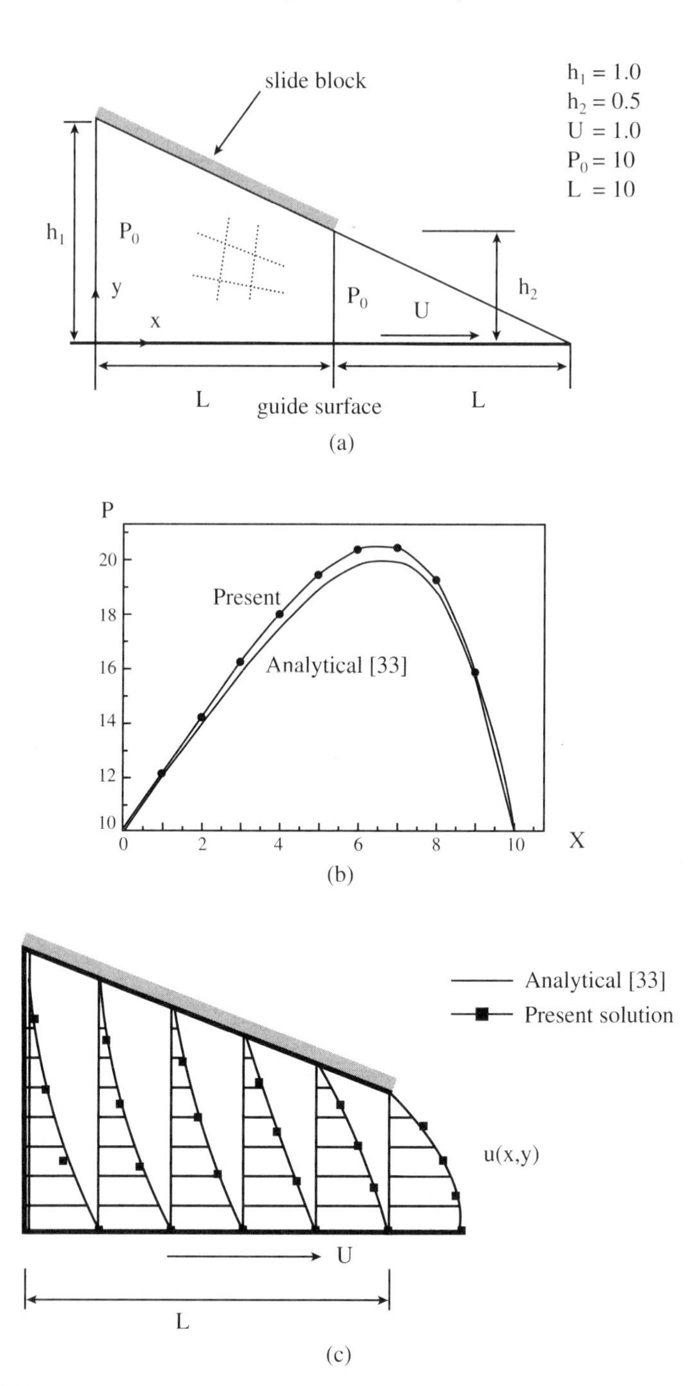

FIGURE 7.20
Lubrication in bearing. (a) Flow in the wedge between the slide block and the guide surface (10×4 meshed); (b) pressure distribution over the slide block; (c) velocity profiles in bearing.

join the system discrete equations, and pure displacement formulations are finally available.

A practical strategy has been suggested to suppress the zero-energy displacement modes, ZEM($\boldsymbol{u}$), stemming from the bilinear term of Reference 4. This strategy entails the deviatoric stress and the deviatoric strain taking the same shape functions in element formulations.

Another advantage of this model is that it is able to unify the computations for both solids and fluids. The proposed hybrid element can be directly used in Stokes flow calculations, and those troublesome numerical stability problems [27] will no longer appear.

References

1. Nagtegaal, J.C., Park, D.M., and Rice, J.R., On numerically accurate finite element solutions in the fully plastic range, *Comp. Methods Appl. Mech. Eng.*, 4, 153, 1974.
2. Argyris, J.H. and Dunne, P.C., Improved displacement finite elements for incompressible materials, in *Energy Method in Finite Element Analysis*, edited by R. Glowinski, E.Y. Rodin, and O.C. Zienkiewicz, 1979, pp. 225–241.
3. Nagtegaal, J.C. et al., On numerically accurate finite element solutions in the fully plastic range, *Comp. Methods Appl. Mech. Eng.*, 4, 153, 1978.
4. Zienkiewicz, O.C., *The Finite Element Method*, 3rd ed., McGraw-Hill, New York, 1977.
5. Needleman, A. and Shih, C.F., Finite element method for plane strain deformations of incompressible solids, *Comp. Methods Appl. Mech. Eng.*, 15, 223, 1978.
6. Thompson, E.G. and Hangue, M.I., A high order finite element for completely incompressible creeping flow, *Int. J. Numer. Methods Eng.*, 6, 315, 1973.
7. Casciaro, R. and Cascini, L., A mixed formulation and mixed finite element for limit analysis, *Int. J. Numer. Methods Eng.*, 18, 211, 1982.
8. Gao, Y., On the Complementary Principles for Elasto-Plastic Systems and Pan-penalty Finite Element Method, Ph.D. thesis, Tsinghua University, Beijing, China, 1985.
9. Bathe, K.J. and Wilson, E.L., *Numerical Methods in Finite Element Analysis*, Prentice-Hall, Englewood Cliffs, NJ, 1976.
10. Wu, C.C., Huang, M.G., and Pian, T.H.H., Convergence criteria of incompatible elements, general formulation of incompatible functions and its application, *Comput. Struct.*, 27, 639, 1987.
11. Wilson, E.L. et al., Incompatible displacement models, in *Numerical and Computer Methods in Structural Mechanics*, Fenves, S.J., Ed., Academic Press, New York, 1973.
12. Herrmann, L.R., Elasticity equations for incompressible and nearly incompressible materials by variational theorem, *AIAA J.*, 3, 1896, 1965.
13. Taylor, R.L., Pister, K.S., and Herrmann, L.R., On a variational theorem for incompressible and nearly incompressible elasticity, *Int. J. Solids Struct.*, 4, 875, 1968.
14. Key, S.W., A variational principle for incompressible and nearly incompressible anisotropic elasticity, *Int. J. Solids Struct.*, 5, 455, 1969.

15. Murakawa, H. and Atluri, S.N., Finite elasticity solutions using hybrid finite elements based on a complementary energy principle. Part 2. Incompressible materials, *J. Appl. Mech.*, 46, 71, 1979.
16. Gadala, M.S., Numerical solutions of nonlinear problems of continua. II. Survey of incompressibility constraints and software aspects, *Comput. Struct.*, 22, 841, 1986.
17. Malkus, D.S. and Hughes, T.J.R., Mixed finite element methods: reduced and selective integration techniques: a unification of concepts, *Comput. Methods Appl. Mech. Eng.*, 15, 63, 1978.
18. Zienkiewicz, O.C. and Taylor, R.L., *The Finite Element Method*, 4th ed., Vol. 1, McGraw-Hill, U.K., 1989.
19. Cook, R.D., *Concepts and Applications of Finite Element Analysis*, 2nd ed., Wiley, New York, 1981.
20. Wu, C.C., Liu, X.Y., and Pian, T.H.H., Incompressible-incompatible deformation modes and plane strain plastic element, *Comput. Struct.*, 41, 449, 1991.
21. Wu, C.C., Yuan, L., and Furukawa, T., Deviating hybrid model and multivariable elimination at element level for incompressible medium, *Int. J. Numer. Methods Eng.*, 46, 729–745, 1999.
22. Zhang, W. and Li, X., On the generalized displacement method in the finite element analysis of incompressible materials, *J. Dalan Inst. Technol.*, 20, 1, 1981 (in Chinese).
23. Li, X.K., Computerized generalized displacement method for finite element analysis of incompressible material, *Eng. Comput.*, 11, 335, 1984.
24. Oden, J.T., *Finite Element of Nonlinear Continua*, McGraw-Hill, New York, 1972.
25. Thompson, E.C., Average and complete incompressibility in the finite element method, *Int. J. Numer. Methods Eng.*, 9, 925, 1975.
26. Wu, C.C., Dual zero energy modes in mixed/hybrid elements: definition, analysis and control, *Comput. Methods Appl. Mech. Eng.*, 81, 39, 1990.
27. Girault, V. and Raviart, P.A., *Finite Element Methods for Navier-Stokes Equations*, Springer-Verlag, Berlin, 1986.
28. Simo, J.C., Taylor, R.L., and Pister, K.S., Variational and projection methods for the volume constraint in finite deformation plasticity, *Comput. Methods Appl. Mech. Eng.*, 51, 177, 1985.
29. Pian, T.H.H. and Leem, S.W., Notes on finite element for nearly incompressible materials, *AIAA J.*, 4, 824, 1976.
30. Spilker, R.L., Improved hybrid stress axisymmetric elements including behavior for nearly incompressible materials, *Int. J. Numer. Methods Eng.*, 17, 483, 1981.
31. Tong, P., An assumed stress hybrid finite element method for an incompressible and nearly incompressible materials, *Int. J. Solids Struct.*, 5, 455, 1969.
32. Bratianu, C. and Atluri, S.N., A hybrid finite method for Stokes flow: Part I. Formulation and numerical studies, *Comput. Methods Appl. Mech. Eng.*, 36, 23, 1983.
33. Hu, H.C., *Variational Principles in Elastic Mechanics and Applications*, Science Press, Beijing, 1981.
34. Pian, T.H.H. and Wu, C.C., A rational approach for choosing stress terms for hybrid finite element formulations, *Int. J. Numer. Methods Eng.*, 26, 2331, 1988.
35. Wu, C.C. and Bufler, H., Multivariable finite element: consistency and optimization, *Sci. China A*, 9, 946, 1990 (in Chinese) or 34, 284 1991 (in English).
36. Strange, G. and Fix, G.J., *An Analysis of the Finite Element Method*, Prentice-Hall, Englewood Cliffs, NJ, 1973.

37. Taylor, R.L., Beresford, P.J., and Wilson, E.L., A non-conforming element for stress analysis, *Int. J. Numer. Methods Eng.*, 10, 1211, 1976.
38. Wu, C.C. and Cheung, Y.K., The patch test condition in curvilinear coordinates: formulation and application, *Sci. China A*, 8, 849, 1992 (in Chinese) or 36, 62, 1993 (in English).
39. Wu, C.C. and Pian, T.H.H., *Incompatible Numerical Analysis and Hybrid Element Method*, Science Press, Beijing, 1997.
40. Wu, C.C., Huang, M.G., and Pian, T.H.H., Consistency condition and convergence criteria of incompressible elements: general formulation of incompatible functions and its application, *Comput. Struct.*, 27, 639, 1987.
41. Pian, T.H.H. and Chen, D.P., On the suppression of zero-energy deformation modes, *Int. J. Numer. Methods Eng.*, 19, 1741, 1983.
42. Cheung, Y.K. and Wu, C.C., A study on the stability of 3-field finite elements by the theory of zero energy modes, *Int. J. Solids Struct.*, 29, 215, 1992.
43. Simo, J.C. and Rifai, M.S., A class of mixed assumed strain methods and the method of incompatible modes, *Int. J. Numer. Methods Eng.*, 29, 1595, 1990.
44. Prager, W. and Hodge, P.G., *Theory of Perfactly Plastic Solids*, John Wiley, 1951.

8

Computational Fracture

8.1 Introduction

In computational fracture mechanics, the estimation of the upper and lower bound for fracture parameters becomes significant, as one cannot obtain an accurate or reliable solution due to the complexity of fracture problems, no matter what experimental or numerical method is used. The bound problem consists of three aspects: (a) theoretically, the existence of an approximate upper and lower bound for fracture parameters; (b) if it is true, numerically, the evaluation approach; and (c) the error measure should be considerable once the bound solutions are obtained.

Rice's J integral has the meaning of the release rate of the strain energy $\Pi(u_i)$ with respect to the crack area [1]. Hence, the bound of the numerical solutions for the J integral, if it exists, may be estimated using the assumed displacement finite element method that is founded on $\Pi(u_i)$. However, the I^* integral proposed as the dual counterpart of the J integral has also been shown to be the release rate of the complementary energy $\Pi_c(\sigma_{ij})$ with respect to the crack area [2]. A natural conjecture is that the bound of the numerical solutions for the I^* integral may be established using the assumed-stress finite element method that is founded on $\Pi_c(\sigma_{ij})$.

8.2 Dual Path-Independent Integral and Bound Theorem

8.2.1 Dual Path-Independent Integral

For a given crack system with actual status of stresses, strains, and displacements $(\sigma_{ij}, \varepsilon_{ij}, u_i)$, the J integral can be defined as:

$$J = \int_{\Gamma} [W(\varepsilon_{ij})dx_2 - \sigma_{ij}n_j(\frac{\partial u_i}{\partial x_1})ds] \tag{8.1}$$

To find a dual integral of J, we introduce the Legendre transformation

$$W(u_i)+B(\sigma_{ij})=\sigma_{ij}\varepsilon_{ij} \tag{8.2}$$

into Equation 8.1. Then an alternative path integral can be derived:

$$\begin{aligned} I^* &= \int_\Gamma \left\{[\sigma_{ij}\varepsilon_{ij}-\beta(\sigma_{ij})]dx_2-\sigma_{ij}n_j\frac{\partial u_i}{\partial x_1}ds\right\} \\ &= \int_\Gamma [-B(\sigma_{ij})dx_2+\sigma_{ij}\varepsilon_{ij}dx_2-\sigma_{i1}\frac{\partial u_i}{\partial x_1}dx_2+\sigma_{i2}\frac{\partial u_i}{\partial x_1}dx_1] \\ &= \int_\Gamma [-B(\sigma_{ij})dx_2+\sigma_{i2}\frac{\partial u_i}{\partial x_j}dx_j] \end{aligned} \tag{8.3}$$

It is easy to verify that I^* is a path independent integral. The I^* integral can also be defined as a complementary energy release rate [2]:

$$I^*=\int_\Gamma [-B(\sigma_{ij})dx_2+u_i\frac{\partial \sigma_{ij}}{\partial x_1}n_j ds+\frac{\partial}{\partial x_j}(u_i\sigma_{i2})dx_j] \tag{8.4}$$

For an equilibrium stress field, $\sigma_{ij,j}=0$, it can be verified that Equation 8.3 is equivalent to Equation 8.4. In numerical calculations, the expression in Equation 8.3 is especially recommended due to its simplicity and the absence of a derivative of stresses.

8.2.2 Bound Theorem

Corresponding to the dual integrals J and I^*, the following bound theorems can be established for a certain linear or nonlinear elasticity crack system with a homogeneous displacement boundary constraint:

Lower Bound Theorem for J: For the given cracked system, if u_i and $\tilde{u}_i$ are, respectively, the exact and the approximate displacement based on the minimum potential energy principle, the approximate J integral will take a lower bound to the exact J integral [2, 31, 44]:

$$J(\tilde{u}_i)\le J(u_i) \tag{8.5}$$

and the lower bound of J can be obtained by the displacement compatible elements.

Upper Bound Theorem for I*: For the given cracked system, if σ_{ij} and $\tilde{\sigma}_{ij}$ are, respectively, the exact and the approximate stresses based on the minimum complementary energy principle, the approximate I^* integral will take an upper bound to the exact I^* integral:

$$I^*(\tilde{\sigma}_{ij}) \geq I^*(\sigma_{ij}) \tag{8.6}$$

and the upper bound can be obtained by stress equilibrium elements.

In the case of linear elasticity, a proof for the above theorems has been presented by Wu et al. [2]. In the case of nonlinear elasticity, including the deformation theory–based plasticity, the upper bound theorem still holds. As an example, the lower bound theorem for J will be proved here.

For the given nonlinear crack system with homogeneous displacement boundary constraints, it can be verified for the actual solutions (u_i, σ_{ij}) that:

$$\Pi_c(\sigma_{ij}) = V(\sigma_{ij}) \tag{8.7}$$

$$\Pi(u_i) = -kU(u_i) \tag{8.8}$$

where $V(\sigma_{ij})$ and $U(u_i)$ are the system complementary energy and the system strain energy, respectively, and the finite constant

$$k = V(\sigma_{ij})/U(u_i) > 0 \tag{8.9}$$

In the case of linear elasticity, k = 1. Let $\tilde{u}_i = u_i + \delta u_i$; then we have

$$\Pi(\tilde{u}_i) = \Pi(u_i) + \delta\Pi + \delta^2\Pi(\delta u_i) \tag{8.10}$$

As for the displacement finite element that is based on the potential energy principle, the first and second variations in Equation 8.10 are, respectively,

$$\delta\Pi = 0$$

and

$$\delta^2\Pi(\delta u_i) = \int_V A(\delta u_i)dV$$

Hence Equation 8.10 becomes

$$\Pi(\tilde{u}_i) - \Pi(u_i) = \int_V A(\delta u_i)dV$$

In accordance with the definition of J,

$$J(\tilde{u}_i) - J(u_i) = -\frac{d}{da}(\Pi(\tilde{u}_i) - \Pi(u_i)) = -\frac{d}{da}\int_V A(\delta u_i)dV \tag{8.10a}$$

Considering an actual status of the given system, $J(u_i) \geq 0$, and Equation 8.8 and Equation 8.9 must be true. Thus

$$J(u_i) = -\frac{d}{da}\Pi(u_i) = -\frac{d}{da}\left[-kU(u_i)\right] = k\frac{d}{da}\int_V A(\delta u_i)dV \geq 0 \tag{8.10b}$$

Observing the strain energy to be positive definite, the comparison of Equation 8.10a and Equation 8.10b results in

$$-\frac{d}{da}\int_V A(\delta u_i)dV \leq 0 \tag{8.11}$$

Thus the inequality (Equation 8.5) holds.

8.3 Numerical Strategy and Error Measure

8.3.1 Numerical Strategy

As for J, its lower bound can easily be obtained by using the conventional isoparametric elements. For I^*, however, its upper bound should be estimated by stress equilibrium elements. Unfortunately it is hard to get a reliable equilibrium model for two-dimensional (2-D) and 3-D problems since some numerical problems, such as rank deficiency, displacement indeterminacy, etc., cannot be avoided. One faces the problem of how to implement the upper bound theorem for I^*.

The stress equilibrium element is based on the complementary energy formulation $\Pi_c(\sigma)$ while the stress hybrid element is based on the Reissner formulation $\Pi_R(\sigma, u)$. However, $\Pi_R(\sigma, u)$ is identical to $\Pi_c(\sigma)$ when the element stress σ meets the equilibrium equation. Thus a hybrid model may degenerate into an equilibrium model when the stress equilibrium equations are enforced to the hybrid model.

For an individual element, let

$$\Pi_R^e = \int_{V^e}\left[\sigma^T(\mathbf{Du}) - \frac{1}{2}\sigma^T\mathbf{S}\sigma\right]dV \tag{8.12}$$

A generalized functional can be created in the manner of Reference [3]:

$$\Pi_{RG}^{e} = \Pi_{R}^{e} - \int_{V^e} (\mathbf{D}^T\sigma)^T(\mathbf{D}^T\sigma)dV \tag{8.13}$$

In Equation 8.13 the penalty factor $\alpha > 0$ is taken to be a large constant such that the homogeneous stress equilibrium condition $\mathbf{D}^T\sigma = \mathbf{0}$ will be enforced to the element in the least-squares sense.

Recalling the four-node plane hybrid stress element P-S, proposed by Pian and Sumihara [4], the assumed element stress trial functions can be expressed, in terms of the element coordinates (ξ, η), as

$$\sigma = \begin{Bmatrix} \sigma_{11} \\ \sigma_{22} \\ \sigma_{12} \end{Bmatrix} = \begin{bmatrix} 1 & 0 & 0 & a_1^2\eta & a_3^2\xi \\ 0 & 1 & 0 & b_1^2\eta & b_3^2\xi \\ 0 & 0 & 1 & a_1b_1\eta & a_3b_3\xi \end{bmatrix} \begin{Bmatrix} \beta_1 \\ \vdots \\ \beta_5 \end{Bmatrix} = \Phi\beta \tag{8.14}$$

where the coefficients a_i and b_i are related to the element nodal coordinates (x_i, y_i) in the manner:

$$\begin{bmatrix} a_1 & b_1 \\ a_2 & b_2 \\ a_3 & b_3 \end{bmatrix} = \frac{1}{4}\begin{bmatrix} -1 & 1 & 1 & -1 \\ 1 & -1 & 1 & -1 \\ -1 & -1 & 1 & 1 \end{bmatrix} \begin{bmatrix} x_1 & y_1 \\ x_2 & y_2 \\ x_3 & y_3 \\ x_4 & y_4 \end{bmatrix} \tag{8.15}$$

By substituting the stress (Equation 8.14) and the bilinear displacements $\mathbf{u}_q = \mathbf{N}(\xi,\eta)\mathbf{q}$ into the functional (Equation 8.13), we have

$$\Pi_{RG}^{e}(\beta, \mathbf{q}) = \beta^T\mathbf{G}\mathbf{q} - \frac{1}{2}\beta^T(\mathbf{H} + \frac{\alpha}{E}\mathbf{H}_p)\beta \tag{8.16}$$

After condensing β, the element stiffness matrix can be formulated as

$$\boldsymbol{K}^e = \boldsymbol{G}^T(\mathbf{H} + \frac{\alpha}{E}\mathbf{H}_p)^{-1}\mathbf{G} \tag{8.17}$$

in which the matrices $\boldsymbol{G}$ and $\boldsymbol{H}$ are identical to those of the P-S element, while the additional penalty matrix

$$\boldsymbol{H}_p = \int_{V^e} (\mathbf{D}^T\Phi)^T(\mathbf{D}^T\Phi)dV \tag{8.18}$$

In such a way, the P-S hybrid element is developed into a penalty-equilibrating model, P-S(α), in which the stress equilibrium equation is imposed in a mining of penalty.

Xiao et al. [6] presented a further study of the dual analysis of crack in mode I/II and the related penalty hybrid element technique.

8.3.2 Error Measure

Let $\delta u_i = \tilde{u}_i - u_i$ be the displacement error induced by using the assumed displacement finite elements, in accordance with the lower bound theorem (Equation 8.5); the relative error for the J integral must be

$$J(\delta u_i) = J(\tilde{u}_i) - J(u_i) \leq 0 \tag{8.19}$$

and the absolute error,

$$\left|J(\delta u_i)\right| = J(u_i) - J(\tilde{u}_i) \tag{8.20}$$

Let $\delta\sigma_{ij} = \tilde{\sigma}_{ij} - \sigma_{ij}$ be the stress error induced by using the assumed-stress finite elements, in accordance with the upper bound theorem (Equation 8.6); the relative error for the I^* integral must be

$$I^*(\delta\sigma_{ij}) = I^*(\tilde{\sigma}_{ij}) - I^*(\sigma_{ij}) \geq 0 \tag{8.21}$$

and

$$\left|I^*(\delta\sigma_{ij})\right| = I^*(\tilde{\sigma}_{ij}) - I^*(\sigma_{ij}) \tag{8.22}$$

Corresponding to Equation 8.20 and Equation 8.22, the relative errors can, respectively, be expressed as

$$\left|J(\delta u_i)\right| / J(u_i) \quad \text{and} \quad \left|I^*(\delta\sigma_{ij})\right| / I^*(\sigma_{ij}) \tag{8.23}$$

The strength estimation for a crack structure can be carried out once the upper and lower bounds are yielded, and the approximate strength is usually taken to be a combination of the bound solutions. In order to measure the error of the fracture parameters given by finite element methods, a relative dual error for J and I^* is defined as

$$\Delta_{J-I^*} = \frac{\left|J(\delta u_i)\right| + \left|I^*(\delta\sigma_{ij})\right|}{J(u_i) + I^*(\sigma_{ij})} \tag{8.24}$$

In Equation 8.24, the sum of reference solutions

$$J(u_i) + I^*(\sigma_{ij}) = J(\tilde{u}_i) + I^*(\tilde{\sigma}_{ij}) - \left[J(\delta u_i) + I^*(\delta\sigma_{ij})\right] \tag{8.25}$$

Because the small quantities $J(\delta u_i) \le 0$, whereas $I^*(\delta\sigma_{ij}) \ge 0$, the last term in Equation 8.24 can be ignored, and we have

$$J(u_i) + I^*(\sigma_{ij}) \approx J(\tilde{u}_i) + I^*(\tilde{\sigma}_{ij}) \tag{8.26}$$

Substitute Equation 8.20, Equation 8.22, and Equation 8.26 into Equation 8.23 and note that the actual J and I^* are identical in value. We finally obtain

$$\tilde{\Delta}_{J-I^*} = \frac{I^*(\tilde{\sigma}_{ij}) - J(\tilde{u}_i)}{I^*(\tilde{\sigma}_{ij}) + J(\tilde{u}_i)} \tag{8.27}$$

The above relative dual-error formula depends on the approximate solutions of J and I^* only, so as to be easily used in nonlinear fracture estimations. The error will vanish when the adopted finite element meshes become more and more fine.

8.4 Numerical Tests of Crack Estimation

In numerical calculations, the well-known four-node isoparametric element Q4 and the penalty-equilibrium element P-S(α) are employed to estimate J and I^*, respectively.

The center cracked panel (CCP) with uniform stretching load σ_∞ (Figure 8.1a) and the single-edge cracked panel single-edge cracked panel (SECP) with uniform stretching load σ_∞ (Figure 8.2b) are calculated in this section. Only a quarter or half of a specimen needs to be considered due to symmetry. Three finite element meshes and two integral paths are shown in Figure 8.2abc. The specimen material consists of Young modulus E = 1.0 and Poisson ratio $\nu = 0.3$. The distributed stretching load $\sigma_\infty = 1.0$. For linear elastic crack problems, all the solutions of J or I^* will be transferred to the stress intensity factor K_I for convenient comparisons. The reference solutions of K_I are offered by Ewalds and Wanhill [5].

To inspect the convergence behavior of the solutions of J and I^*, three meshes with different densities and two independent integral paths are simultaneously considered for each specimen. From the results shown in Figure 8.3a,b and Figure 8.4a,b, it can be seen that the solutions of J by Q4 always converge to the exact solution from a certain lower bound. On the contrary, the solutions of I^* by P-S(α) always converge to the exact solution

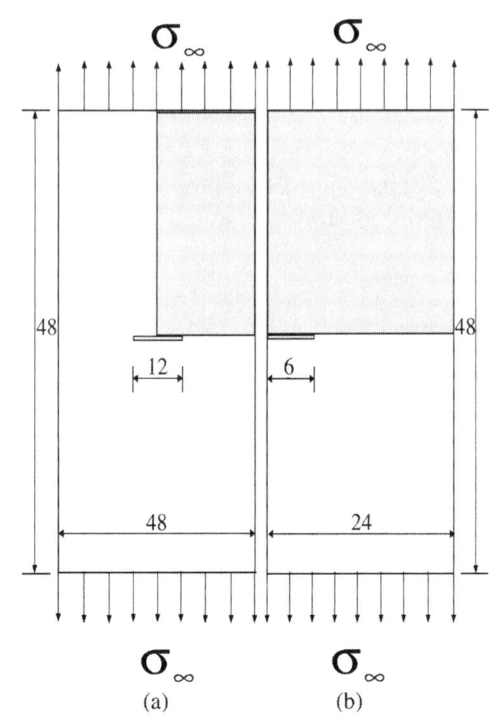

FIGURE 8.1
Specimen (a) CCP and (b) SECP.

from a certain upper bound. All the numerical solutions demonstrate the bound theorems presented in this section.

The error formula $\tilde{\Delta}_{J-I^*}$ in Equation 8.27 is implemented to measure the relative error of the bound solutions for CCP and SECP. The results are listed in Table 8.1 through Table 8.4. The results given by the formula Δ_{J-I^*} in Equation 8.24 are also shown in parentheses in the tables for comparison. Independent of the selection of meshes and paths, both $\tilde{\Delta}_{J-I^*}$ and Δ_{J-I^*} always offer almost the same results. These numerical tests exhibit the efficiency of the approximate error formula (Equation 8.27).

8.5 Incompatible Numerical Simulation of an Axisymmetric Cracked Body

8.5.1 Introduction

Various singular finite element methods have been suggested for the solution of linear elastic fracture problems. In comparison with standard elements, the singular elements that exhibit the square root singularity represent the

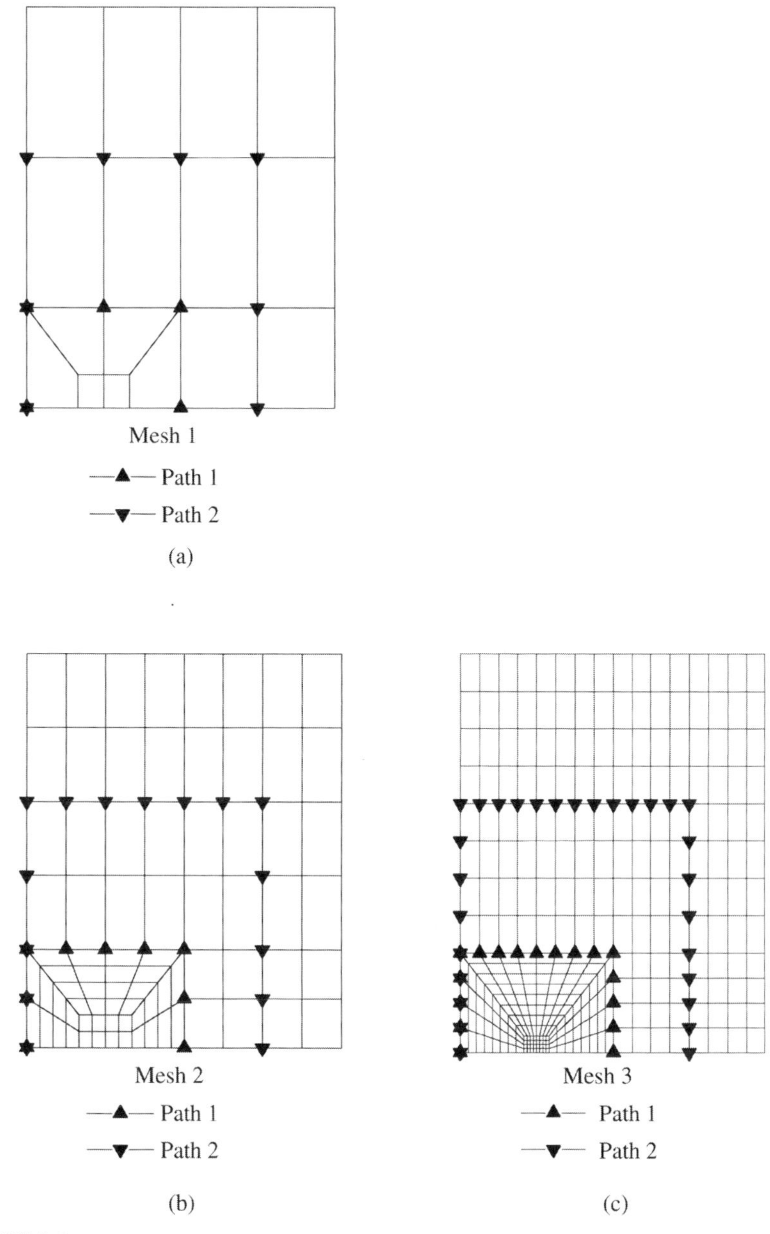

FIGURE 8.2
The employed finite element meshes and the selected integral paths.

crack tip singularity more naturally and provide more accurate solutions of the stress intensity factors [7]. For the plastic fracture problems, however, the majority of singular elements are no longer efficient because the stress and strain fields near the tip of the crack become very complex, and the assumption of square root singularity will no longer hold. Other endeavors

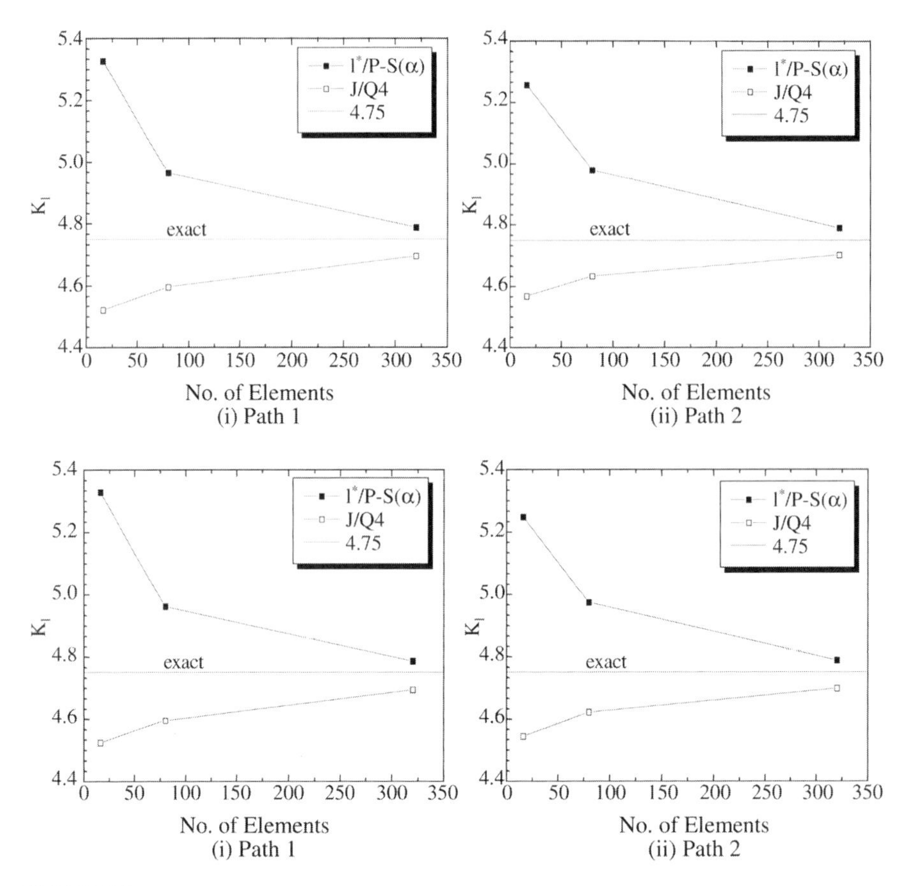

FIGURE 8.3
The bound behavior of the solutions of J and I^* (transferred to K_I) in three meshes with different densities and two independent integral paths of CCP.

to develop plastic crack elements with the HRR (Hutchinson, Rice, and Rosengren) asymptotic field have not been found to be very attractive.

The standard elements, in particular the isoparametric ones, can be conveniently used for the solution of ductile fractures under plane stress conditions. However, such compatible elements cannot be employed to solve the axisymmetric plastic crack problems. It seems that the incompressibility of plastic material provides axisymmetric isoparametric elements with such strong deformation constraints that they are unable to approximate the stress and strain fields in the plastic domain near the crack tip. In this case the refinement of the element meshes in the plastic domain is useless because the assembly of compatible elements usually strengthens the plastic deformation constraints even further [8].

The incompressible finite element method [9] has been used to determine the fully plastic solutions of 2-D crack samples [10]. This method, however, cannot be used to determine the limit loads of plastic structures except by

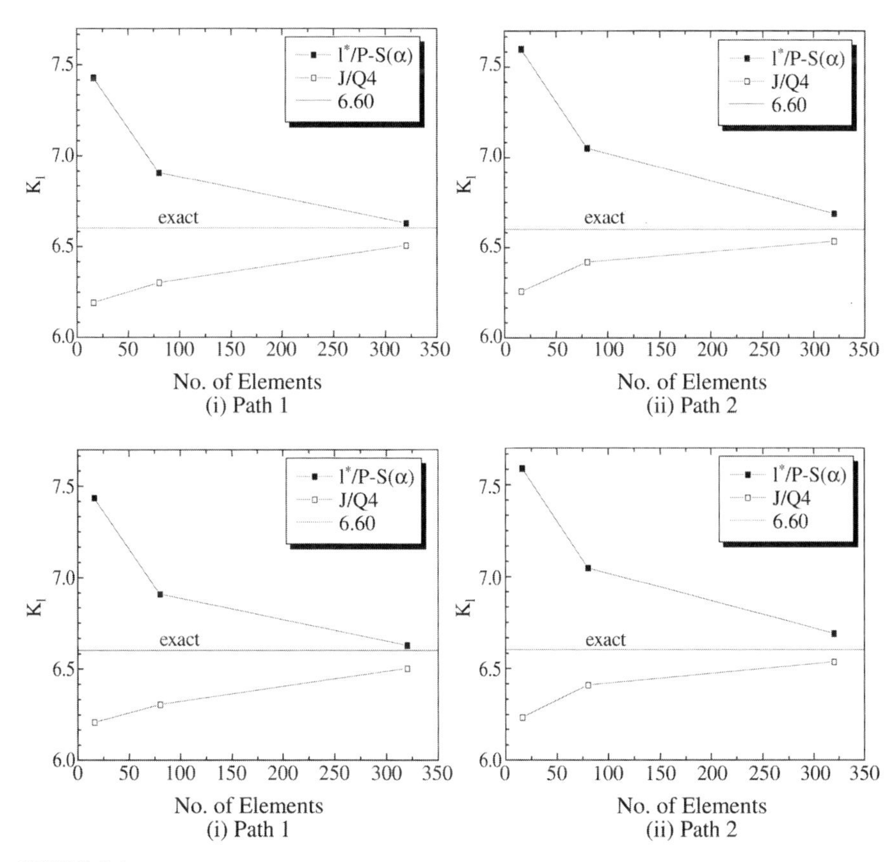

FIGURE 8.4
The bound behavior of the solutions of J and I^* (transferred to K_I) in three meshes with dfferent densities and two independent integral paths of SECP.

TABLE 8.1
Relative Error $\tilde{\Delta}_{J-I^*}$ and Δ_{J-I^*} for CCP (Plane Strain)

Mesh	1	2	3
Path 1	8.187 (8.487)	3.855 (3.880)	0.965 (0.963)
Path 2	7.015 (7.225)	3.597 (3.639)	0.922 (0.921)

TABLE 8.2
Relative Error $\tilde{\Delta}_{J-I^*}$ and Δ_{J-I^*} for CCP (Plane Stress)

Mesh	1	2	3
Path 1	8.157 (8.459)	3.828 (3.851)	0.971 (0.969)
Path 2	7.815 (7.405)	3.661 (3.698)	0.942 (0.941)

TABLE 8.3

Relative Error $\tilde{\Delta}_{J-J^*}$ and Δ_{J-J^*} for SECP (Plane Strain)

Mesh	1	2	3
Path 1	9.084 (9.373)	4.578 (4.581)	0.926 (0.922)
Path 2	9.681 (10.16)	4.681 (4.777)	1.159 (1.162)

TABLE 8.4

Relative Error $\tilde{\Delta}_{J-J^*}$ and Δ_{J-J^*} for SECP (Plane Stress)

Mesh	1	2	3
Path 1	8.991 (9.294)	4.558 (4.563)	0.952 (0.947)
Path 2	9.801 (10.26)	4.741 (4.834)	1.155 (1.157)

letting the strain hardening parameter $n \to \infty$, which makes the approach impossible in numerical implementation [11].

The solution of an elastic-plastic cracked body is a nonlinear moving-boundary problem. As the loading is increased, the plastic domain at the crack tip is extended step by step, and set after set of elastic elements (usually compressible) become plastic ones (incompressible). Therefore, as a rational elastic-plastic discrete model, the element should be capable of being used for both compressible and incompressible materials and for domains both near and far from the tip of the crack.

It will be demonstrated that a current isoparametric-axisymmetric element can be changed to an incompatible one by the addition of local displacements. The element is universal for both elastic and plastic computation and has been employed to solve the axisymmetric problem of a circular edge crack on the inner surface of a long cylinder made of ideally elastic-plastic material. The cylinder is loaded axially by a uniformly distributed force.

8.5.2 Incompatible Axisymmetric Element AQ6

The displacement trial function of the four-node isoparametric element, Axi-Q4, is a bilinear interpolation with eight nodal-displacement parameters that correspond to one rigid body motion and seven independent deformation modes of the element. However, under the following zero volumetric condition in plastic flow theory,

$$\varepsilon = \frac{\partial u}{\partial r} + \frac{\partial v}{\partial z} = 0 \tag{8.28}$$

the Axi-Q4 element has only two deformation modes, which are independent of each other and can be expressed as

$$\begin{Bmatrix} u \\ v \end{Bmatrix} = \begin{bmatrix} r & 0 \\ -2z & r \end{bmatrix} \begin{Bmatrix} c_1 \\ c_2 \end{Bmatrix} \tag{8.29}$$

The related stretch mode and shear mode are shown in Figure 8.5. Such an element with only two basic modes is unable to approximate the complex plastic deformations near the crack tip. In order to improve the element performance, we add a set of incompatible displacements, $\mathbf{u}_\lambda = \begin{bmatrix} u_\lambda & v_\lambda \end{bmatrix}$, to the bilinear displacement of the Axi-Q4 element. It turns out that some locked deformations, for example, the linear and quadratic bending modes, may be released from the incompressible condition (Equation 8.28). However, the introduction of an incompatible displacement $\mathbf{u}_\lambda$ possibly destroys the convergence of discrete solutions. According to the incompatible element theory [12], the incompatible displacement $\mathbf{u}_\lambda$ must satisfy the constant stress patch test [13]. Fortunately, the axisymmetric form of the patch test condition (PTC) has been established [14] for an individual element, and it takes the form

$$\oint r \begin{bmatrix} n_r \\ n_z \end{bmatrix} u_\lambda dS = 0, \tag{8.30a}$$

or equivalently,

$$\int\int_{A^e} \begin{bmatrix} \dfrac{1}{r} + \dfrac{\partial}{\partial r} \\ \dfrac{\partial}{\partial z} \end{bmatrix} u_\lambda r dA \tag{8.30b}$$

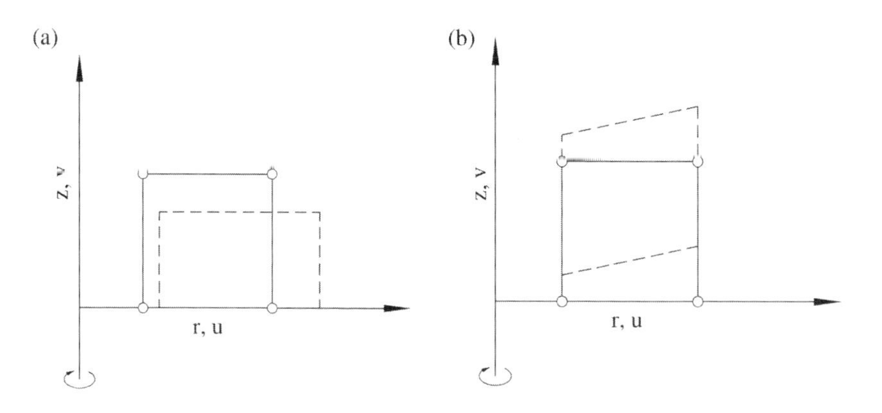

FIGURE 8.5
Incompressible deformation modes of Axi-Q4.

where the curvilinear integral is carried out along the element edges in the $r-z$ cross section of the axisymmetric element; n_r and n_z are the components of the unit outward normal on the element edges. In accordance with Equation 8.30, additional displacements can be chosen in the form of

$$u_\lambda = \begin{bmatrix} u_\lambda & v_\lambda \end{bmatrix} = \begin{bmatrix} \xi^2 & \eta^2 \end{bmatrix} \begin{bmatrix} \lambda_1 & \lambda_3 \\ \lambda_2 & \lambda_4 \end{bmatrix} \tag{8.31}$$

where ξ and η are the element isoparametric coordinates, and $\lambda_{1,2,3,4}$ are the local parameters that can be statistically condensed at the element level. After that, the desired incompatible displacements that satisfied the convergence condition (Equation 8.30) can easily be derived by means of the general incompatible formulation in Reference 13. Thus the displacements (Equation 8.31) are modified as

$$u_\lambda^* = \begin{bmatrix} u_\lambda & v_\lambda \end{bmatrix} = \begin{bmatrix} \xi^2 - a\xi - c\eta & \eta^2 - b\xi - d\eta \end{bmatrix} \begin{bmatrix} \lambda_1 & \lambda_3 \\ \lambda_2 & \lambda_4 \end{bmatrix} \tag{8.32}$$

where the coefficients

$$a = (p_{11}q_{22} - p_{21}q_{12}) / e$$

$$b = (p_{12}q_{22} - p_{22}q_{12}) / e$$

$$c = (p_{21}q_{11} - p_{11}q_{21}) / e$$

$$d = (p_{22}q_{11} - p_{12}q_{21}) / e$$

$$e = q_{11}q_{22} - q_{12}q_{21}$$

$$p_{11} = 2b_2a_4 + 3b_3a_1 - b_1a_3 \qquad q_{11} = 3b_3a_4 + 2b_2a_1 - b_1a_2$$

$$p_{12} = -2b_2a_4 + b_3a_1 - 3b_1a_3 \qquad q_{12} = -3b_1a_4 - 2b_2a_3 + b_3a_2$$

$$p_{21} = -2a_2a_4 - 2a_3a_1 \qquad q_{21} = -3a_3a_4 - a_1a_2$$

$$p_{22} = -p_{21} \qquad q_{22} = 3a_1a_4 + a_2a_3$$

and

$$\begin{bmatrix} a_1 & b_1 \\ a_2 & b_2 \\ a_3 & b_3 \\ a_4 & b_4 \end{bmatrix} = \frac{1}{4} \begin{bmatrix} -1 & 1 & 1 & -1 \\ 1 & -1 & 1 & -1 \\ -1 & -1 & 1 & 1 \\ 1 & 1 & 1 & 1 \end{bmatrix} \begin{bmatrix} r_1 & z_1 \\ \vdots & \vdots \\ r_4 & z_4 \end{bmatrix}$$

The total element displacements are then given as a sum of two parts:

$$\mathbf{u} = \mathbf{u}_q + \mathbf{u}_\lambda^* \tag{8.33}$$

where $\mathbf{u}_q$ is the standard bilinear displacement functions of the Axi-Q4 element. In terms of Equation 8.33, the new axisymmetric element, AQ6, can be formulated conveniently by using the potential energy principle. The resulting incompatible element is complete in quadratic terms in the isoparametric system and possesses six independent deformation modes even if the incompressible condition (Equation 8.28) is rigidly enforced.

8.5.3 Elastic Solution

For linear elastic plane-strain mode I problems, there exists the following relationship between the J integral and the stress intensity factor K_I [15]:

$$J = (1-v^2)K_1^2/E \tag{8.34}$$

As mentioned in References 16 and 17, the stress and strain fields near the crack tip in the axisymmetric problem correspond to the fields of the plane strain ones. Thus we can use Equation 8.34 to calculate the stress intensity factor in terms of the J integral. An advantage of using J is the unification of both the linear and nonlinear fracture calculations. The axisymmetric formulation of the J integral is presented in References 18 and 19. Referring to Figure 8.6, it can be expressed as

$$J = \frac{1}{R_i + a}\left\{\int_\Gamma r\left[Wdz - (T_r\frac{\partial u}{\partial r} + T_z\frac{\partial v}{\partial r})ds\right] - \int\int_A (W - \sigma_\theta\frac{u}{r})dA\right\} \tag{8.35}$$

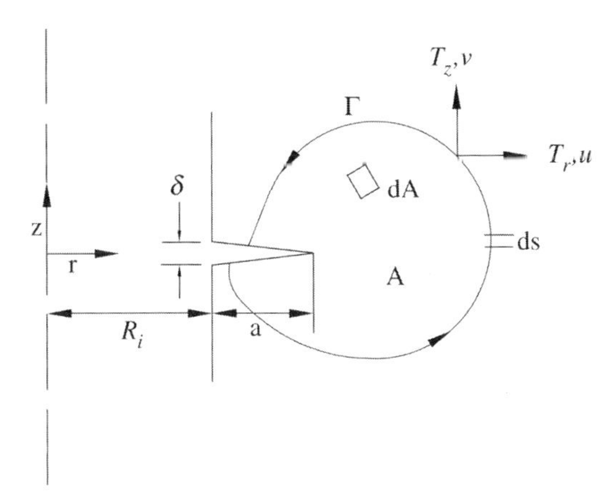

FIGURE 8.6
Definition of the J integral for axisymmetric crack.

where

$$W = \frac{1}{2}(\sigma_r \varepsilon_r + \sigma_z \varepsilon_z + \sigma_\theta \varepsilon_\theta + \tau_{rz} \gamma_{rz})$$

is the strain energy density; T_r and T_z are the boundary force components on the integral path Γ (Figure 8.6).

We consider the thick-walled cylinder with a circumferential crack in Figure 8.7. In the case of crack depth $a = b/4$ and $b/R_i = 1/10$ or $1/5$ (b is the wall thickness and R_i is the inner radius). In Table 8.5, a series of rational numerical solutions are provided by the incompatible element method with much less effort than is required in Reference 20. Additionally, an incompressible numerical test has confirmed that when the Poisson's ratio $v \to 0.5$, for instance $v = 0.49$, the solutions of the Axi-Q4 become very bad and

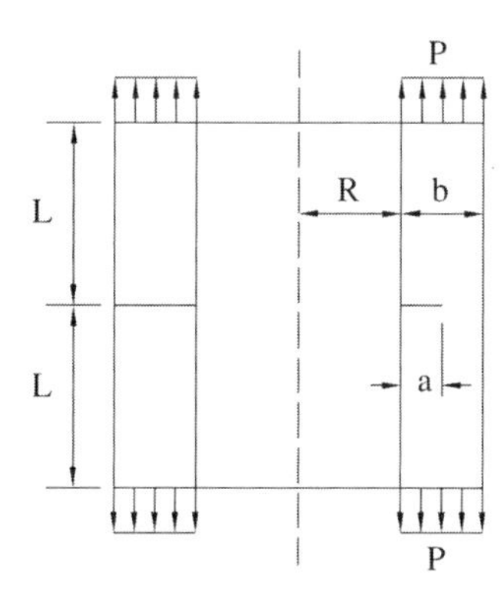

FIGURE 8.7
A stretched cylinder with a circumferential crack.

TABLE 8.5
Elastic Solutions of the Cracked Cylinder (Figure 8.7 and Figure 8.8)

$\frac{b}{R_i}$	Method	n = 0.3		n = 0.49	
		AQ6	**Ref. 21**	**AQ6**	**Axi-Q4**
$\frac{1}{10}$	d¢	1.76	1.81	1.78	1.30
	D¢p	0.28	—	0.28	0.20
	Kđ(1)	1.34	1.34	1.36	1.03
	Kđ(2)	1.35	1.34	1.36	1.18
	Kđ(3)	1.34	1.34	1.36	1.21
$\frac{1}{5}$	d¢	1.62	1.70	1.65	1.22
	D¢p	0.24	—	0.25	0.18
	Kđ(1)	1.25	1.26	1.27	0.97
	Kđ(2)	1.25	1.26	1.27	1.10
	Kđ(3)	1.25	1.26	1.27	1.14

unacceptable, while the incompatible element AQ6 is able to provide the rational solutions as shown in Table 8.5.

8.5.4 Elastic-Plastic Solution

It is assumed that the cylinder in Figure 8.7 [23] is made of an ideally elastic-plastic material obeying von Mises' yield criterion, and the yield strength is $\sigma_s = 200$ KN/mm^2. The discrete mesh in Figure 8.8 is adopted once again. The incremental tangential stiffness method is used in iterative calculations, in which the residual forces, that is, the differences between equivalent nodal forces and external applied nodal loads, are taken as the control parameters for the iterative convergence.

The plastic zone starting from the tip of the crack is developed gradually as the loading is increased. A series of elastic-plastic interfaces are described in Figure 8.9. In the cross section of the stretched cylinder, the plastic zone near the crack tip has a similar shape to the Mode I problem in the plane strain case. However, it is assumed in Reference 22 that for the cylinder under axial loading, a small plastic zone develops at the crack tip, having the shape of a thin in-plane layer annulus. Our experience has shown that the stress and strain states near the crack tip for the axisymmetric problem are consistent with those for the plane strain problem, and not the plane stress problem as implied in Reference 22.

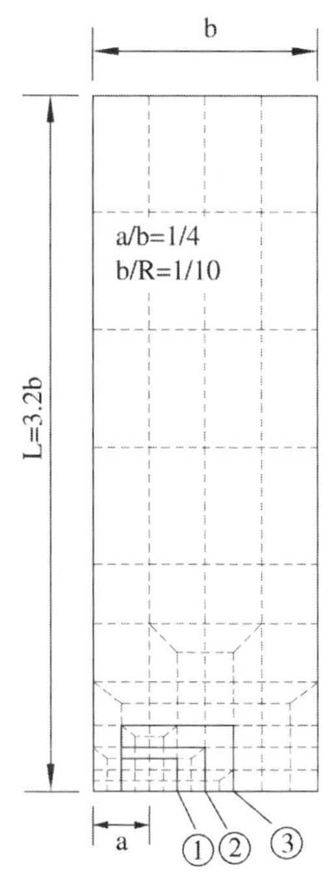

FIGURE 8.8
Mesh division (for a quarter of a cross section) and J integral paths.

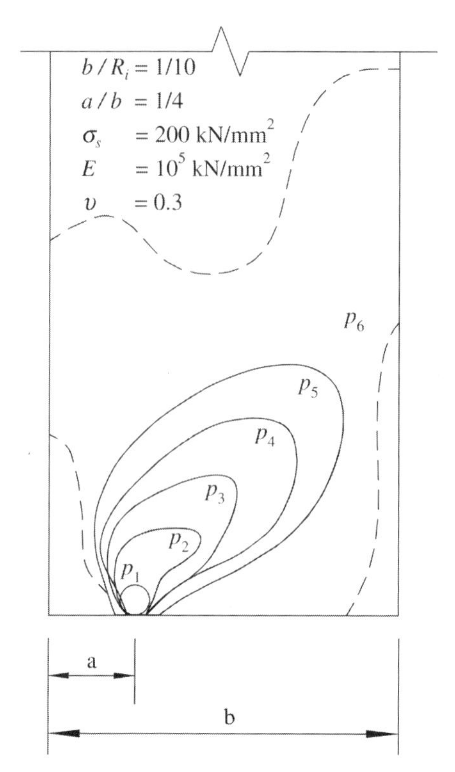

FIGURE 8.9
Extension of the plastic domains near the crack tip.

For the small-deformation static crack problems, the J integral based on deformation theory has been widely used as a crack control parameter in 2-D elastic-plastic fractures without the limitation of small-scale yielding. Our investigation further verifies that for the axisymmetric crack problem based on plastic flow theory, the J integral defined by Equation 8.35 is still path independent in the case of a small- or even a finite-scale yielding, with finite-scale meaning that the radius of the plastic zone around the crack tip r_s is less than the crack depth a.

When the incompatible element is used to evaluate the axisymmetric J integral (Equation 8.35), a simple numerical treatment for stresses and strains is suggested. The computed stress and strain values at the Gauss points near the selected integral path are used as the basis value for a bilinear extrapolation to this path. In this way, we obtain three sets of J values: $J(i)$, $i = 1$, 2, and 3, referring to the paths 1, 2, and 3, respectively. The results in Table 8.6 indicate that in the case of finite-scale yielding, the relative differences between $J(i)$, denoted by $\Delta J/J(i)$ in Table 8.6, are not more than 1.5%, and therefore the J integral can be regarded as path independent. However, when the loading level $p > 0.7\,p_s$ (p_s is the plastic limit load), or $r_s > a$ (see Figure 8.9), the J integral is no longer path independent. The above limits for J path independence are also suitable

TABLE 8.6

Inspection of the *J* Path Independence (Figure 8.7 and Figure 8.8; a/b = 1/4)

p (kN/mm²)	$b/R_i = 1/10(p_s = 178)$				$b/R_i = 1/5(p_s = 180)$			
	J(1)	J(2)	J(3)	ΔJ/J(i)	J(1)	J(2)	J(3)	ΔJ/J(i)
40	33.2	33.7	33.5	1.5%	28.7	29.1	29.0	1.4%
80	132.9	134.8	134.1	1.4%	114.9	116.6	116.0	1.5%
100	213.8	216.9	215.8	1.4%	182.2	184.8	183.0	1.4%
120	325.6	328.7	327.5	0.9%	269.7	273.5	272.4	1.4%
140	486.8	513.6	513.3	5.2%	393.7	411.9	410.0	4.4%
160	849.4	922.6	931.7	8.8%	616.6	648.3	663.7	7.1%
170	1379	1491	1504	8.3%	891.8	952.1	960.4	7.1%

for the cracked cylinder of a/b = 1/2(b/R_i = 1/10 or 1/5), but not for the case of $a/b > 1/2$ [17]. It is therefore suggested that under the conditions

$$p < 0.7p_s \quad r_s < a \le b/2 \tag{8.36}$$

the *J* integral defined by Equation 8.35 is path independent and can be taken as a primary control parameter for axisymmetric elastic-plastic crack problems.

8.5.5 Plastic Limit Analysis

When the external applied load p reaches a certain limit value p_s, plastic flow of the structure will occur, which is characterized by a large displacement even though the loading increment Δp is very small. Because of this, the numerical iteration process may be broken off too early, and a trick has to be introduced to carry on the iteration to obtain an ideal plastic solution. The trick involves halving the current increment Δp and repeating the iteration; such repeated iterations should continue automatically until Δp is reduced to a certain small value defined *a priori*.

For a cylinder of $a/b = 1/4$ and $b/R_i = 1/10$, the load-displacement curves, $p - \Delta_p$ and $p - \delta$, are as shown in Figure 8.10 and Figure 8.11, respectively, in which the results for both the incompatible axisymmetric element Axi-Q6 and the compatible Axi-Q4 element are presented. For the incompatible element, the loading-point displacement Δ_p as well as the opening-mouth displacement δ suddenly become very large. At that point the axial stretch load p reaches $p_s = 178$ and the iteration process terminates. That value is nothing but the plastic limit load of the cracked cylinder. However, no limit value of p exists for the isoparametric element. A number of plastic limit loads solved by the Axi-Q6 incompatible approach are listed in Table 8.7. They are in agreement with the lower bound of the limit loads provided by Reference 20.

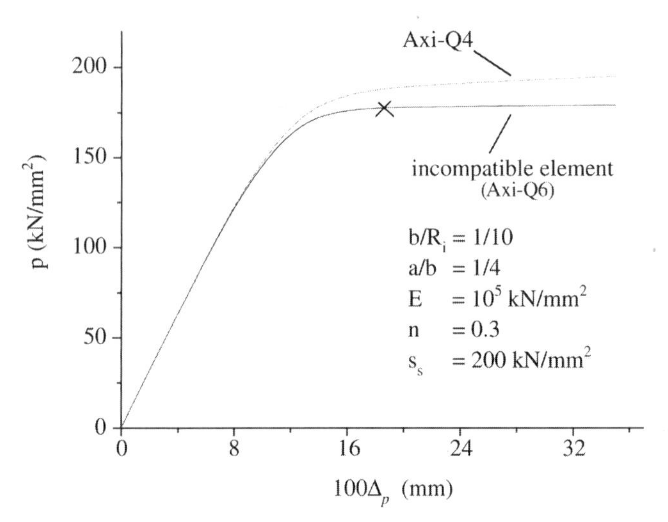

FIGURE 8.10
Load-loading-point displacement curves.

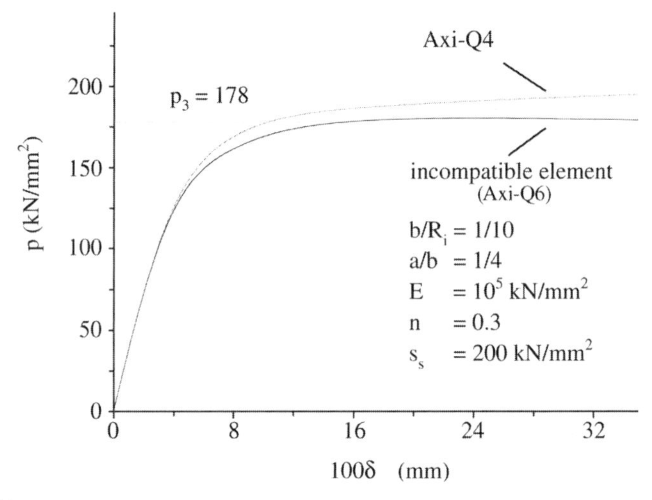

FIGURE 8.11
Load-opening displacement of crack mouth curves.

TABLE 8.7
Plastic Limit Loads (kN/mm²) of the Cracked Cylinder (Figure 8.7)

a/b	b/Ri	Axi-Q4	Axi-Q6	Ref. 14 (lower bound solution)	Material
1/4	1/10	No	178	175.3	E = 105 kN/mm²
	1/5	No	180	177.1	v = 0.3
1/2	1/10	No	120	118.2	ss = 200 kN/mm²
	1/5	No	126	120.7	H = 0.0

8.5.6 Conclusion

We have proposed an incompatible axisymmetric element Axi-Q6 that can overcome the difficulties experienced by isoparametric elements in axisymmetric plastic problems with incompressible materials. The proposed incompatible axisymmetric element can be used conveniently both in the plastic region near the crack tip and in the exterior elastic region. The element can be readily incorporated into many widely available codes, since apart from an additional static condensation of a 4×4 matrix involving the internal parameters, it is otherwise similar to a standard isoparametric element.

The proposed incompatible element was applied to a systematic study of the Mode I problem of an ideal elastic-plastic cylinder with annular cracks. From the numerical results the following can be concluded:

1. The shape of the plastic region near the crack tip for the axisymmetric problem is the same as that for the plane strain case.
2. The J integral for the axisymmetric problem is path independent even under finite-scale yielding situations.
3. As an elastic-plastic crack control parameter for axisymmetric problems, the J integral can be used within the limits of $p < 0.7p_s$ and $r_s < a \leq b/2$.
4. The plastic limit load can be computed accurately by the proposed element but not by the axisymmetric isoparametric elements.

8.6 Extension of J to Dynamic Fracture of a Functional Graded Material

8.6.1 Introduction

Functional graded materials (FGMs) have been widely used in technological applications. Their mechanical behaviors, especially their fracture behaviors, have been extensively studied in recent years. A comprehensive review has been presented recently by Suresh and Mortensen [24]. For the fracture of an FGM, Jin and Noda [25] have shown that the singularity and the angular distribution of the near-tip stress and displacement fields are the same as those of homogeneous materials. Erdogan and Wu [26] have presented an analytical result for a surface crack in a plate. Chan et al. [27] and Jin and Paulino [28] have shown the stress intensity factors for FGMs in some important cases. It is also known that for cracks in the gradient direction, the J integral becomes path dependent. Honein and Herrmann [29] have derived an extension of the J integral for FGMs that is path independent. Jian et al. [30] have also presented a modified J integral for FGMs. Wu et al. [2, 31] have set up a dual and bound theorem and developed the corresponding finite element methodology. However, the extension of the J integral to the dynamic fracture of FGMs has not appeared in the literature. It is well known that the crack tip energy release rate formula can be derived conveniently from a variational form of momentum balances (see, for example, Moran and Shih [32]). In this section, an extended dynamic J integral for FGMs will be introduced following the procedure of Moran and Shih [32].

For numerical implementation of the extended dynamic J integral, a suitable choice would be the meshless methods suggested in References 33 to 35. The element-free Galerkin method (EFGM) proposed by Belytschko et al. [33, 34] is a meshless method in terms of the moving least-squares (MLS) approximation based only on the finite discrete nodes. Since no element connectivity data are needed, the extension of the crack is treated naturally by the growth of the surfaces of the crack. This method provides a higher resolution of localized derivations of strains and stresses. It can also adopt the material properties at the integration points to simulate the variation of material properties.

However, in EFGM, the MLS interpolants do not pass through the nodal parameter values, the imposition of boundary conditions on the dependent variables is quite awkward, and the computational cost is very burdensome. To solve the problem, the EFGM is coupled to the finite element method by a collocation method [36]. The basic idea is to evaluate the real value of the node at the interface between the finite element and the EFG regions using the MLS method for EFG and then assign this value to the finite element nodal. This method is easily implemented by converting the finite element matrices to the matrices relevant to the EFG virtual-nodal values.

This coupled method is used to calculate the dynamic J integral for a stationary propagating crack. The EFG is only used near the crack tip, where great versatility and high resolution are needed, while finite elements are applied in the remaining part. The boundary conditions are treated by finite elements; in the meantime, the computation efficiency is greatly improved. Since the background finite element is used as an integral mesh in the EFGM, so that it is very convenient to the numerically implemented method.

8.6.2 The Extended Dynamic J Integral for FGMs

The dynamic J integral for cracked homogeneous linear elastic materials can be formulated as [32]:

$$J_t = \lim \int_\Gamma [(W + L)\delta_{1j} - \sigma_{ij} u_{i,1}] n_j d\Gamma \tag{8.37}$$

where σ_{ij} and u_i are the stress and displacement, respectively. δ_{ij} is the Kronecker symbol. n_j is the outward unit normal to the integration path Γ. The stress work density W and kinetic energy density L, in a linear elastic medium, are given by

$$W = \frac{1}{2}\sigma_{ij}\varepsilon_{ij} \qquad L = \frac{1}{2}\rho\dot{u}_i^2 \tag{8.38}$$

where ρ and $\dot{u}$ are the mass density and the velocity, respectively.

Consider a 2-D body with a crack oriented along the negative x_1. S is a closed domain surrounding the crack tip (see Figure 8.12).

The domain integral representation of J is more suited for numerical computation. According to Moran [32], we introduce a weighting function q, which has a value of unity on Γ_1 and zero on Γ_2. Within the area S, q is an arbitrary smooth function of x_1 and x_2 with values ranging from zero to one. Thus Equation 8.37 can be rewritten as

$$J = \oint [-(W+L)\delta_{1j} - \sigma_{ij}u_{i,1}]n_j q d\Gamma - \int_{BC+DA} \sigma_{i2}u_{i,1}n_2 q d\Gamma \tag{8.39}$$

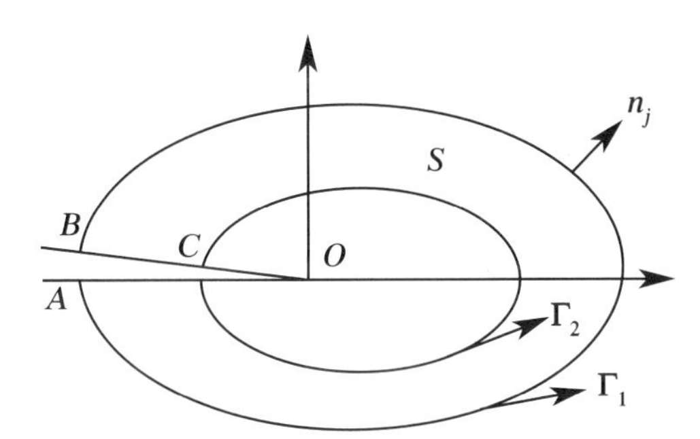

FIGURE 8.12
Conventions at the crack tip.

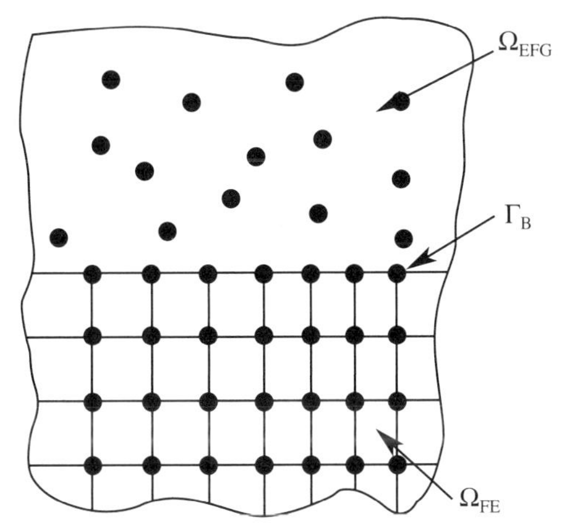

FIGURE 8.13
Interface Γ_B between the domain Ω_{EFG} and the domain Ω_{FE}.

Applying the divergence theorem and, without loss of generality, taking the crack faces to be traction free, we obtain

$$J = -\int\int_S \{[(W+L)\delta_{1j} - \sigma_{ij}u_{i,1}]q_j + [(W+L)_{,1} - (\sigma_{ij}u_{i,1})_{,j}]q\}dS \tag{8.40}$$

where $W_{,1}$ and $L_{,1}$ can be rearranged as follows:

$$W_{,1} = (\frac{1}{2}\sigma_{ij}\varepsilon_{ij})_{,1} = (\frac{1}{2}\varepsilon_{ij}D_{ijkl}\varepsilon_{kl})_{,1} = \sigma_{ij}\varepsilon_{ij,1} + \frac{1}{2}\varepsilon_{ij}D_{ijkl,1}\varepsilon_{kl} \tag{8.41}$$

In the absence of body forces, the equation of motion is written as $\sigma_{ij,j} = \rho\ddot{u}$. The first term on the right-hand side in Equation 8.41 can be expressed as

$$\sigma_{ij}\varepsilon_{ij,1} = (\sigma_{ij}u_{i,1})_{,j} - \sigma_{ij,j}u_{i,1} = (\sigma_{ij}u_{i,1})_{,j} - \rho\ddot{u}u_{i,1} \tag{8.42}$$

With the use of Equation 8.42, $W_{,1}$ can be rewritten as

$$W_{,1} = (\sigma_{ij}u_{i,1})_{,j} - \rho\ddot{u}u_{i,1} + \frac{1}{2}\varepsilon_{ij}D_{ijkl,1}\varepsilon_{kl} \tag{8.43}$$

$$L_{,1} = (\frac{1}{2}\rho\dot{u}_i^2)_{,1} = \frac{1}{2}\rho_{,1}\dot{u}_i^2 + \rho\dot{u}_i\dot{u}_{i,1} \tag{8.44}$$

Inserting Equation 8.43 and Equation 8.44 into Equation 8.39, an extended dynamic J integral for FGMs is available as follows

$$\begin{aligned} J_t = \int\int_S &[-(W+L)q_{,1} + \sigma_{ij}u_{i,1}q_j + \rho(\ddot{u}u_{i,1} - \dot{u}_i\dot{u}_{i,1})q \\ &- \frac{1}{2}(\varepsilon_{ij}D_{ijkl,1}\varepsilon_{kl} + \rho_{,1}\dot{u}_i^2)q]dS \end{aligned} \tag{8.45}$$

In Equation 8.45 the last term

$$\int\int_S \frac{1}{2}(\varepsilon_{ij}D_{ijkl,1}\varepsilon_{kl} + \rho_{,1}\dot{u}_i^2)qdS$$

exists for FGMs only. For homogeneous materials, since $D_{ijkl,1} = 0$ and $\rho_{,1} = 0$, it vanishes. Equation 8.45 reduces to Equation 8.46, which is the same as the one given by Moran et al. [32]:

$$J = \int\int_S [-(W+L)q_{,1} + \sigma_{ij}u_{i,1}q_j + \rho(\ddot{u}u_{i,1} - \dot{u}_i\dot{u}_{i,1})q]dS \tag{8.46}$$

In the calculation, D_{ijkl} and ρ adopt the values at the integral points, so that a high precision is available.

8.6.3 The Element-Free Galerkin Method

The most important difference between the EFGM and the finite element method is the construction of the shape functions and test functions. In the EFGM, the field variable is approximated by MLS approximations, and unlike in the finite element method, no element connection data are needed. We briefly describe the EFG methodology in the following.

In Ω, the MLS interpolation $\mathbf{u}^*$ is given by

$$\mathbf{u}^h(\mathbf{x}) = \mathbf{p}^T(\mathbf{x})\mathbf{a}(\mathbf{x}) = \sum_{j=1}^{m} p_j(\mathbf{x})a_j(\mathbf{x}) \tag{8.47}$$

Here, $\mathbf{p}(x)$ is a complete polynomial basis of arbitrary order, and $\mathbf{a}(x)$ are coefficients that are functions of the space coordinates. We have used a linear polynomial basis in all the calculations in the following, so $\mathbf{p}(x) = \begin{bmatrix} 1 & x & y \end{bmatrix}$, $m = 3$. Equation 8.47 is the global approximation; the corresponding local approximation is

$$\mathbf{u}^h(\mathbf{x}, \bar{\mathbf{x}}) = \mathbf{p}^T(\bar{\mathbf{x}})\mathbf{a}(\mathbf{x}) = \sum_{j=1}^{m} p_j(\bar{\mathbf{x}})a_j(\mathbf{x}) \tag{8.48}$$

The coefficients $\mathbf{a}(x)$ are obtained at any point by minimizing the following:

$$J = \sum_{I=1}^{n} w(\mathbf{x} - \mathbf{x}_I)[\mathbf{p}^T(\mathbf{x}_I)\mathbf{a}(\mathbf{x}) - \mathbf{u}^*(\mathbf{x}_I)]^2 \tag{8.49}$$

Here, $w(\mathbf{x} - \mathbf{x}_I)$ is a weighting function with compact support and n is the number of nodes in the neighborhood, or the domain of influence of x. In the domain of influence, $w > 0$; otherwise $w = 0$; $\mathbf{u}_I^* = \mathbf{u}^*(\mathbf{x}_I)$.

Equation 8.49 can be rewritten in the matrix form

$$J = (\mathbf{Pa} - \mathbf{u}^*)^T \mathbf{W}(\mathbf{x})(\mathbf{Pa} - \mathbf{u}^*) \tag{8.50a}$$

where

$$\mathbf{P} = \begin{bmatrix} p_1(\mathbf{x}_1) & p_2(\mathbf{x}_1) & \dots & p_m(\mathbf{x}_1) \\ p_1(\mathbf{x}_2) & p_2(\mathbf{x}_2) & \dots & p_m(\mathbf{x}_2) \\ \dots & \dots & & \dots \\ p_1(\mathbf{x}_n) & p_2(\mathbf{x}_n) & \dots & p_m(\mathbf{x}_n) \end{bmatrix} \tag{8.50b}$$

$$\mathbf{W}(\mathbf{x}) = \begin{bmatrix} w(\mathbf{x}-\mathbf{x}_1) & 0 & \cdots & 0 \\ 0 & w(\mathbf{x}-\mathbf{x}_2) & \cdots & 0 \\ \cdots & \cdots & \ddots & \cdots \\ 0 & 0 & \cdots & w(\mathbf{x}-\mathbf{x}_n) \end{bmatrix} \tag{8.50c}$$

$$\mathbf{u}^* = (\mathbf{u}_1^*, \mathbf{u}_2^*, \ldots, \mathbf{u}_n^*) \tag{8.50d}$$

To find the coefficients **a(x)**, we calculate the extremum of J, i.e., $\partial J / \partial a = 0$ and have

$$\mathbf{A}(\mathbf{x})\mathbf{a}(\mathbf{x}) - \mathbf{B}(\mathbf{x})\mathbf{u}^* = \mathbf{0} \tag{8.51a}$$

and

$$\mathbf{A}(\mathbf{x}) = \mathbf{P}^T \mathbf{W}(\mathbf{x})\mathbf{P}, \quad \mathbf{B}(\mathbf{x}) = \mathbf{P}^T \mathbf{W}(\mathbf{x}) \tag{8.51b}$$

From Equation 8.51a, we have

$$\mathbf{a}(\mathbf{x}) = \mathbf{A}^{-1}(\mathbf{x})\mathbf{B}(\mathbf{x})\mathbf{u}^* \tag{8.52}$$

Inserting Equation 8.52 into Equation 8.48, we have

$$\mathbf{u}^h(\mathbf{x}) = \mathbf{p}^T(\mathbf{x})\mathbf{A}^{-1}(\mathbf{x})\mathbf{B}(\mathbf{x})\mathbf{u}^* = \Phi(\mathbf{x})\mathbf{u}^* \tag{8.53a}$$

where

$$\Phi(\mathbf{x}) = \mathbf{p}^T(\mathbf{x})\mathbf{A}^{-1}(\mathbf{x})\mathbf{B}(\mathbf{x}) = [\varphi_1 \ \ \varphi_2 \ \ \ldots \ \ \varphi_n] \tag{8.53b}$$

So the shape functions $\Phi(\mathbf{x})$ can be obtained by the MLS method and then its derivatives can be formulated as

$$\begin{aligned} \Phi(x)_{,i} &= [\mathbf{p}^T(\mathbf{x})\mathbf{A}^{-1}(\mathbf{x})\mathbf{B}(\mathbf{x})]_{,i} \\ &= \mathbf{p}^T(\mathbf{x})_{,i}\mathbf{A}^{-1}(\mathbf{x})\mathbf{B}(\mathbf{x}) + \mathbf{p}^T(\mathbf{x})\mathbf{A}^{-1}(\mathbf{x})_{,i}\mathbf{B}(\mathbf{x}) + \mathbf{p}^T(\mathbf{x})\mathbf{A}^{-1}(\mathbf{x})\mathbf{B}(\mathbf{x})_{,i} \end{aligned} \tag{8.54}$$

where

$$\mathbf{A}^{-1}(\mathbf{x})_{,i} = -\mathbf{A}^{-1}(\mathbf{x})\mathbf{A}(\mathbf{x})_{,i}\mathbf{A}^{-1}(\mathbf{x}) \tag{8.55}$$

8.6.4 Collocation Method for Coupling the Finite Element and the EFG Methods

As shown in Figure 8.13, in the meshless domain Ω_{EFG}, we adapt the MLS method to construct the trial functions of point $\mathbf{x}_{el}$:

$$\mathbf{u}^h(\mathbf{x}_{el}) = \Phi(\mathbf{x})\breve{\mathbf{u}} \tag{8.56}$$

For an element in the finite element domain Ω_{FE}, which is not adjacent to the interface Γ_b between the subdomains Ω_{FE} and Ω_{EFG}, we construct the trial functions of point x_{fl} in the usual way:

$$\mathbf{u}^h(\mathbf{x}_{fl}) = \sum_{i=1}^{ND} N_i \mathbf{u}_i \tag{8.57}$$

where ND is the number of nodes in the element and $\mathbf{u}_i$ is the element nodal values.

For a finite element with node $\mathbf{x}_{bl}$ on the interface Γ_b using the MLS method we have

$$\mathbf{u}^h(\mathbf{x}_{bl}) = \Phi(\mathbf{x})\breve{\mathbf{u}} \tag{8.58}$$

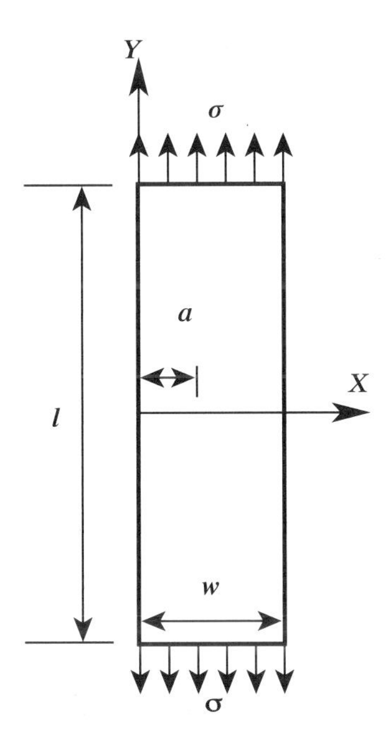

FIGURE 8.14
Single-edge cracked FGM panel.

Then the trial functions of the finite element can be expressed as

$$u^h(x_{fl}) = \sum_{i=1}^{ND} N_i u_i(x_i) \ , \ u_i = \begin{cases} u_i \ , & x_i \in \Omega_{FE} \\ \Phi(x)\breve{u} \ , & x_i \in \Gamma_b \end{cases} \tag{8.59}$$

The prescribed displacements in Ω_{FE} can be enforced directly. It should be noted that the collocation method can be used to couple any MLS-based meshless methods with displacement-based compatible, incompatible, and hybrid finite elements.

8.6.5 Numerical Implementation of the Boundary Value Problem

The energy functional of the 2-D boundary value problem in elasticity can be formulated as

$$\int_{\Omega} \delta(\nabla_S u^T) : \Phi \, d\dot{\Omega} - \int_{\Omega} \delta u^T \cdot f - \int_{\Gamma_s} \delta u^T \cdot \bar{t} d\Gamma = 0 \quad \forall u, \delta u \in H^1 \tag{8.60}$$

where $\nabla_s u^T$ is the symmetry part of ∇u^T, u is the test function, and H^1 is the 1-D Sobolev space. The stiffness matrix of the EFGM can be obtained as follows.

With the MLS method, the displacement u of a point x can be approximated by the displacement u^* in its influence domain

$$u = \Phi u^* \tag{8.61a}$$

where

$$u = \begin{bmatrix} u \\ v \end{bmatrix} \tag{8.61b}$$

$$\phi = \begin{bmatrix} \varphi_1 & 0 & \varphi_2 & 0 & \cdots & \varphi_n & 0 \\ 0 & \varphi_1 & 0 & \varphi_2 & \cdots & 0 & \varphi_n \end{bmatrix} \tag{8.61c}$$

$$u^* = \begin{bmatrix} u_1^* & v_1^* & u_2^* & v_2^* & \cdots & u_n^* & v_n^* \end{bmatrix}^T \tag{8.61d}$$

So we have the strain at the point x

$$\varepsilon = \mathrm{B}u^* \tag{8.62a}$$

and/or

$$\varepsilon = \begin{bmatrix} \varepsilon_x & \varepsilon_y & \gamma_{xy} \end{bmatrix} \tag{8.62b}$$

$$B = \begin{bmatrix} \varphi_{1,x} & 0 & \varphi_{2,x} & 0 & \cdots & \varphi_{n,x} & 0 \\ 0 & \varphi_{1,y} & 0 & \varphi_{2,y} & \cdots & 0 & \varphi_{n,y} \\ \varphi_{1,y} & \varphi_{1,x} & \varphi_{2,y} & \varphi_{2,x} & \cdots & \varphi_{n,y} & \varphi_{n,x} \end{bmatrix} \tag{8.62c}$$

We have the stiffness matrix

$$K = \int_{\Omega_E} B^T D B d\Omega \tag{8.63}$$

where D is the matrix of the elastic materials and is expressed as

$$D = \frac{E(x,y)}{1-\nu^2(x,y)} \begin{bmatrix} 1 & \nu(x,y) & 0 \\ \nu(x,y) & 1 & 0 \\ 0 & 0 & \frac{1-\nu(x,y)}{2} \end{bmatrix} \tag{8.64}$$

in which Young's modulus $E(x,y)$ and Poisson's ratio $v(x,y)$ adopt the values of the integration points in finite elements and background finite elements.

Finally, we can get the corresponding discrete equation of the 2-D boundary value problem:

$$Ku^* = F \tag{8.65a}$$

where

$$F = \int_{\Omega} \Phi^T \mathbf{f} d\Omega + \int_{\Gamma_u} \Phi^T \bar{\mathbf{t}} d\Gamma \tag{8.65b}$$

In the implementation, the domain is divided into finite element method and EFGM subdomains first, and the boundary (Γ_u) is overridden by the finite element method subdomain. Then K and F are formed by numerical integrations. Solving Equation 8.60, we can get the numerical results of u. Then we can obtain the global displacement field of the 2-D elastic problem through Equation 8.56.

In the EFGM, a crack is modeled by its effect on the domain of influence of nodes. It is regarded as an inner boundary, and thus sampling points are not included in the domain of influence of nodes that are on the other side of the crack because the domain of influence does not extend across boundaries. So the growth of the crack can be modeled by introducing new surfaces, and no additional computation is required.

8.6.6 Numerical Examples

An SECP of unit thickness in an elastic plane stress condition is considered in Figure 8.14. l and w are the length and the width of the plate. a is the length of the edge crack. Poisson's ratio v and the mass density ρ are constant. Young's modulus $E(x)$ is given by the following expression:

$$E(x) = E_1 \exp\left(\frac{x}{h}\ln\frac{E_2}{E_1}\right) \tag{8.66}$$

In this section, l=4, w=1, a=0.25, $E_1 = 1.11\times10^{11}$, $E_2 = 2.22\times10^{11}$, $v = 0.3$, $\rho = 7800$, and $\sigma = 1.0$.

The distribution of nodes is shown in Figure 8.15. The plate is divided into 30×40 elements, including the background elements for quadrature. The number of nodes is 1271. The nodes of the EFGM are only distributed near the crack tip. The meshless domain is divided into 30×4 background elements. The rest of the plate is divided into 30×36 elements. The nearer to the crack surface, the closer together the nodes are arranged along Y. Otherwise, the material properties vary along X, so nodes are arranged more finely along X than along Y.

In terms of Equation 8.45, for a stationary crack, we evaluate the dynamic J integral for FGMs in two different integral domains (see Figure 8.16) with the increasing of the time step. A fixed time step equal to 4×10^{-6} was employed for a total of 170 time steps. The numerical results are plotted in Figure 8.17.

The curves in Figure 8.17 show that before the stress waves reach the crack surface, the J integral is zero. After the stress waves reach the crack surface, the J integral begins to increase. Because the velocity of the stress waves varies with the Young's modulus, a little difference between the J integral in domain

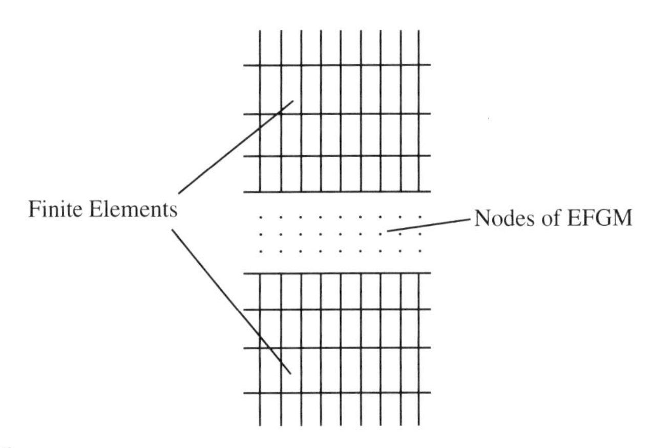

FIGURE 8.15
Node distribution.

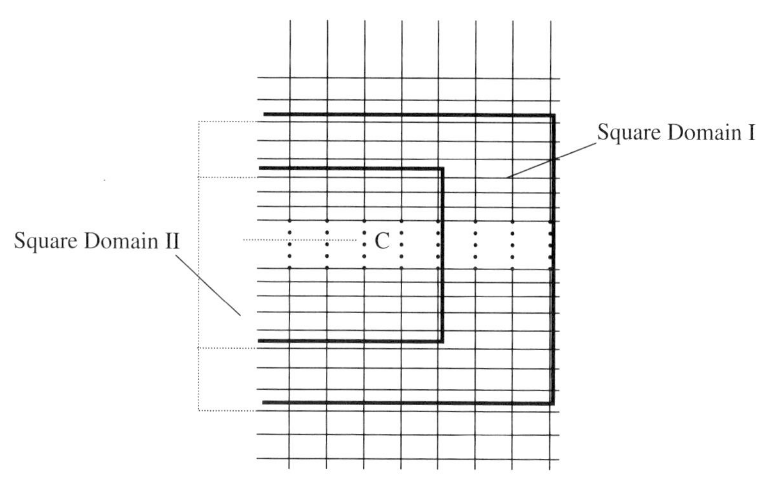

FIGURE 8.16
Discrete model (SECP) and square domain with the crack tip "C."

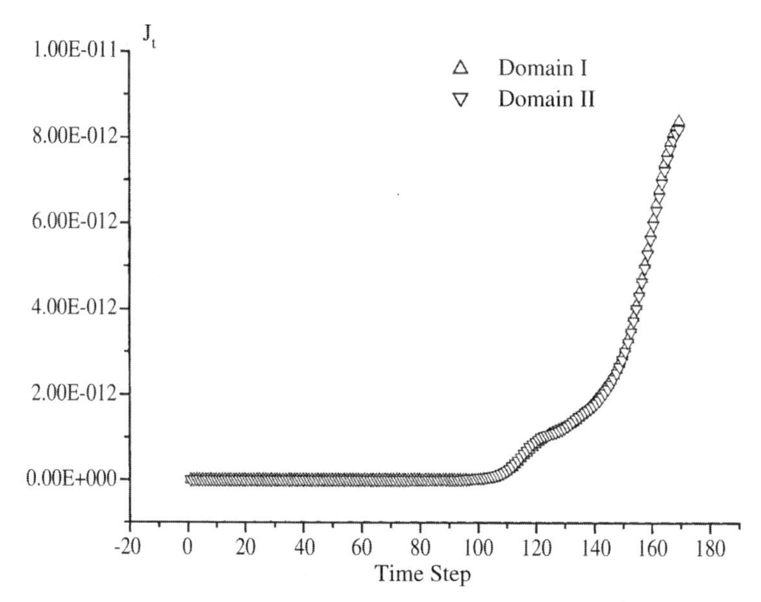

FIGURE 8.17
Variation of J integral with time step for stationary crack.

I and J in domain II exists. However, after the stress waves pass through the whole area for evaluating the J integral, the J integrals in the two domains are almost the same. Therefore, the J integral is path independent.

For the crack that starts propagating with $v = 1400$, as soon as the stress wave front reaches its planes, the dynamic J integrals are calculated in the same domains. A fixed time step equal to 4×10^{-6} was employed for a total of 120 time steps. The crack moved from $x = 0.25$ to 0.37. The numerical results are shown in Figure 8.18.

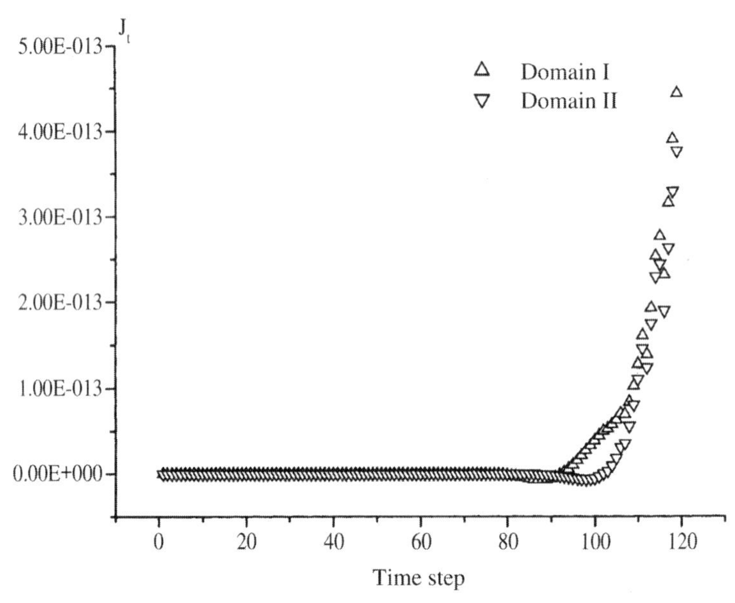

FIGURE 8.18
Variation of J integral with time step for propagating crack.

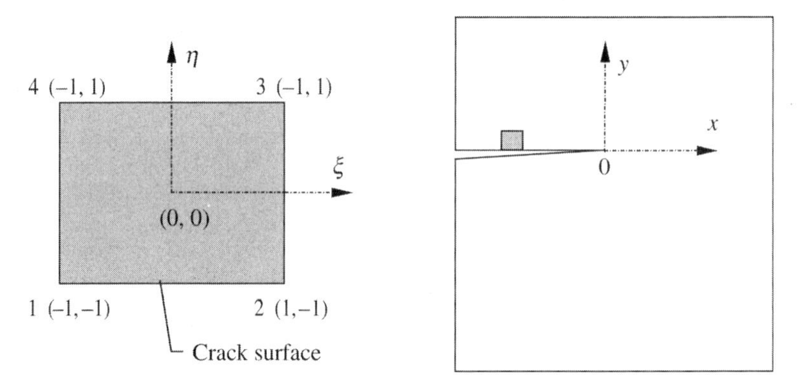

FIGURE 8.19
Crack surface element.

The curves in Figure 8.17 exhibit slight oscillations, but the J integral in the two domains tends to be equal with the time. The two numerical examples show that the extended J integral is path independent. Its trend of variation is similar to that in homogeneous materials [34]. The results also show that the coupled finite element–EFG method is efficient for FGMs.

8.6.7 Conclusion

In this section, Rice's J integral has been extended to the dynamic fracture of FGMs by properly taking into account the nonhomogeneity. Numerical results show that the extended integral varies with time and crack propagation as in homogeneous materials.

In analyzing the dynamic fracture of FGMs and computing the extended J integral, the EFGM is coupled with a finite element for enforcing boundary conditions directly and improving the computation efficiency. Numerical results show that the technique is efficient. In forming mass and stiffness matrices and evaluating the extended integral, we adopt the material properties at Gauss integration points for FGM, so that only a small number of nodes are needed. The coupled finite element–EFG method facilitates the modeling of a growing crack problem and provides accurate results. It is powerful in dealing with dynamic crack propagation in the FGM.

8.7 Evaluation of Electromechanical Crack Systems

8.7.1 Introduction

As an electromechanical coupling material, piezoelectric ceramics are brittle and likely to crack at all scales from domains to devices. Under mechanical and electrical loading, these materials can fail prematurely due to the propagation of flaws or defects induced during the manufacturing process and by in-service electromechanical loading. Hence, it is necessary to understand and be able to analyze the fracture characteristics of piezoelectric materials so that reliable service life prediction of the electromechanical coupling system can be conducted. Electromechanical modeling of piezoelectric fracture is complicated by the piezoelectric materials exhibiting electro-elastic coupling behavior as well as anisotropy behavior. Among the theoretical studies of cracked piezoelectric bodies, permeable and impermeable conditions at the crack face by Parton [37] and Deeg [38], respectively, are mostly adopted. Whichever condition at the crack face is adopted, it is hard to get the exact solutions except in few cases, and the numerical method is often used [39].

In the finite element analysis, Allik and Hughes presented a piezoelectric element [40] for the linear electromechanical materials and structures, in which case the basic variables, both the displacement and the electric potential, are taken to be a linear interpolation, so that it is an isoparametric-compatible model. A further discussion of the piezoelectric model has been presented by Landis [41]. In recent years, the hybrid finite element has shown ideal numerical behavior in the nonlinear electromechanical coupling analysis [42]. To simulate the characteristic singularity at the crack tip, Wu et al. [43] present a piezoelectric hybrid element, and with it a series of singular fields at the crack tip are simulated.

For the conventional elastic-plastic fracture problem, a bound analysis for the path-independent integrals has been suggested by Wu et al. [2]. In this section, the upper and lower bound approach will be extended to the piezoelectric fracture. To this end, the following subjects should be considered for the electromechanical coupling system: (a) dual path-independent integrals for the piezoelectric crack; (b) bound theorems for the dual crack parameters; (c) dual piezoelectric finite elements, with which the upper and lower bound theorems can be implemented; and (d) error measure for the obtained numerical solutions.

8.7.2 Dual Integrals for Electromechanical Systems

The J integral method suggested by Rice et al. [1] can be developed into the electromechanical crack analysis. A path-independent integral for a piezoelectric medium has been suggested by Pak [45] as

$$J(u_i,\ \varphi) = \int_S \left[H(\varepsilon_{ij}, E_i)dx_2 - (\sigma_{ij}n_j \frac{\partial u_i}{\partial x_1} + D_j n_j \frac{\partial \varphi}{\partial x_1})ds \right] \tag{8.67}$$

where D_i and φ are the electric displacement and the electric potential, respectively, and the electric enthalpy

$$H(\varepsilon_{ij}, E_i) = \frac{1}{2} c^E_{ijkl}\varepsilon_{ij}\varepsilon_{kl} - \frac{1}{2} \in^{\varepsilon}_{ij} E_i E_j - e_{ikl}\varepsilon_{kl}E_i \tag{8.68}$$

with the strain $\varepsilon_{ij} = (u_{ij} + u_{ji})$ and the electric field strength $E_i = -\varphi,_i$. The related piezoelectric potential energy (without body force and bulk charge) can be

$$\Pi_p(u_i, \varphi) = \int_v H(\varepsilon_{ij}, E_i)dV - \int_{S\sigma} \bar{T}_i u_i ds + \int_{Sw} \bar{q}_s \varphi ds \tag{8.69}$$

The integral (Equation 8.67) can be defined as $J(u_i, \varphi) = -\ d\Pi_p/da$, that is, the potential energy release rate for the piezoelectric crack system.

In order to build the dual of $J(u_i, \varphi)$, initially we consider the dual of Rice's J, that is, the complementary energy release rate suggested by Wu et al. [2, 44]:

$$I^*(\sigma_{ij}) = \int_S [-B(\sigma_{ij})dx_2 + \sigma_{i2} \frac{\partial u_i}{\partial x_j} dx_j] \tag{8.70}$$

The path-independent integral I^* is defined for the conventional elastic materials but not for the piezoelectric one. However it can easily be developed as a piezoelectric crack parameter, and only it is needed to add the electric energy contribution induced by the electric loading D_i and φ to the I^* integral (Equation 8.70). In this way, the desired piezoelectric crack parameter can be formulated as

$$I(\sigma_{ij}, D_i) = \int_S [-G(\sigma_{ij}, D_i)dx_2 + (\sigma_{i2}\frac{\partial u_i}{\partial x_j} + D_2\frac{\partial \varphi}{\partial x_j})dx_j\ ] \tag{8.71}$$

where the complementary electric enthalpy can be formulated as follows

$$G(\sigma_{ij}, D_i) = \frac{1}{2}s_{ijkl}^D\sigma_{ij}\sigma_{kl} - \frac{1}{2}\beta_{ij}^\sigma D_i D_j + g_{ijk}\sigma_{ij}D_k \tag{8.71a}$$

For a linear piezoelectric crack system, $I(\sigma_{ij}, D_i) = J(u_i, \varphi)$. Indeed, $I(\sigma_{ij}, D_i)$ takes the meaning of the piezoelectric complementary energy release rate; that is,

$$I(\sigma_{ij}, D_i) = d\Pi_c/da \tag{8.72}$$

where the system complementary energy can be formulated as

$$\Pi_c(\sigma_{ij}, D_i) = \int_v G(\sigma_{ij}, D_i)dv - \int_{S_u} \bar{u}_i\sigma_{ij}n_j ds - \int_{S_\varphi} \bar{\varphi} D_j n_j ds \tag{8.72a}$$

where S_u denotes the boundary on which the prescribed displacement $\bar{u}_i$ acting and S_φ denotes the boundary on which the prescribed electric potential $\bar{\varphi}$ is acting.

8.7.3 Bound Theorem

For conventional homogeneous materials, elastic or elastic-plastic, a bound analysis for crack estimation has been suggested by Wu et al. [2]. The bound method can be extended to the electromechanical systems. It is well known that various upper and lower bound analyses depend on the positive-definite property of the given mathematical and physical problem. For the piezoelectric system, the related electric enthalpy and the complementary enthalpy are all assumed to be positive definite, and we have the preconditional: $H(\varepsilon_{ij}, E_i) > 0$ and $G(\sigma_{ij}, D_i) > 0$.

8.7.3.1 Lower Bound Theorem for $J(u_i, \varphi)$

For a given piezoelectric crack system with the homogeneous boundary constraints, $\bar{u}_i|_{Su} = 0$, $\bar{\varphi}|_{S\varphi} = 0$, the approximate J integral based on the potential energy principle takes the lower bound to the actual one:

$$J(\tilde{u}_i,\tilde{\varphi}) \leq J(u_i,\varphi) \tag{8.73}$$

Here (u_i, φ) and $(\tilde{u}_i,\tilde{\varphi})$ are, respectively, the actual solution and that given by the displacement electric potential compatible-finite element method.

Proof

Let $\tilde{u}_i = u_i + \delta u_i$, $\tilde{\varphi} = \varphi + \delta\varphi$, δu_i, and $\delta\varphi$ be, respectively, the virtual displacement and the virtual electric potential that satisfy the u_i / φ homogeneous boundary condition. Thus the approximate potential energy can be expressed as

$$\Pi_p(\tilde{u}_i,\tilde{\varphi}) = \Pi_p(u_i,\varphi) + \delta\Pi_p + \delta^2\Pi_p \tag{8.74}$$

in which $\delta\Pi_p = 0$ as the stationary condition of the system potential energy, and

$$\delta^2\Pi_p = \Pi_p(\tilde{u}_i,\tilde{\varphi}) - \Pi_p(u_i,\varphi) = \int_V H(\delta u_i,\delta\varphi)dV \geq 0 \tag{8.75}$$

In accordance with the definition of the J integral,

$$J(\tilde{u}_i,\tilde{\varphi}) = -\frac{d}{da}\Pi_p(\tilde{u}_i,\tilde{\varphi}) = -\frac{d}{da}(\Pi_p(u,\varphi) + \delta^2\Pi_p) = J(u_i,\varphi) + \delta^2 J \tag{8.76}$$

and

$$J(u_i,\varphi) = -\frac{d}{da}\Pi_p(u_i,\varphi) = \frac{d}{da}\int_v H(u_i,\varphi)dv \geq 0 \tag{8.77}$$

$$\delta^2 J = -\frac{d}{da}\delta^2\Pi_p = -\frac{d}{da}\int_v H(\delta u_i,\delta\varphi)dv \tag{8.78}$$

Observing that $H(u_i,\varphi)$ and $H(\delta u_i,\delta\varphi)$ possess the same function configuration, the comparison of Equation 8.77 and Equation 8.78 indicates that $\delta^2 J \leq 0,$ and the inequality (Equation 8.73) holds.

8.7.3.2 Upper Bound Theorem for $I(\sigma_{ij}, D_i)$

For a given piezoelectric crack system with homogeneous boundary constraints, $\bar{u}_i|_{Su} = 0,$ $\bar{\varphi}|_{S\varphi} = 0,$ the approximate solution to the I-integral based on the complementary energy principle takes the upper bound to the actual one:

$$I(\tilde{\sigma}_{ij},\tilde{D}_i) \geq I(\sigma_{ij},D_i) \tag{8.79}$$

where (σ_{ij}, D_i) and $(\tilde{\sigma}_{ij}, \tilde{D}_i)$ are, respectively, the actual solution and that given by the stress-electric displacement equilibrium finite element method.

Proof

Let $\tilde{\sigma}_{ij} = \sigma_{ij} + \delta\sigma_{ij}$ and $\delta\sigma_{ij}$ be the virtual stresses and $\tilde{D}_i = D_i + \delta D_i$, δD_i be the virtual electric displacement, and we have

$$\Pi_c(\tilde{\sigma}_{ij}, \tilde{D}_i) = \Pi_c(\sigma_{ij}, D_i) + \delta\Pi_c + \delta^2\Pi_c \tag{8.80}$$

in which $\delta\Pi_c = 0$ is the functional stationary condition. With the homogeneous constraints $\bar{u}_i|_{Su} = 0$, $\bar{\varphi}|_{S\varphi} = 0$, the complimentary energy in Equation 8.80 is then (see Equation 8.72a)

$$\Pi_c(\sigma_{ij}, D_i) = \int_v G(\sigma_{ij}, D_i) dV \tag{8.81}$$

$$\delta^2\Pi_c = \frac{d}{da}\int_v G(\delta\sigma_{ij,}\delta D_i) dV \tag{8.82}$$

Equation 8.80 through Equation 8.82 give

$$I(\tilde{\sigma}_{ij}, \tilde{D}_i) = \frac{d}{da}\Pi_c(\tilde{\sigma}_{ij}, \tilde{D}_i) = I(\sigma_{ij}, D_i) + \delta^2 I \tag{8.83}$$

$$I(\sigma_{ij}, D_i) = \frac{d}{da}\Pi_c(\sigma_{ij}, D_i) = \frac{d}{da}\int_v G(\sigma_{ij,} D_i) dv \geq 0 \tag{8.84}$$

$$\delta^2 I = \frac{d}{da}\delta^2\Pi_c = \frac{d}{da}\int_v G(\delta\sigma_{ij,}\delta D_i) dv \tag{8.85}$$

Because both $G(\sigma_{ij,} D_i)$ and $G(\delta\sigma_{ij,}\delta D_i)$ possess the same function configurations, the comparison of Equation 8.84 and Equation 8.85 results in $\delta^2 I \geq 0$, and the inequality (Equation 8.79) is true.

8.7.4 A Limitation on the Bound Analysis

The above bound theorems are conditional for piezoelectric fracture estimations. The lower bound theorem for J depends on the positive definiteness of the electric enthalpy: $H(\varepsilon_{ij,} E_i) > 0$, and the upper theorem for I is dependent on the positive definiteness of the complementary electric enthalpy: $G(\sigma_{ij}, D_i) > 0$. The complexity lies in that the positive definiteness of H and G is dependent on the electric loading. In the case of large electric loading

(E_i and/or D_i) the electric enthalpy or the complementary enthalpy is likely to lose the positive definiteness, so that the bound theorems no longer hold.

Taking the crack of model I as an example, the energy release rate can be expressed as [46]

$$G = \frac{1}{2}\pi[A\sigma_\infty^2 + B\sigma_\infty D_\infty - CD_\infty^2] \tag{8.86}$$

in which σ_∞ and D_∞ are, respectively, the mechanical loading and the electric loading far from the crack tip. Then the nonnegative condition, $G > 0$, leads to the following requirement for the loading rate $\rho = D_\infty / \sigma_\infty$:

$$A + B\rho - C\rho^2 > 0 \tag{8.87}$$

The inequality (Equation 8.87) shows a limitation on the bound theorems.

8.7.5 Piezoelectric Finite Elements

As for the plane fracture problem, the four-node piezoelectric isoparametric element (PZT-Q4) can easily be formulated in terms of the assumed bilinear displacement and the electric potential. The resulting compatible model can be employed to estimate the lower bound of $J(u_i, \varphi_i)$. For $I(\sigma_{ij}, D_i)$, the upper bound should be estimated by the stress-electric displacement equilibrium element. Unfortunately it is hard to obtain a reliable equilibrium model for 2D/3D problems because some numerical difficulties, such as the element rank deficiency, cannot be avoided. One faces the problem of how to implement the upper bound theorem for $I(\sigma_{ij}, D_i)$ [43, 48].

The $\sigma_{ij} \sim D_i$ equilibrium element that is based on the complementary energy functional $\Pi_c(\sigma_{ij}, D_i)$ in Equation 8.72a should be equivalent to the hybrid element that is based on Reissner functional $\Pi_R(u_i, \sigma_{ij}, \varphi, D_i)$ [48] since under the equilibrium constraints ($\sigma_{ij,j} = 0,\ D_{i,i} = 0$), $\Pi_R(u_i, \sigma_{ij}, \varphi, D_i)$ is identical to $\Pi_c(\sigma_{ij}, D_i)$. Therefore, the hybrid model based on $\Pi_R(u_i, \sigma_{ij}, \varphi, D_i)$ should be equivalent to the equilibrium model based on $\Pi_c(\sigma_{ij}, D_i)$ when the $\sigma_{ij} \sim D_i$ equilibrium equations are enforced to the element.

In this section, a four-node piezoelectric hybrid element will be developed for the upper bound estimation. The element displacements and the element electric potential are assumed as the bilinear interpolations:

$$u = \begin{Bmatrix} u \\ v \\ \varphi \end{Bmatrix} = \frac{1}{4}\sum_{i=1}^{4}(1+\xi_i\xi)(1+\eta_i\eta)\begin{Bmatrix} u_i \\ v_i \\ \varphi_i \end{Bmatrix} = \boldsymbol{N}\boldsymbol{q},$$

$$\mathbf{q} = \{u_1, v_1, \varphi_1, \ldots, u_4, v_4, \varphi_4\}^T \tag{8.88}$$

For the element stresses, the Pian–Sumihara stress mode [47] is adopted as follows:

$$\sigma = \begin{Bmatrix} \sigma_x \\ \sigma_y \\ \tau_{xy} \end{Bmatrix} = \begin{bmatrix} 1 & 0 & 0 & a_3^2\xi & a_1^2\eta \\ 0 & 1 & 0 & b_3^2\xi & b_1^2\eta \\ 0 & 0 & 1 & a_3 b_3\xi & a_1 b_1\eta \end{bmatrix} \begin{Bmatrix} \beta_1 \\ \vdots \\ \beta_5 \end{Bmatrix} = \boldsymbol{P}_m\beta \tag{8.89}$$

where the element geometric parameter is

$$\begin{bmatrix} a_1 & b_1 \\ a_2 & b_2 \\ a_3 & b_3 \end{bmatrix} = \frac{1}{4}\begin{bmatrix} -1 & 1 & 1 & -1 \\ 1 & -1 & 1 & -1 \\ -1 & -1 & 1 & 1 \end{bmatrix} \begin{bmatrix} x_1 & y_1 \\ x_2 & y_2 \\ x_3 & y_3 \\ x_4 & y_4 \end{bmatrix}$$

The element electric displacement is initially assumed as a linear distribution in the element $\xi-\eta$ coordinates. After introducing an energy consistency condition [49], it can be optimized as

$$\mathbf{D} = \begin{Bmatrix} D_x \\ D_y \end{Bmatrix} = \begin{bmatrix} 1 & 0 & a_3\xi & a_1\eta \\ 0 & 1 & b_3\xi & b_1\eta \end{bmatrix} \begin{Bmatrix} \beta_1 \\ \vdots \\ \beta_4 \end{Bmatrix} = \mathbf{P}_e\beta \tag{8.90}$$

In terms of the element trial function (Equation 8.88 through Equation 8.90), the Reissner energy functional [47] for the individual element can be formulated as

$$\Pi_R^e(\mathbf{u},\varphi,\sigma,\mathbf{D}) = \int_{V^e} [G(\sigma,\mathbf{D}) - \sigma^T(\partial\mathbf{u}) - \mathbf{D}^T(\nabla\varphi)]dV \tag{8.91}$$

where $G(\sigma, D)$ is the complementary electric enthalpy and $\partial\mathbf{u}$ is the element strain. Substitution of the element trial function (Equation 8.88 through Equation 8.90) into Equation 8.91 results in

$$\Pi_R^e(\beta,\mathbf{q}) = \beta^T\mathbf{G}\mathbf{q} - \frac{1}{2}\beta^T\mathbf{H}\beta \tag{8.92}$$

$$\mathbf{G} = \int_{V^e} \mathbf{P}^T(\partial\mathbf{N})\,dv, \quad \mathbf{H} = \int_{V^e} \mathbf{P}^T\mathbf{S}\mathbf{P}\,dv$$

and

$$\mathbf{P} = \begin{bmatrix} \mathbf{P}_m & \mathbf{0}_{3\times4} \\ \mathbf{0}_{2\times5} & \mathbf{P}_e \end{bmatrix}^T \tag{8.93}$$

where $\partial\mathbf{N}$ is the element strain matrix, and $\mathbf{S}$ is the piezoelectric constitution matrix. By the functional stationary condition $\partial\Pi_R^e(\beta,\mathbf{q})=0$ results in $\mathbf{H}\beta$ = $\mathbf{G}q$, (after condensing β, the functional (Equation 8.92) can be reformulated as

$$\Pi_R^e=\frac{1}{2}\mathbf{q}^T\mathbf{K}^e\,\mathbf{q} \tag{8.94}$$

and the element stiffness matrix

$$\mathbf{K}^e=\mathbf{G}^T\mathbf{H}^{-1}\mathbf{G} \tag{8.95}$$

Because it is difficult to build a stress-equilibrium element directly, the penalty-equilibrium hybrid element presented by Wu and Cheung [3] is suggested here. Let the functional $\Pi_R^e(u_i,\sigma_{ij},\varphi,D_i)$ be generalized as

$$\Pi_{RG}^e=\Pi_R^e-\alpha\left(\int_{V^e}\sigma_{ij,j}\sigma_{ij,j}dv+\int_{V^e}D_{i,j}D_{i,j}dv\right) \tag{8.96}$$

where the penalty factor α is taken to be a large constant (e.g., 10^4). In terms of the assumed element trial functions (Equation 8.88 through Equation 8.90), the functional in Equation 8.96 can be formulated as

$$\Pi_{RG}^e(\beta,\mathbf{q})=\beta^T\mathbf{G}\mathbf{q}-\frac{1}{2}\beta^T\mathbf{H}\beta-\alpha\beta^T\mathbf{H}_p\beta \tag{8.97}$$

where the penalty element equilibrium matrix

$$\mathbf{H}_p=\iint_{v^e}(\partial^T\mathbf{P})^T(\partial^T\mathbf{P})dv,\ (\partial^T\text{ is the equilibrium differential operator}). \tag{8.98}$$

In such a way, the equilibrium equations $\sigma_{ij,j}=0$ and $D_{i,j}=0$ will be enforced within the element in the meaning of penalty, so that the functional $\Pi_R^e(u_i,\sigma_{ij},\varphi,D_i)\Rightarrow\Pi_c^e(\sigma_{ij},D_i)$, and then the complementary energy functional is available. The hybrid element based on Π_R^e will degenerate into an equilibrium model, which is based on Π_c^e and able to implement the upper bound theorem. The resulting penalty-equilibrium element is PZT-PS (α) [3].

In the numerical calculations the singular stresses and electric displacement in the crack tip domain are hard to simulate by the conventional isoparametric element, and the numerical solutions are very sensitive since the traction-free and electric displacement–free conditions cannot be imposed on the crack surface. Fortunately, the homogeneous traction–electric displacement condition can easily be imposed on the crack surface by the hybrid element since the element stress and the element electric displacement are independently assumed in the element formulation on the crack surface.

8.7.6 Dual Error Measure

A dual bound error formula for Rice's *J*-kind of crack parameters has been presented by Wu, Xiao, and Li [44]. Here the formula will be developed to the electromechanical crack problem.

Let $\delta u_i = \tilde{u}_i - u_i$ be the displacement error and let $\delta\varphi = \tilde{\varphi} - \varphi$ be the electric potential error induced by using the assumed displacement-electric potential finite element. Then, in accordance with the lower bound theorem (Equation 8.73), the relative error for the *J* integral must be

$$J(\delta u_i, \delta\varphi) = J(\tilde{u}_i, \tilde{\varphi}) - J(u_i, \varphi) \leq 0 \tag{8.99}$$

and the absolute error is then

$$|J(\delta u_i, \delta\varphi)| = J(u_i, \varphi) - J(\tilde{u}_i, \tilde{\varphi}) \tag{8.100}$$

Let $\delta\sigma_{ij} = \tilde{\sigma}_{ij} - \sigma_{ij}$ be the stress error and $\delta D_i = \tilde{D}_i - D_i$ be the electric-displacement error. In accordance with the upper bound theorem (Equation 8.79), the relative error for the *I*-integral should be

$$I(\delta\sigma_{ij}, \delta D_i) = I(\tilde{\sigma}_{ij}, \tilde{D}_i) - I(\sigma_{ij}, D_i) \geq 0 \tag{8.101}$$

and the absolute error,

$$|I(\delta\sigma_{ij}, \delta D_i)| = I(\tilde{\sigma}_{ij}, \tilde{D}_i) - I(\sigma_{ij}, D_i) \geq 0 \tag{8.102}$$

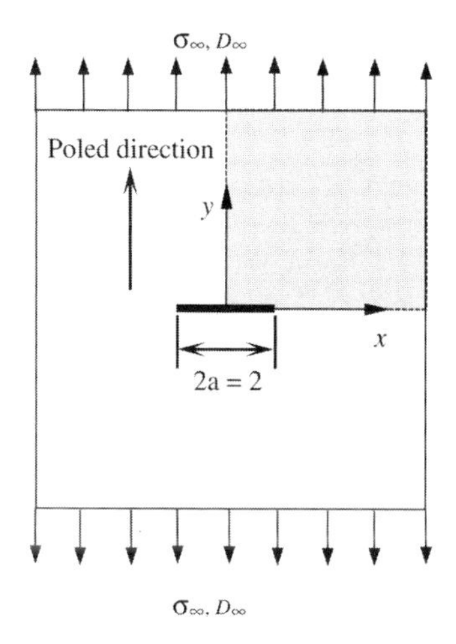

FIGURE 8.20
8 × 8 CCP specimen.

Corresponding to Equation 8.100 and Equation 8.102, the relative errors can be respectively expressed as

$$\left|J(\delta u_i,\delta\varphi\right|/\,J(u_i,\varphi) \text{ and } \left|I(\delta\sigma_{ij},\delta D_i)\right|\big/I(\sigma_{ij},D_i) \quad (8.103\text{a and } 8.103\text{b})$$

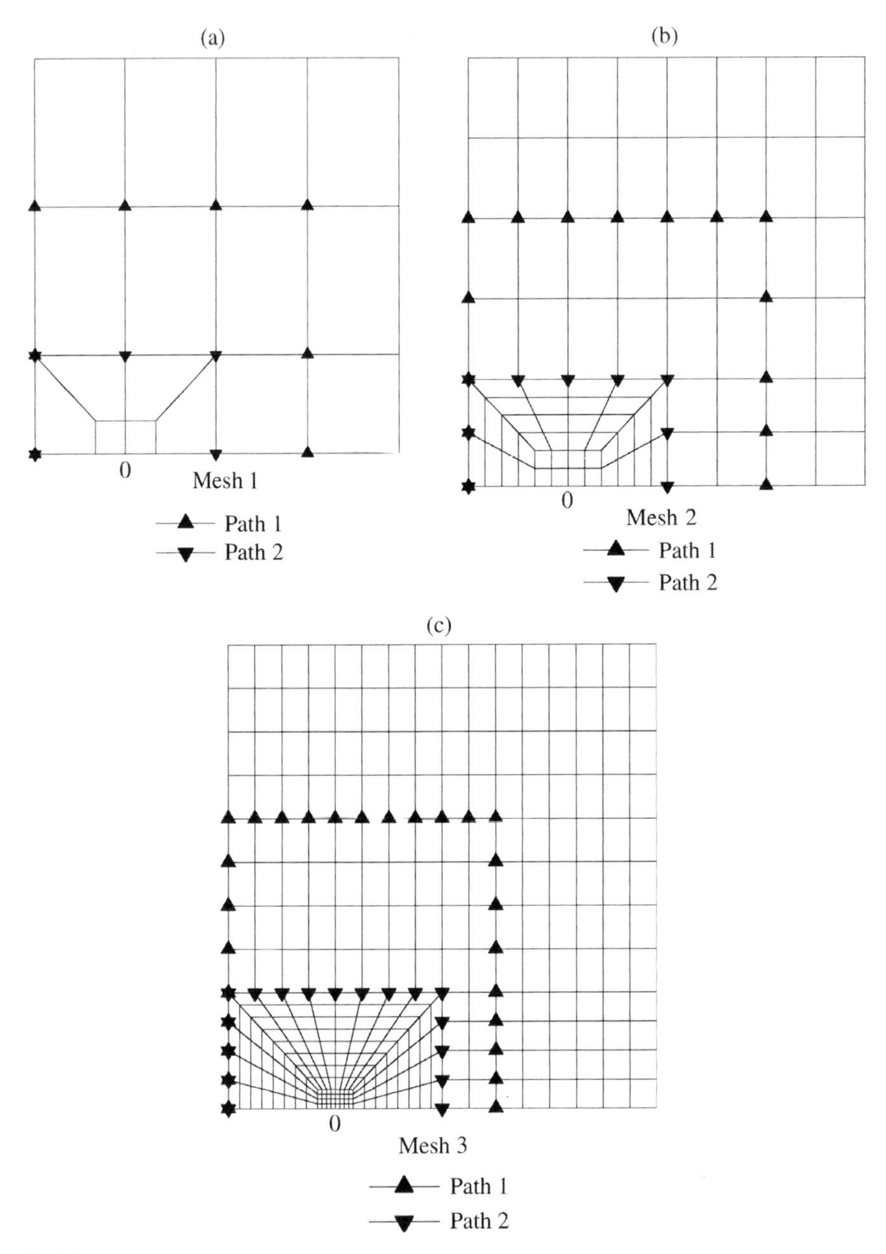

FIGURE 8.21
Employed finite element meshes and the selected integral paths for CCP.

For a given crack component, the strength estimation can be carried out once the upper and lower bounds are yielded. The approximate strength is usually taken to be a combination of the bound solutions. In order to measure

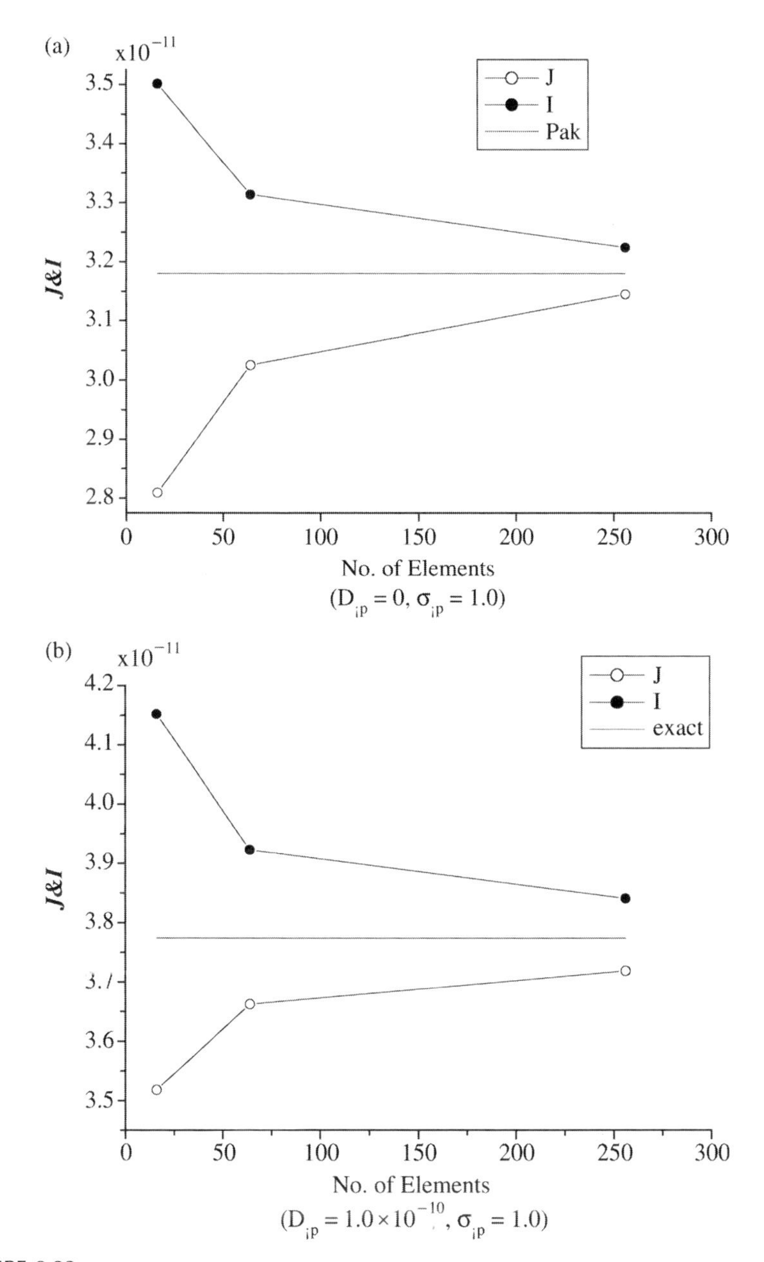

FIGURE 8.22
Bound behavior of the solutions of *J* and *I* in three meshes with different densities (PZT-4).

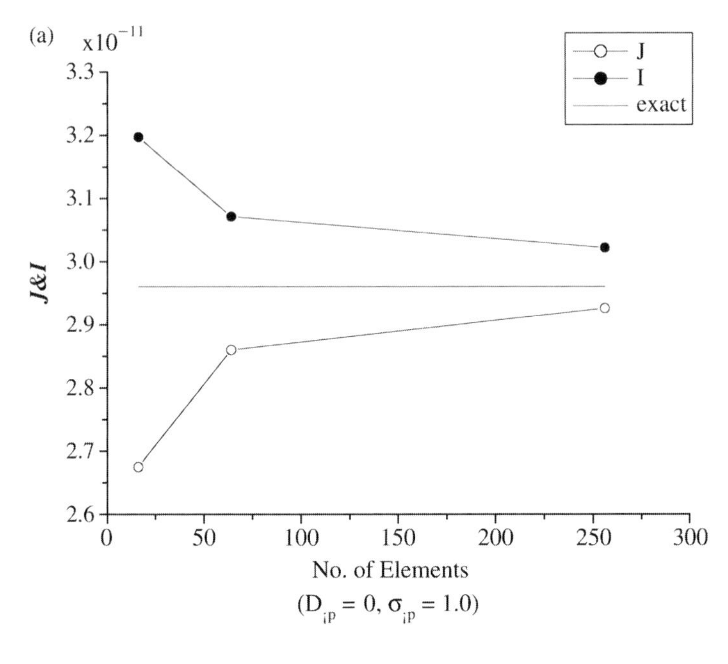

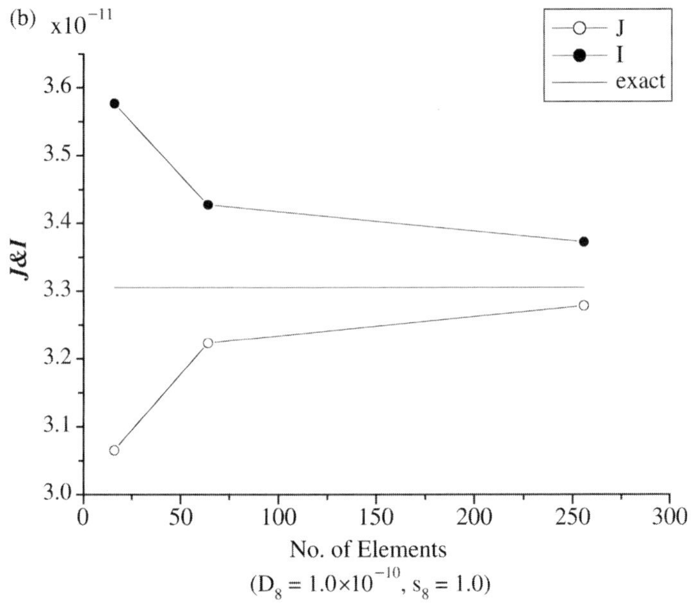

FIGURE 8.23
Bound behavior of the solutions of J and I in three meshes with different densities (PZT-5H).

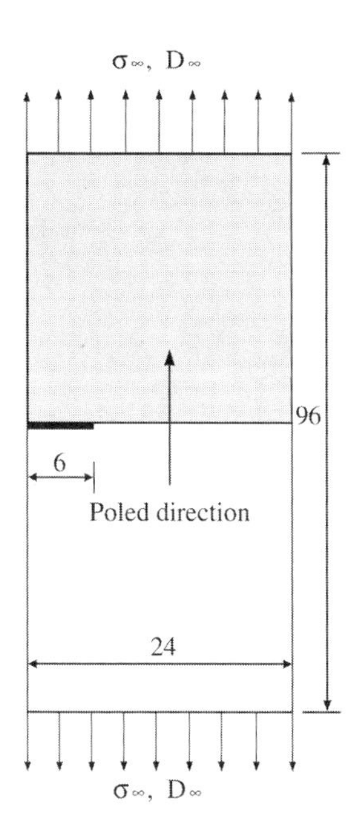

FIGURE 8.24
SECP specimen.

TABLE 8.8

Material Constants for PZT-4 and PZT-5H Where C_{ij}'s Are in 10 GNm^{-2}, e_{ij}'s Are in Cm^{-2} and $\in_{ij}$ Are μCV^{-1} m^{-1}

	C_{11}	C_{12}	C_{13}	C_{33}	C_{44}	e_{31}	e_{33}	e_{15}	$\in_{11}$	$\in_{33}$
PZT-4	13.9	7.78	7.43	11.3	2.56	−6.98	13.84	13.44	6.00	5.47
PZT-5H	12.6	5.5	5.3	11.7	3.53	−6.5	23.3	17.0	15.1	13.0

TABLE 8.9A

Relative Error Δ_{J-J^*} *(%)* for CCP (PZT-4)

Mesh	Mesh 1 (16 Elements)	Mesh 2 (64 Elements)	Mesh 3 (256 Elements)
Load 1: $D_\infty = 0$, $\sigma_\infty = 1.0$	10.968	4.549	1.248
Load 2: $D_\infty = 2.2 \times 10^{-10}$, $\sigma_\infty = 1.0$	8.261	3.438	1.614

TABLE 8.9B

Relative Error Δ_{J-J^*} *(%)* for CCP (PZT-5H)

Mesh	Mesh 1 (16 Elements)	Mesh 2 (64 Elements)	Mesh 3 (256 Elements)
Load 1: $D_\infty = 0$, $\sigma_\infty = 1.0$	8.901	3.560	1.606
Load 2: $D_\infty = 1.0 \times 10^{-10}$, $\sigma_\infty = 1.0$	7.694	3.068	1.409

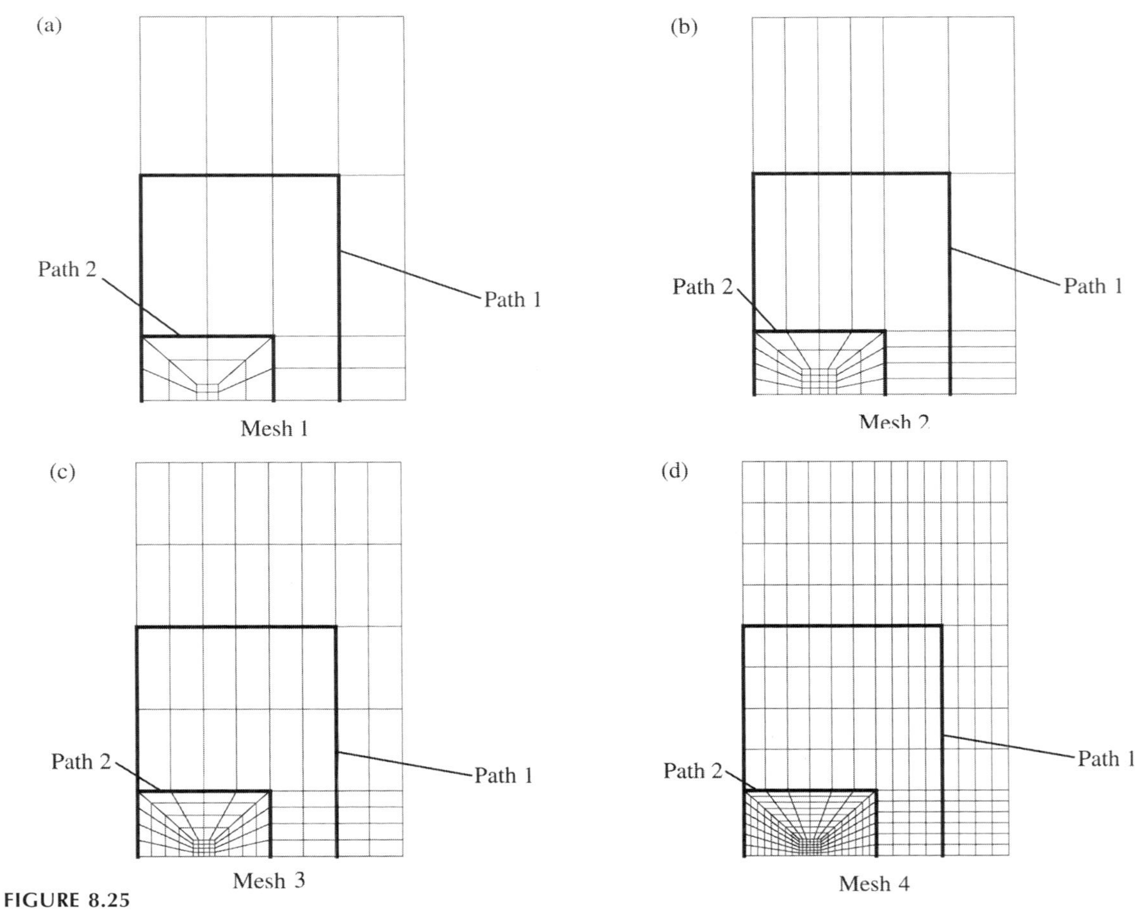

FIGURE 8.25
Employed finite element meshes and selected integral paths for SECP.

TABLE 8.10A

Relative Error Δ_{J-I^*} *(%)* for SECP (PZT-4)

Mesh	Mesh 1 (28 Elements)	Mesh 2 (60 Elements)	Mesh 3 (112 Elements)	Mesh 4 (340 Elements)
Load 1: $D_\infty = 0$, $\sigma_\infty = 1.0$	10.206	3.0173	1.5501	1.0329
Load 2: $D_\infty = 1.0 \times 10^{-10}$, $\sigma_\infty = 1.0$	8.8523	2.3245	1.2483	0.8935

TABLE 8.10B

Relative Error Δ_{J-I^*} *(%)* for SECP (PZT-5H)

Mesh	Mesh 1 (28 Elements)	Mesh 2 (60 Elements)	Mesh 3 (112 Elements)	Mesh 4 (340 Elements)
Load 1: $D_\infty = 0$, $\sigma_\infty = 1.0$	8.1892	1.6126	0.8014	0.5853
Load 2: $D_\infty = 1.0 \times 10^{-10}$, $\sigma_\infty = 1.0$	7.4638	1.2829	0.7884	0.5125

TABLE 8.11

Relative Error Δ_{J-I^*} *(%)* for DECP (PZT-5H)

Mesh	Mesh 1 (60 Elements)	Mesh 2 (60 Elements)	Mesh 3 (564 Elements)	Mesh 4 (844 Elements)
Load 1: $D_\infty = 0$, $\sigma_\infty = 1.0$	4.45892	0.640265	0.22452	0.19732
Load 2: $D_\infty = 1.0 \times 10^{-10}$, $\sigma_\infty = 1.0$	3.48058	0.634605	0.37866	0.10618

the numerical error of the crack parameters, the dual relative error formula is defined as [44]

$$\Delta_{J-I} = \frac{|J(\delta u_i, \delta\varphi)| + |I(\delta\sigma_{ij}, \delta D_i)|}{J(u_i, \varphi) + I(\sigma_{ij}, D_i)} \tag{8.104}$$

where the reference solution is defined as

$$J(u_i, \varphi) + I(\sigma_{ij}, D_i) = J(\tilde{u}_i, \tilde{\varphi}) + I(\tilde{\sigma}_{ij}, \tilde{D}) - \left[J(\delta u_i, \delta\varphi) + I(\delta\sigma_{ij}, \delta D_i)\right] \tag{8.105}$$

Observing the small quantities $J(\delta u_i, \delta\varphi) \leq 0$, while $I(\delta\sigma_{ij}, \delta D_i) \geq 0$, the last term in Equation 8.105 can be ignored, so Equation 8.105 becomes

$$J(u_i, \varphi) + I(\sigma_{ij}, D_i) \approx J(\tilde{u}_i, \tilde{\varphi}) + I(\tilde{\sigma}_{ij}, \tilde{D}_i) \tag{8.106}$$

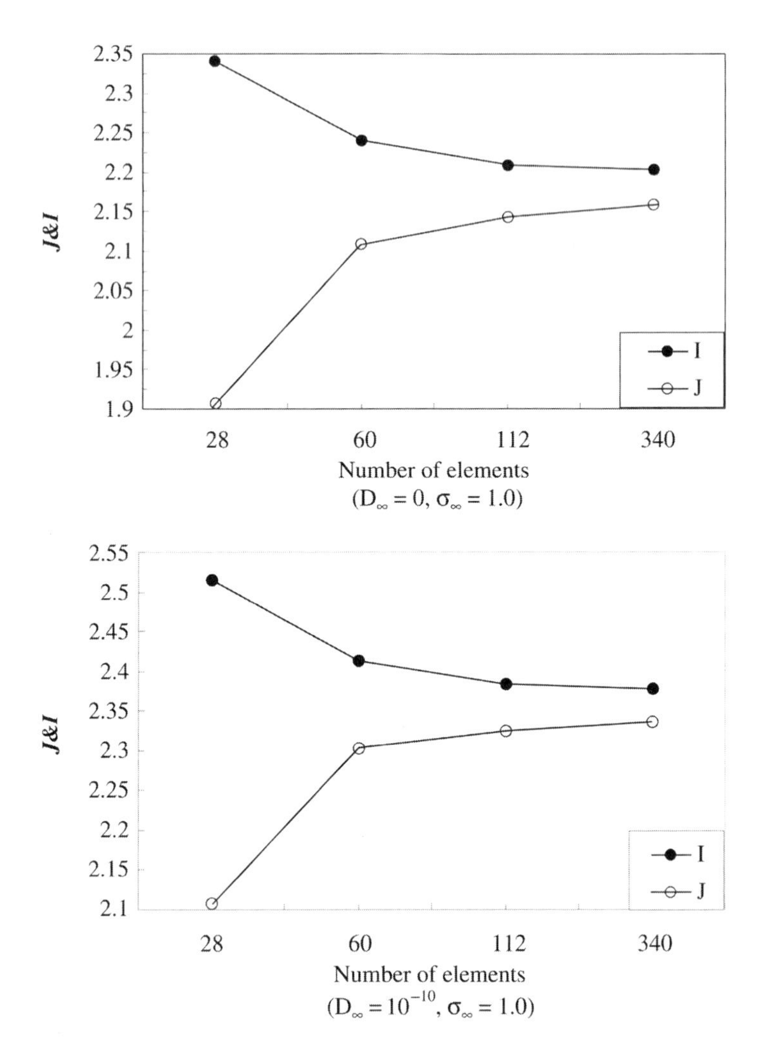

FIGURE 8.26
Bound solutions of J and I ($\times 10^{-9}$) of SECP (PZT-4).

Substitution of Equation 8.100, Equation 8.102, and Equation 8.105 into Equation 8.104 gives the exact solutions, and $J(u_i,\varphi)-I(\sigma_{ij},D_i)=0$ holds, so that the relative error formula (Equation 8.104) becomes

$$\Delta_{J-I}=\frac{I(\tilde{\sigma}_{ij},\tilde{D}_i)-J(\tilde{u}_i,\tilde{\varphi})}{I(\sigma_{ij},D_i)+J(u_i,\varphi)}\approx\frac{I(\tilde{\sigma}_{ij},\tilde{D}_i)-J(\tilde{u}_i,\tilde{\varphi})}{I(\tilde{\sigma}_{ij},\tilde{D}_i)+J(\tilde{u}_i,\tilde{\varphi})} \tag{8.107}$$

It has been verified that the above relative error $\Delta_{J-I}\to 0$ when $h\to 0$ (i.e., the finite element size tends to zero). The dual error measure formula

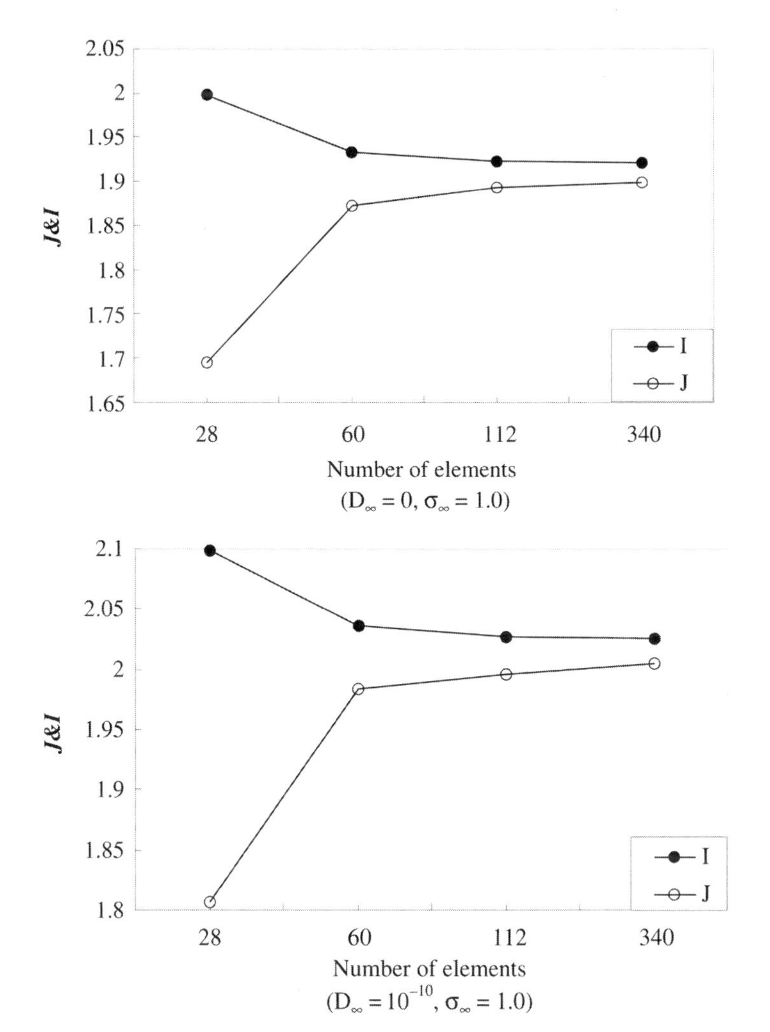

FIGURE 8.27.
Bound solutions of J and I ($\times 10^{-9}$) of SECP (PZT-5H).

(Equation 8.107) depends on th e numerical solutions to J and I only, so it is easy to implement in the crack assessment.

In the case of pure mechanical loading, the electric displacement $\tilde{D}$ and the electric potential $\tilde{\varphi}$ will disappear from the formula (Equation 8.107), and one then has

$$\Delta_{J-I} = \frac{I(\tilde{\sigma}_{ij}) - J(\tilde{u}_i)}{I(\tilde{\sigma}_{ij}) + J(\tilde{u}_i)} \tag{8.108}$$

Equation 8.108 is Wu et al.'s dual error formula for the pure mechanical crack evaluation [44, 2, 31].

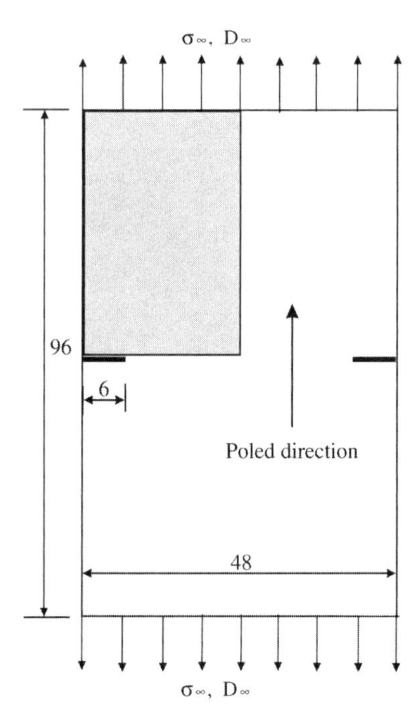

FIGURE 8.28
Double-edged crack panel specimen (DECP).

8.7.7 Numerical Example

In numerical calculations, the piezoelectric isoparametric element PZT-Q4 and the piezoelectric penalty-equilibrium element PZT-PS(α) are employed to estimate J and I, respectively. A CCP (Figure 8.20), an SECP (Figure 8.24), and a double-edged crack panel (DECP) (Figure 8.28) with pure mechanical loading and mechanical-electrical mixed loading are considered. For the symmetry, only a quarter of the CCP and DECP specimens and half of the SECP specimen need to be considered. Three finite-element meshes and two integral paths are shown in Figure 8.21a–c, Figure 8.25a–d, and Figure 8.29a–d for CCP, SECP, and DECP, respectively. The material properties of PZT-4 and PZT-5H are listed in Table 8.8.

To inspect the convergence behavior of the solutions to J and I, meshes with different densities and independent integral paths are simultaneously considered for each specimen. The value of J and I is taken to be an average of that from the neighboring elements. Numerical results of CCP are shown in Figure 8.22a,b and Figure 8.23a,b; numerical results of SECP are shown in Figure 8.26a,b and Figure 8.27a,b; numerical results of DECP are shown in Figure 8.30a,b. For different mechanical and electrical loading, the J-solutions of the PZT-Q4 compatible element always converge to the exact one from a certain lower bound. On the contrary, the I-solutions of the PZT-PS(α) penalty-equilibrium element always converge to the exact

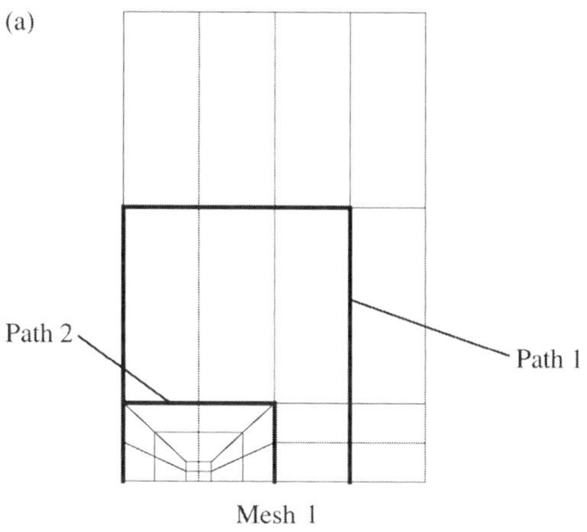

Mesh 1

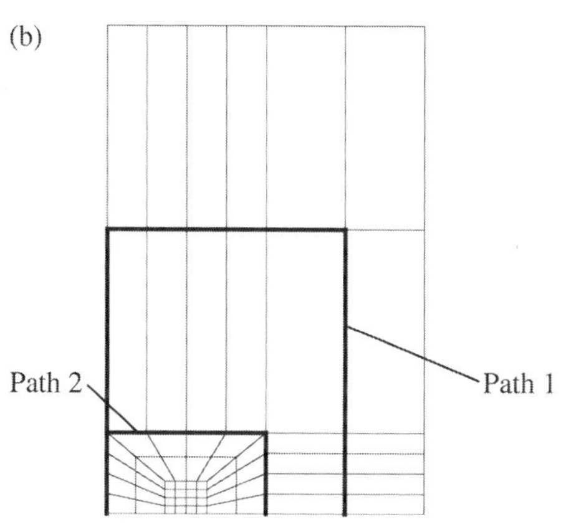

Mesh 2

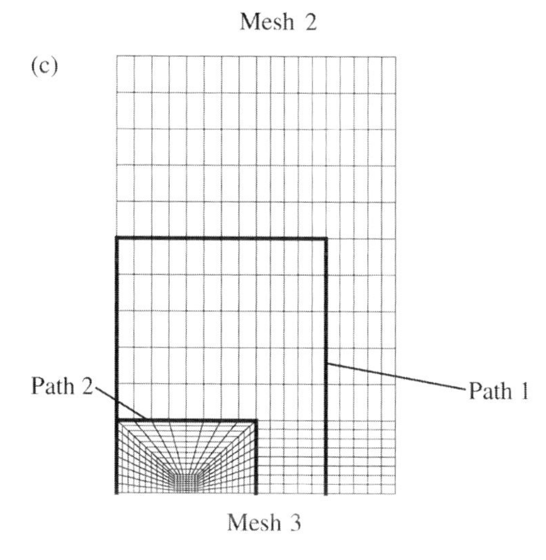

Mesh 3

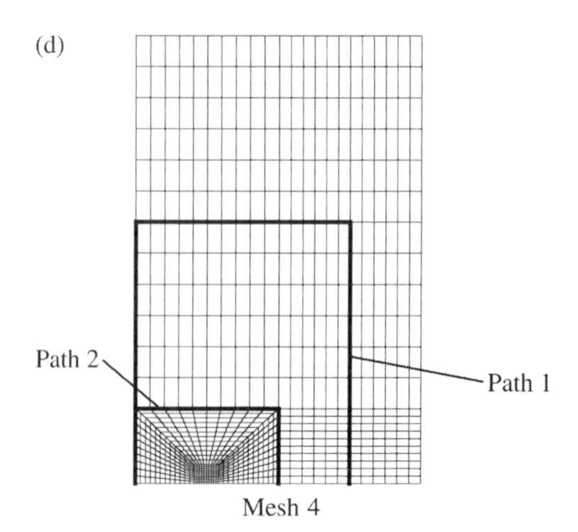

FIGURE 8.29
Employed finite element meshes and selected integral paths for DECP.

one from a certain upper bound. All the numerical solutions demonstrate the bound theorems.

The error formula Δ_{J-I} in Equation 8.107 is implemented to measure the relative error of the bound solutions to the piezoelectric CCP, SECP, and DECP specimens. The results are listed in Table 8.9A and Tables 8.9B and 8.10A and Tables 8.10B and 8.11, respectively. Independent of the selection of finite element meshes and the integral paths, the relative error Δ_{J-I}, offers similar results. These numerical tests exhibit the efficiency of the suggested approximate error formula. The best feature is that no reference solutions are needed. A quantitative error measure can be carried out, and then an error control to the crack parameter is available.

8.7.8 Conclusion

- The suggested I integral, as the dual of the J integral, makes dual analysis possible for piezoelectric fractures. In general, one has

$$J(\tilde{u}_i,\tilde{\varphi}) \leq J(u_i,\varphi) = I(\sigma_{ij},D_i) \leq I(\tilde{\sigma}_{ij},\tilde{D}_i)$$

- The lower bound of J can be estimated by the piezoelectric compatible element PZT-Q4.
- The upper bound of I can be estimated by the piezoelectric penalty-equilibrium hybrid element PZT-PS (α).
- The suggested dual error formula makes the fracture assessment more reliable, and the numerical accuracy can easily be controlled.

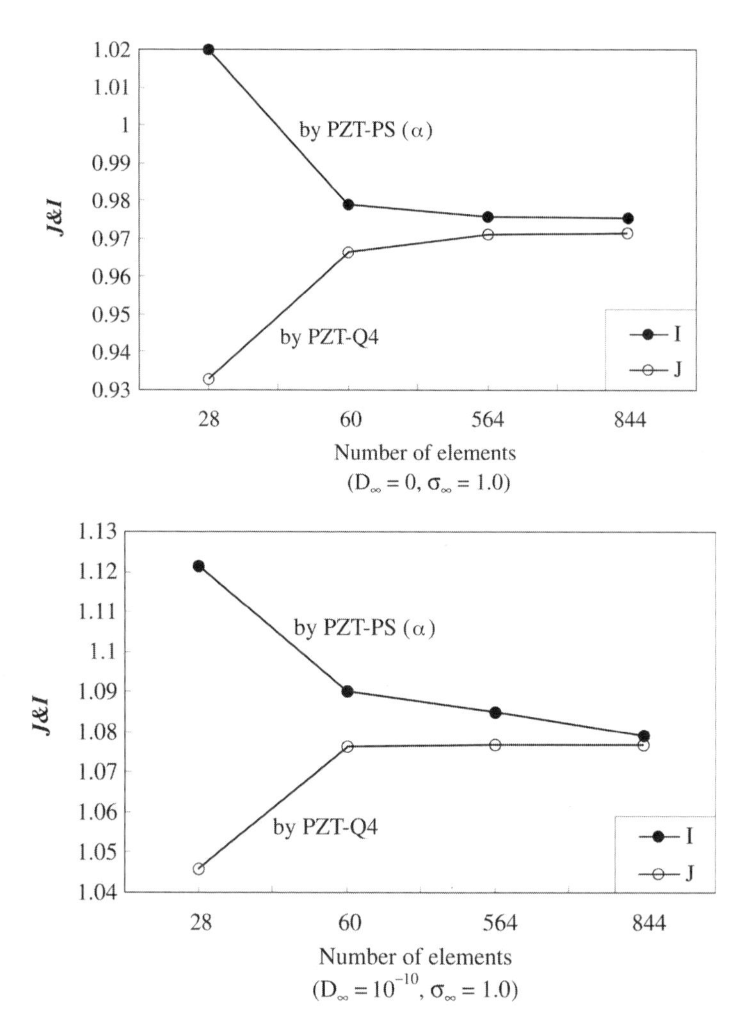

FIGURE 8.30
Bound solutions of J and I ($\times 10^{-9}$) of DECP (PZT-5H).

References

1. Rice, J.R., Mathematical analysis in the mechanics of fracture, in *Fracture: An Advanced Treatise,* Vol. II, Liebowitz, H., Ed., Academic Press, New York, 1968, pp. 191–308.
2. Wu, C.C., Xiao, Q.Z., and Yagawa, G., Dual analysis for path integrals and bounds for crack parameter, *Int. J. Solids Struct.,* 35, 1635, 1998.
3. Wu, C.C. and Cheung, Y.K., On optimization approaches of hybrid stress element, *Finite Elements Anal. Des.,* 21, 111, 1995.

4. Pian, T.H.H. and Sumihara, K., Rational approach for assumed stress finite elements, *Int. J. Numer. Methods Eng.*, 20, 1685, 1984.
5. Ewalds, H.L. and. Wanhill, R.J.H., *Fracture Mechanics*, Edward Arnold, London, 1984.
6. Xiao, Q.Z., Karihaloo, B.L., and Williams, F.W., Application of penalty-equilibrium hybrid stress element method to crack problems, *Eng. Fract. Mech.*, 63, 1, 1999.
7. Liebowitz, H. and Moyer, E.T., Finite element methods in fracture mechanics, *Comput. Struct.*, 31, 1, 1989.
8. Nagtegaal, J.C., Parks, D.M., and Rice, J.R., On numerically accurate finite element solutions in the fully plastic range, *Comput. Methods Appl. Mech. Eng.*, 4, 153, 1978.
9. Needleman, A. and Shih, C.H., Finite element method for plane strain deformations of incompressible solids, *Comput. Methods Appl. Mech. Eng.*, 5, 223, 1978.
10. Kumar, V., de Lorenzi, H.G., Andrews, W.K., Shih, C.F., German, M.D., and Mowbray, D.F., *4th Semiannual Report to EPRI*, Contract No. RP1237-1, General Electric Company, Schenectady, NY (June 1, 1980–January 31, 1981).
11. Huang, Y.Q., Wu, C.C., and Huang, M.G., Incompatible numerical analysis for plane strain fully plastic fracture, in *Theories and Applications for Computational Mechanics*, Science Press, Beijing, 1992, pp. 25–28.
12. Strang, G. and Fix, G.J., *An Analysis of the Finite Element Method*, Prentice-Hall, Englewood Cliffs, NJ, 1973.
13. Wu, C.C., Huang, M.G., and Pian, T.H.H., Consistency condition and convergence criteria of incompatible elements: general formulation of incompatible functions and its application, *Comput. Struct.*, 27, 639, 1987.
14. Cheung, Y.K. and Wu, C.C., The patch test condition in curvilinear coordinate system: formulation and application, *Sci. China A*, 36, 62, 1993.
15. Hwang, K.C. and Yu, S.W., *Elastic-Plastic Fracture*, Tsinghua University Press, Beijing, China, 1985.
16. Rice, J.R., Mathematical analysis in the mechanics of fracture, in *Fracture*, Vol. II, Liebowitz, H., Ed., Academic Press, New York, 1968, pp. 191–311.
17. Sneddon, I.N. and Lowengrub, M., *Crack Problems in Classical Theory of Elasticity*, John Wiley, New York, 1969.
18. Astiz, M.A., Elices, M., and Galvez, V.S., On the energy release rates in axisymmetrical problems, in *Fracture*, The 4th International Conference on Fracture Mechanics, Waterloo, Canada (June 19–24, 1977, pp. 395–400.
19. Bergkvist, H. and Huong, G.-L.L., J-integral related quantities in axi-symmetric cases, *Int. J. Fracture*, 13, 556, 1977.
20. Kumar, V., German, M.D., and Shih, C.F., An engineering approach for elastic-plastic fracture analysis, EPRI NP-1931, New York, 1981.
21. Nied, H.F. and Erdogan, F., The elasticity problem for a thick-walled cylinder containing a circumferential crack, *Int. J. Fracture*, 22, 277, 1983.
22. Singh, B.M., Cardou, A., and Au, M.C., Plastic zone correction in a stretched thick-walled cylinder with an internal circumferential crack, *Eng. Fracture Mech.*, 29, 503, 1988.
23. Huang, Y.C., A Finite Element Analysis for Elastic-Plastic Cracked Cylinder, B.S. thesis, The University of Science & Technology of China, Hefei, China, 1991.
24. Suresh, S. and Mortensen, A., *Fundamentals of Functionally Graded Materials*, Cambridge University Press, Cambridge, U.K., 1998.

25. Jin, Z.-h. and Noda, N., Crack-tip singular fields in nonhomogeneous materials, *ASME J. Appl. Mech.*, 61, 738, 1994.
26. Erdogan, F. and Wu, B.H., The surface crack problem for a plate with functionally graded properties, *J. Appl. Mech.*, 64, 449, 1997.
27. Chan, Y., Paulino, G.H., and Fannjiang, A.C., The crack problem for nonhomogeneous materials under antiplane shear loading: a displacement based formulation, *Int. J. Solids Struct.*, 38, 1989, 2001.
28. Jin, Z.H. and Paulino, G.H., Transient thermal stress analysis of an edge crack in a functionally graded material, *Int. J. Fracture*, 107, 73, 2001.
29. Honein, T. and Herrmann, G., Conversion laws in nonhomogeneous plane elastostatics, *J. Mech. Phys. Solids*, 45, 789, 1997.
30. Jian, C., Linzhi, W., and Shanyi, D., A modified J integral for functional graded materials, *Mech. Res. Commun.*, 27, 301, 2000.
31. Wu, C.C., Xiao, Q.Z., and Yagawa, G., Finite element methodology for path integrals in fracture mechanics, *Int. J. Numer. Methods Eng.*, 43, 69, 1998.
32. Moran, B. and Shih, C.F., Crack tip and associated domain integrals from momentum and energy balance, *Eng. Fracture Mech.*, 27, 615, 1987.
33. Belytschko, T., Lu, Y.Y., and Gu, L., Element-free Galerkin methods, *Int. J. Numer. Methods Eng.*, 37, 229, 1994.
34. Lu, Y.Y., Belytschko, T., and Tabbara, M., Element-free Galerkin method for wave propagation and dynamic fracture, *Comput. Methods Appl. Mech. Eng.*, 126, 131, 1995.
35. Furukawa, T., Yang, C., Yagawa, G., and Wu, C.C., Quadrilateral approaches for accurate free mesh method, *Int. J. Numer. Methods Eng.*, 47, 1445, 2000.
36. Xiao, Q.Z., Boyd, P., and Dhanasekar, M., Collocation method for coupling EFG and FE, Presented at 5th International Conference on Computational Structures Technology, Belgium, September 2000.
37. Parton, V.Z., Fracture mechanics of piezoelectric materials, *Acta Astronaut.*, 3, 671, 1976.
38. Deeg, W.F., The Analysis of Dislocation, Crack and Inclusion Problems in Piezoelectric Solids, Ph.D. dissertation, Stanford University, Stanford, CA.
39. Kumar, S. and Singh, R.N., Crack propagation in piezoelectric materials under combined mechanical and electrical loadings, *Acta Mater.*, 44, 173, 1996.
40. Allik, H. and Hughes, T.J.R., Finite element method for piezoelectric vibration, *Int. J. Numer. Methods Eng.*, 2, 151, 1970.
41. Landis, C.M., A new finite-element formulation for electromechanical boundary value problems, *Int. J. Numer. Methods Eng.*, 55, 613, 2002.
42. Ghandi, K. and Hagood, N.W., A hybrid finite-element model for phase transition in nonlinear electro-mechanically coupled material, in *Smart Structures and Materials 1997*, Proc. SPIE, Vol. 3039, Varadan, V.V. and Chandra, J., Eds, SPIE, Washington, DC, 1997, pp. 97–112.
43. Wu, C.C., Sze, K.Y., and Huang, Y.Q., Numerical solutions on fracture of piezoelectric materials by hybrid elements, *Int. J. Solids Struct.*, 38, 4315–4329, 2001.
44. Wu, C.C., Xiao, Q.Z., and Li, Z.-R., Fracture estimation: bound theorem and numerical strategy, in *Computational Mechanics in Structural Engineering*, Cheng, F.Y. and Gu, Y., Elsevier, Amsterdam, 1999, 349-360. (Initially presented at the second Sino–US Joint Symposium/Workshop on Recent Advancement of Computational Mechanics in Structural Engineering, May 25–28, 1998, Dalian, China.)

45. Pak, Y.E., Crack extension force in a piezoelectric material, *J. Appl. Mech.*, 57, 647, 1990.
46. Suo, Z., Kuo, C.-M., Barnett, D.M., and Willis, J.R., Fracture mechanics in piezoelectric material, *J. Mech. Phys. Solids*, 40, 739, 1992.
47. EerNisse, E.P., Variational method for electro vibration analysis, *IEEE Trans. J. Sonics Ultrason.*, 14, 59, 1983.
48. Liu, M. and Wu, C.C., Numerical analysis of crack-tip fields of piezoelectric media by hybrid finite element, *Chin. Sci. Bull.*, 44, 995, 1999.

9

Computational Materials

9.1 Hybrid Element Analysis of Composite Laminated Plates

9.1.1 Introduction

Understanding the transverse and interlaminar stresses in laminated composite plates and shells is important in the initiation and propagation of delaminations between layers. Many analytical and numerical approaches have been proposed to simulate the nonlinear distribution of the interlaminar stresses along the thickness direction. The finite element method appears to be most attractive of these in view of its versatility in handling complex boundaries and loads. However, the conventional finite element method, in which only the displacements are independently assumed, suffers from numerical difficulties such as the shear locking phenomenon when the plate thickness becomes very thin. It is also difficult to capture the nonlinear distribution of the transverse stresses along the thickness direction, even with a higher-order laminate theory [1]. Further, it is almost impossible to satisfy the interlaminar continuity requirements in the conventional finite element method. The relevant survey and comments can be found in References 2 and 3.

The hybrid stress element method, in which both the displacements and stresses are independently assumed, shows excellent flexibility and numerical performance in strain analysis. The numerical accuracy of the stresses can be improved by using the hybrid element without increasing the global degrees of freedom since the stress parameters are condensed on the element level. The hybrid stress element has been applied to predict the nonlinear distribution of the interlaminar stresses for laminated structures. Mau et al. [4] were the first to apply hybrid stress elements to the analysis of thick laminated plates. Subsequently, Spilker [5–7] published a series of papers on both thin and thick multilayer laminates. In these papers, both the transverse and in-plane stress components within the element are independently assumed. The interpolation functions appear to be very complicated in order to enforce the interlaminar continuity conditions. Starting from the continuous inplane strain field in the transverse direction, Spilker [8] and Pian and Li [9] determined the inplane stresses via the constitutive relationship and calculated the transverse stresses for each layer by introducing the equilibrium equations and the interface continuity requirements. The resulting stress field has a fixed number of stress parameters independent of the total number of layers. However, recalculations are needed when different ply configurations are used. Therefore, further extensions are limited.

Jing and Liao [11] presented a partially hybrid finite element method for laminated plates, in which only the transverse shear stresses are independently assumed. This reduces the degrees of freedom (DOFs) of the hybrid element and greatly improves the computation efficiency, but the interface continuity requirements of the transverse normal stress are ignored. Di et al. [12, 13] numerically analyzed the laminated plates and shells based upon the higher-order theory by the optimized hybrid techniques [14, 15], but the interface continuity requirements were still not met.

In this section, a partially hybrid stress element for laminated plates is set up based on the mixed variational principle of Reissner [16, 17], Moriya [18], and Huang [19], where only the displacement and transverse stress components are used as the basic field variables. It has been verified [20] that the Reissner functional is equivalent to the Hamilton principle when considering the spatial coordinate z as the time measurement. Therefore, the two independently assumed field variables, the displacement and transverse stress components, can be regarded as the state functions in the Hamilton space, or the state space. This model based on the state space precisely satisfies the continuous requirements of the interlaminar stresses on the interfaces, as well as the traction-free conditions on the top and bottom surfaces. To capture the transverse nonlinearity, the shear and normal stress distributions across the thickness direction are assumed to be parabolic and cubic, respectively. The same interpolation functions for the transverse stresses are adopted in each layer. This stress assumption can be further extended to laminated shells without any difficulties. The simply supported three-layer laminated strip under cylindrical bending and rectangular plates under double-sinusoidal load are considered for studying the accuracy, convergence, and capability of the proposed method. The nonlinear distributions of the transverse and normal stresses are also carefully investigated.

9.1.2 The State Space and Energy Formulations of Laminated Plates

Figure 9.1 shows a three-dimensional (3-D) composite laminate, in which h is the plate thickness and n is the total number of layers. The displacement components of the point (x, y, z) can be written as $u = \{u\ v\ w\}^T$, and the inplane and transverse stresses are defined as $\sigma_p = \{\sigma_x\ \sigma_y\ \tau_{xy}\}^T$ and $\sigma_t = \{\tau_{xz}\ \tau_{yz}\ \sigma_z\}^T$, respectively. The inplane and transverse strain-displacement relations are written as [15].

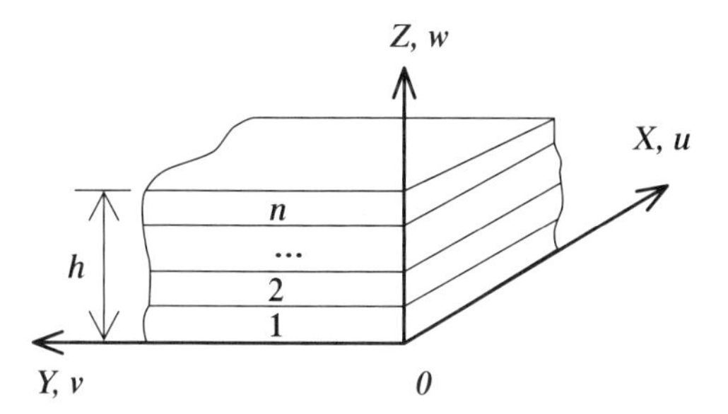

FIGURE 9.1
Configurations of a composite laminated plate.

$$\begin{Bmatrix} \varepsilon_x \\ \varepsilon_y \\ \varepsilon_z \end{Bmatrix} = \begin{bmatrix} \frac{\partial}{\partial x} & 0 & 0 \\ 0 & \frac{\partial}{\partial y} & 0 \\ \frac{\partial}{\partial y} & \frac{\partial}{\partial x} & 0 \end{bmatrix} \begin{Bmatrix} u \\ v \\ w \end{Bmatrix} \quad or \quad \varepsilon_p = \mathbf{D}_p \mathbf{u} \tag{9.1}$$

and

$$\begin{Bmatrix} \gamma_{xz} \\ \gamma_{yz} \\ \varepsilon_z \end{Bmatrix} = \begin{bmatrix} \frac{\partial}{\partial z} & 0 & \frac{\partial}{\partial x} \\ 0 & \frac{\partial}{\partial z} & \frac{\partial}{\partial y} \\ 0 & 0 & 0 \end{bmatrix} \begin{Bmatrix} u \\ v \\ w \end{Bmatrix} \quad or \quad \varepsilon_t = \mathbf{D}_t \mathbf{u} \tag{9.2}$$

The transverse stress continuity and the displacement consistency on the interface are required for laminated plates, while the inplane stress discontinuity between layers is permitted. It is difficult to guarantee the interlayer traction continuity by the traditional displacement elements when the transverse stresses are calculated from the transverse strains via the constitutive relations. Hybrid elements based on the Hellinger–Reissner principle with all six stress components independently assumed can satisfy the interlayer stress continuity, however, it fails to obtain the continuous inplane strains between layers since the inplane strain-displacement relation is satisfied in the sense of the integral. Therefore, a set of mixed-type variables with the displacement and transverse stress components is defined as the state space for the laminated system:

$$\Theta = \{u \; v \; w \,|\, \tau_{xz} \; \tau_{yz} \; \sigma_z\} = \{\mathbf{u} \,|\, \sigma_t\} \tag{9.3}$$

It has been verified [10] that the transverse stress components are the dual variables of the displacement components when considering the z-direction as the timescale in the Hamilton system. Once the variables in the space are solved, the rest of the mechanical variables, such as the inplane strains and stresses, are all available using the geometrical and physical equations of the laminated plate. The stress–strain relationship can be written in the form

$$\begin{Bmatrix} \varepsilon_p \\ \varepsilon_t \end{Bmatrix} = \begin{bmatrix} \mathbf{S}_p & \mathbf{S}_{pt} \\ \mathbf{S}_{pt} & \mathbf{S}_t \end{bmatrix} \begin{Bmatrix} \sigma_p \\ \sigma_t \end{Bmatrix} \tag{9.4}$$

or, equivalently,

$$\begin{Bmatrix} \sigma_p \\ \varepsilon_t \end{Bmatrix} = \begin{bmatrix} \mathbf{C}_p & \mathbf{C}_{pt} \\ -\mathbf{C}_{pt} & \mathbf{C}_t \end{bmatrix} \begin{Bmatrix} \varepsilon_p \\ \sigma_t \end{Bmatrix} \tag{9.5}$$

in which

$$\mathbf{C}_p = \mathbf{S}_p^{-1}, \mathbf{C}_{pt} = -\mathbf{C}_p \mathbf{S}_{pt} = \mathbf{C}_{tp}^{\ T}, \mathbf{S}_t^* = \mathbf{S}_{pt}^{\ T} \mathbf{C}_p \mathbf{S}_{pt}$$

For a given layer (i.e., the ith layer), the Hellinger–Reissner energy functional can be expressed as (ignoring body force and boundary tractions)

$$\Pi_R^i(\sigma, \mathbf{u}) = \int_V \left[-\frac{1}{2} \sigma^T S \sigma + \sigma^T \mathbf{D} \mathbf{u} \right] dV \tag{9.6}$$

or, equivalently,

$$\Pi_R{}^i(\sigma, \mathbf{u}) = \int_V \left[-\frac{1}{2} \begin{Bmatrix} \sigma_p \\ \sigma_t \end{Bmatrix}^T \begin{bmatrix} \mathbf{S}_p & \mathbf{S}_{pt} \\ \mathbf{S}_{pt}^{\ T} & \mathbf{S}_t \end{bmatrix} \begin{Bmatrix} \sigma_p \\ \sigma_t \end{Bmatrix} + \begin{Bmatrix} \sigma_p \\ \sigma_t \end{Bmatrix}^T \begin{Bmatrix} \varepsilon_p \\ \varepsilon_t \end{Bmatrix} \right] dV \tag{9.7}$$

Eliminating the inplane stresses from Equation 9.7 using Equation 9.6, the desired state functional based on the sate space Θ can be formulated as

$$\Pi_\Theta^i(\sigma_t, \mathbf{u}) = \int_V \left[\frac{1}{2} (\mathbf{D}_p \mathbf{u})^T \mathbf{C}_p (\mathbf{D}_p \mathbf{u}) + \sigma_t^{\ T} \mathbf{C}_{tp} (\mathbf{D}_p \mathbf{u}) - \frac{1}{2} \sigma_t^{\ T} \mathbf{S}_t^* \sigma_t + \sigma_t^{\ T} (\mathbf{D}_t \mathbf{u}) \right] dV \tag{9.8}$$

The system energy functional for the laminated plate with n layers is then

$$\Pi_\Theta(\sigma_t, \mathbf{u}) = \sum_{i=1}^{n} \Pi_\Theta{}^i(\sigma_t, \mathbf{u}) \tag{9.9}$$

It has been shown in Reference 10 that the stationary requirement for the functional (Equation 9.9), $\delta \Pi_\Theta(\sigma_t, \mathbf{u}) = 0$, leads to a set of equilibrium equations and geometrical consistency conditions. In addition, the stress continuity requirements on the interfaces and the surface traction-free restrictions for the concerned laminated system can also be obtained. In comparison with the functional $\Pi_R^i(\sigma_p, \sigma_t, \mathbf{u})$, the functional (Equation 9.8) possesses fewer unknown field variables, which makes it possible to reduce computational costs. The major advantage of adopting $\Pi_\Theta(\sigma_t, \mathbf{u})$ is that it allows the interface continuity requirements of the transverse stresses and the traction-free conditions to be imposed easily. In addition, the interlaminar discontinuity requirements of the inplane stresses and the transverse strains are also satisfied.

9.1.3 The Laminate Hybrid Element Based on the State Space

A 16-node hexahedron element with constant thickness hi, as shown in Figure 9.2, has been developed to model a single layer of composite plates. The displacements and transverse stresses are assumed independently within the element [15]. In terms of the arranged nodal displacement parameters, a set of bilinear distributions in the $\Pi_\Theta(\sigma_t, \mathbf{u})$ or $x - y$ plane can be interpolated for the element displacements (u, v, w). Simultaneously, the displacement distributions in the z-direction, that is, the cubic inplane displacements (u, v) and the linear transverse displacement w, can also be defined. Thus the assumed element displacement functions can be expressed as

$$u = Nq \tag{9.10a}$$

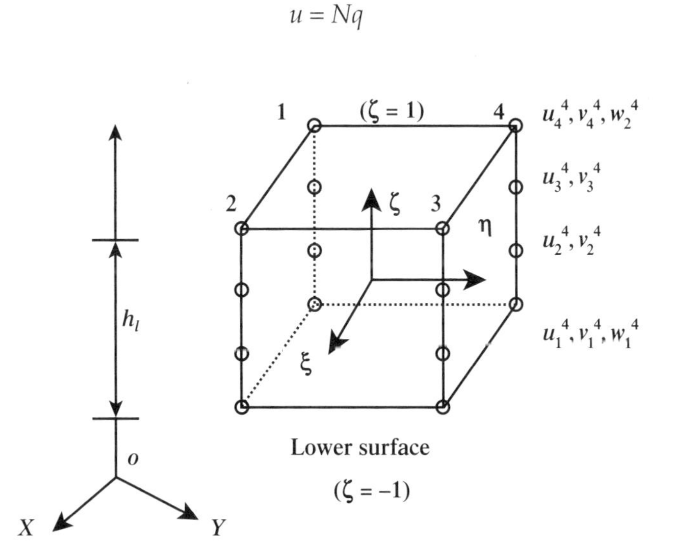

FIGURE 9.2
Definition of a 16-node hexahedron element.

where $N_k^{(1\times1)}(\xi,\eta)$ is a four-node bilinear shape function in the $\Pi_\Theta(\sigma_t,u)$ plane, and $L_k^{(3)}(\zeta)$ and $L_k^{(1)}(\zeta)$ are, respectively, the cubic and linear interpolations in the z-direction. Their detailed expressions are given below:

$$
\begin{aligned}
N_K{}^{(1\times1)}(\xi,\eta) &= \frac{1}{4}(1+\xi\xi_K)(1+\eta\eta_K),(K=1-4)\\
L_1{}^{(3)}(\zeta) &= -\frac{1}{16}(1-3\zeta)(1-\zeta)(1+3\zeta)\\
L_2{}^{(3)}(\zeta) &= \frac{9}{16}(1-\zeta)(1-3\zeta)(1+\zeta)\\
L_3{}^{(3)}(\zeta) &= \frac{9}{16}(1-\zeta)(1+3\zeta)(1+\zeta)\\
L_4{}^{(3)}(\zeta) &= -\frac{1}{16}(1-3\zeta)(1+\zeta)(1+3\zeta)\\
L_1{}^{(1)}(\zeta) &= \frac{1}{2}(1-\zeta)\\
L_2{}^{(1)}(\zeta) &= \frac{1}{2}(1+\zeta)
\end{aligned}
\tag{9.10b}
$$

Numerical tests in Reference 10 show that the inplane displacements should be assumed as a cubic distribution in the thickness direction in order to be consistent with the parabolic distribution of the transverse shear stresses. These play a significant role in delamination and failure analysis of laminated structures with thick and moderate thickness.

Corresponding to the element displacement shown in Equation 9.10a, there exists an isoparametric relationship between the global and local coordinate systems, which takes the form of

$$
\begin{Bmatrix} x \\ y \\ z \end{Bmatrix} = \begin{Bmatrix} \sum_{k=1}^{4} N_k{}^{(1\times1)}(\xi,\eta)x^k \\ \sum_{k=1}^{4} N_k{}^{(1\times1)}(\xi,\eta)y^k \\ \sum_{j=1}^{4} L_j{}^{(1)}(\zeta)z_j \end{Bmatrix} \tag{9.11}
$$

For the convenience of imposing the stress continuity on the interlaminar surfaces, the transverse stresses $(\tau_{xz},\tau_{yz},\sigma_z)$ within the element can be assumed as

$$\tau_{xz} = \left[L_1^{(2)}(\zeta) \quad L_2^{(2)}(\zeta) \quad L_3^{(2)}(\zeta) \right] \begin{Bmatrix} \beta_1 + \xi\beta_2 + \eta\beta_3 \\ \beta_4 + \xi\beta_5 + \eta\beta_6 \\ \beta_7 + \xi\beta_8 + \eta\beta_9 \end{Bmatrix} \tag{9.12a}$$

$$\tau_{yz} = \left[L_1^{(2)}(\zeta) \quad L_2^{(2)}(\zeta) \quad L_3^{(2)}(\zeta) \right] \begin{Bmatrix} \beta_{10} + \xi\beta_{11} + \eta\beta_{12} \\ \beta_{13} + \xi\beta_{14} + \eta\beta_{15} \\ \beta_{16} + \xi\beta_{17} + \eta\beta_{18} \end{Bmatrix} \tag{9.12b}$$

$$\sigma_z = \left[L_1^{(3)}(\zeta) \quad L_2^{(3)}(\zeta) \quad L_3^{(3)}(\zeta) \quad L_4^{(3)}(\zeta) \right] \begin{Bmatrix} \beta_{19} \\ \beta_{20} \\ \beta_{21} \\ \beta_{22} \end{Bmatrix} \tag{9.12c}$$

in which the quadratic basis functions $L_k^{(2)}(\zeta)$ in the z-direction are

$$L_1^{(2)}(\zeta) = 1/2\zeta(\zeta - 1)$$

$$L_2^{(2)}(\zeta) = (1 - \zeta^2)$$

$$L_3^{(2)}(\zeta) = 1/2\zeta(\zeta + 1)$$

The assumed stresses above can be denoted in the matrix form as

$$\sigma_t = \begin{Bmatrix} \tau_{xz} \\ \tau_{yz} \\ \sigma_z \end{Bmatrix} = \Phi_t \beta \tag{9.12d}$$

where $\beta = \{\beta_1, \beta_2, \cdots, \beta_{22}\}^T$ is the element stress vector. In the above assumed stress field, τ_{xz} and τ_{yz} take a linear distribution in the $\xi - \eta$ plane, while they take a parabolic distribution in the z-direction. This approximation is consistent with the classic laminated plate theory.

Substituting the element displacements in Equation 9.10a into Equation 9.1 and Equation 9.2, the element strains can be expressed as

$$\varepsilon_p = \boldsymbol{D}_p \boldsymbol{u} = \boldsymbol{B}_p \boldsymbol{q}, \;\; \boldsymbol{B}_p = \boldsymbol{D}_p N, \tag{9.13}$$

$$\text{where } \; \varepsilon_t = \boldsymbol{D}_t \boldsymbol{u} = \boldsymbol{B}_t \boldsymbol{q}, \text{ where } \boldsymbol{B}_t = \boldsymbol{D}_t N. \tag{9.14}$$

For an individual laminate element i with the nodal displacements $\mathbf{q}^i$ and the stress parameters β_i, the element state functional of Equation 9.8 takes the form of

$$\Pi_\Theta{}^i(q^i,\beta^i)=\frac{1}{2}q^{iT}\mathbf{K}_p{}^i q^i+\beta^{iT}G^i q^i-\frac{1}{2}\beta^{iT}H^i\beta^i \tag{9.15}$$

where

$$H^i=\int_{V^i}\mathbf{Q}_t{}^{iT}\mathbf{S}_t{}^{*i}\Phi_t{}^i dV$$

$$G^i=\int_{V^i}(\Phi_t{}^{iT}\mathbf{C}_{tp}{}^i\mathbf{B}_p{}^i+\Phi_t{}^{iT}\mathbf{B}_t{}^i)dV$$

9.1.4 Implementation of the Interface Stress Continuity

Considering the upper and lower surfaces of an element in Figure 9.2, only the stress parameters relating to $L_3^{(2)}(\zeta)$ and $L_4^{(3)}(\zeta)$ contribute to the stresses on the upper surface, while $L_1^{(2)}(\zeta)$ and $L_1^{(3)}(\zeta)$ contribute to the stresses on the lower surface (see Equations 9.12a–d). In this case, the traction continuity on the interlaminar surfaces and surface traction-free conditions can be easily satisfied when the assumed transverse stresses of Equations 9.12a–d are adopted.

Observing the transverse interface of two adjacent 3-D elements in Figure 9.2 (provided that each layer is discretized by only one element in the thickness direction), the continuity requirements of the interlaminar transverse stresses can be expressed as

$$(\tau_{xz}{}^i \quad \tau_{yz}{}^i \quad \sigma_z{}^i)\,|_{\zeta=1}=(\tau_{xz}{}^{i+1} \quad \tau_{yz}{}^i \quad \sigma_z{}^{i+1})\,|_{\zeta=-1} \tag{9.16}$$

where i denotes the lower layer and $i+1$ denotes the upper layer. In view of the stress modes in Equations 9.12a–d, the continuity conditions shown in Equation 9.16 can also be expressed as

$$\begin{aligned}&\{\beta_7 \quad \beta_8 \quad \beta_9 \quad \beta_{16} \quad \beta_{17} \quad \beta_{18} \quad \beta_{22}\}^{i^T}\\&=\{\beta_1 \quad \beta_2 \quad \beta_3 \quad \beta_{10} \quad \beta_{11} \quad \beta_{12} \quad \beta_{19}\}^{i+1^T}\end{aligned} \tag{9.17a}$$

or

$$\beta_{up}{}^i=\beta_{low}{}^{i+1} \tag{9.17b}$$

For the elements on the upper and lower surfaces of the laminated plates, the surface traction-free conditions can be satisfied by enforcing

$$(\tau^i{}_{xz} \quad \tau^i{}_{yz} \quad \sigma^i{}_z)_{\zeta=1} = 0 \quad \text{or}$$

$$\beta^i{}_{up} = 0 \tag{9.18a}$$

on the upper surface and

$$(\tau^i{}_{xz} \quad \tau^i{}_{yz} \quad \sigma^i{}_z)_{\zeta=-1} = 0 \quad \text{or}$$

$$\beta^i{}_{low} = 0 \tag{9.18b}$$

on the lower surface.

Implementation of the stress constraints in Equation 9.17a to Equation 9.18b will produce the interface element or the surface element, in which the interlayer continuity requirements and the traction-free conditions are satisfied exactly [21]. Moreover, if the conditions in Equation 9.17a to Equation 9.18b are implemented on each interface or surface of a laminated system, one can build a super hybrid element composed of n individual 3-D laminated hybrid elements. Because the stress parameters can be eliminated at the super-element level, the system stiffness matrix for the laminated plate can be directly assembled in the x–y plane in terms of the super-element stiffness matrix. The state functional of a super-element of n laminates can be formulated as

$$\Pi_\Theta(\mathbf{q}_*, \beta_*) = \sum_{i=1}^{n} \Pi_\Theta{}^i(\mathbf{q}^i, \beta^i)$$
$$= \frac{1}{2}\mathbf{q}_*{}^T \mathbf{K}_p \mathbf{q}_* + \beta_*{}^T \mathbf{G} \mathbf{q}_* - \frac{1}{2}\beta_*{}^T \mathbf{H} \beta_*, \tag{9.19}$$

Consequently, the stiffness matrix of the super-element can be formulated as

$$\mathbf{K} = \mathbf{K}_p + \mathbf{G}^T \mathbf{H}^{-1} \mathbf{G} \tag{9.20}$$

where $\mathbf{q}_*$, β_*, $\mathbf{K}_p$, $\mathbf{G}$, and $\mathbf{H}$ are obtained by assembling all the single-layer element matrices $\mathbf{q}^i$, β^i, $\mathbf{K}_p^i$, $\mathbf{G}^i$, and $\mathbf{H}^i$.

9.1.5 Numerical Examples

Three numerical examples [10] are considered in this section to illustrate the numerical performance of the hybrid model that satisfies the interlaminar continuity requirements and the surface traction-free conditions exactly.

The numerical accuracy, the convergence of the method, and the stress distributions in the transverse direction are studied.

The first example is a simply supported three-layer [0°/90°/0°] laminate of infinite length under cylindrical bending (see Figure 9.3). The laminate is subjected to a sinusoidal distributed load, $q(x) = q_0 \sin(\pi x / a)$, on the upper surface, where a is the plate width and h is the thickness. Each layer has the equal thickness, and its material constants are

$$E_T = 10^6 psi$$

$$G_{TT} = 0.2 \times 10^6 psi$$

$$\mu_{TT} = 0.25.$$

This is a plane strain problem and has no strains associated with the y-direction. Owing to the symmetry, only half of the plate strip needs to be modeled. The half strip with unit length in the y-direction is then divided into 20 × 3 elements in the x–z plane. A small span-to-thickness ratio, $S = a/h = 4$, is selected since thick laminates are more prone to delamination or failure resulting from the interlaminar stresses.

Figure 9.4 and Figure 9.5 show the transverse distributions of τ_{xz} on the simply supported edge and σ_x at the center of the plate using the present model. Analytic solutions by Pagano [22] and 2-D finite element results based on the first-order shear deformation theory are also plotted for comparison. Note that all results are normalized as follows:

$$\bar{w} = \frac{100E_T}{q_0 S h^4} w, \; (\bar{\sigma}_x, \bar{\sigma}_y) = \frac{1}{q_0 S^2}(\sigma_x, \sigma_y),$$
$$\bar{\tau}_{xz} = \frac{1}{q_0 S} \tau_{xz} \tag{9.21}$$

Figure 9.4 and Figure 9.5 show that the proposed model presents very consistent results with theoretical solutions for both in-plane and transverse shear stresses. In contrast, the 2-D finite element fails to simulate the nonlinear distribution of sxz and gives wrong predictions of rx at the interlaminar surfaces and the lower and upper surfaces.

The second example is a three-layer [0°/90°/0°] square laminate with its width a and thickness h. The plate is simply supported on all edges and subjected to a double-sinusoidal load, $q(x,y) = q_0 \sin(\pi x / a)\cos(\pi y / a)$, on the upper surface. Again, each layer has equal thickness. The same material constants, span-to-thickness ratio, and normalization as for the first example are used here.

Due to the symmetry in geometry, only a quarter of the plate is considered. An 8×8 element discretization in the x–y plane and a single element in each

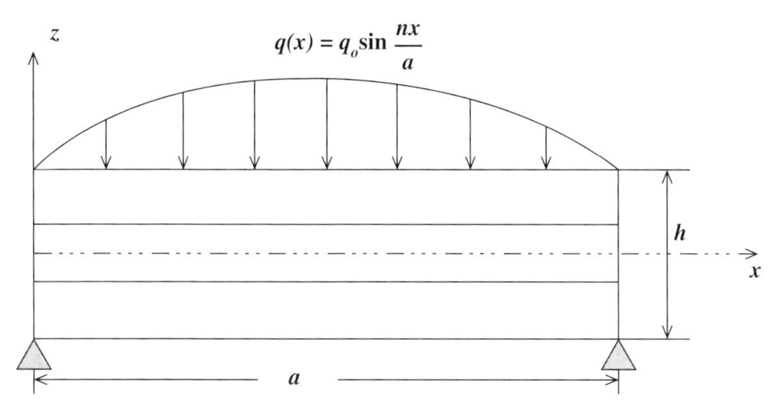

FIGURE 9.3
A simply supported three-layer [0°/90°/0°] laminate under cylindrical bending.

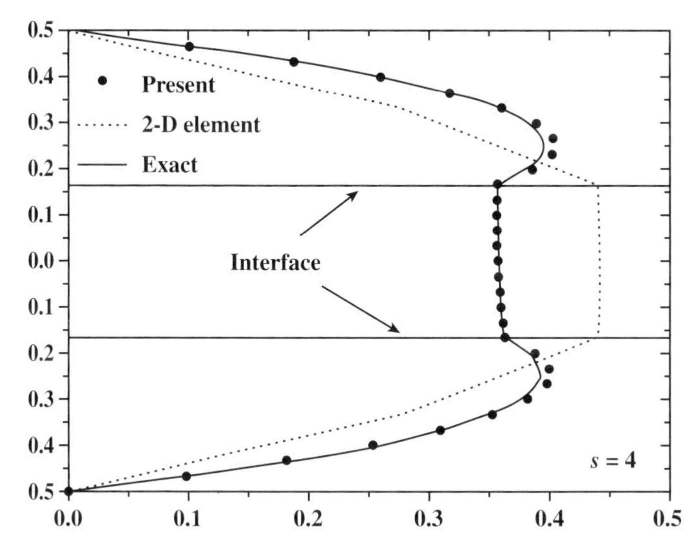

FIGURE 9.4
Distribution of transverse shear stress across the thickness at the edge of the simply supported [0°/90°/0°] laminate under cylindrical bending ($x = 0$, $s = 4$).

layer along the thickness direction are adopted in computation. The transverse distribution of the shear stress τ_{xz} at the middle edge and the normal stress σ_x at the center of the plate are plotted in Figure 9.6 and Figure 9.7, respectively. Analytical solutions by Pagano [23] are also plotted for comparison. From the figures, it is again seen that the numerical results are in excellent agreement with the analytic solutions, especially for the transverse shear stress τ_{xz}.

A three-layer rectangular laminated plate is the final example to study the numerical accuracy and convergence of the transverse displacement and some typical stresses. The plate has the same lamination geometry, materials

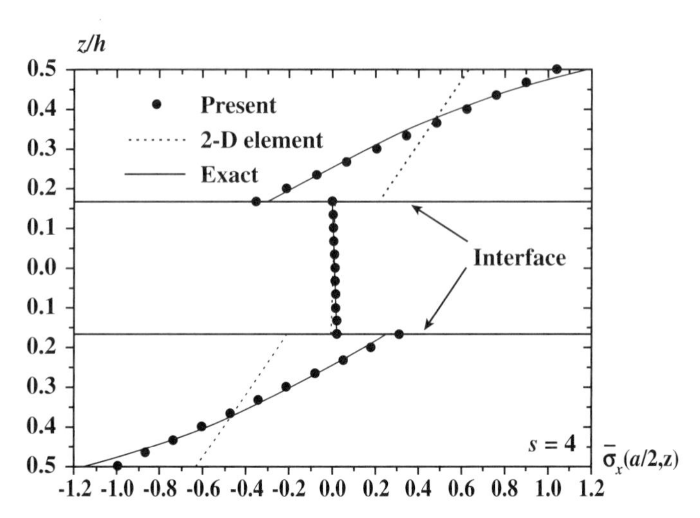

FIGURE 9.5
Distribution of inplane normal stress across the thickness at the center of the simply supported [0°/90°/0°] laminate under cylindrical bending ($x = a/2$, $s = 4$).

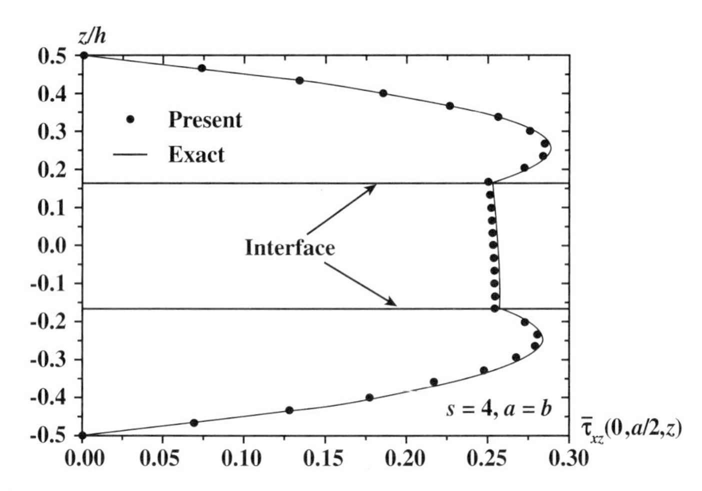

FIGURE 9.6
Transverse shear stress distribution across the thickness at the middle of the edge of the simply supported [0°/90°/0°] square laminate ($x = 0$, $y = a/2$, $s = 4$).

constants, and boundary conditions as the second example, but $b = 3a$, where a and b are the ledge length along the x- and y-axes, respectively. A double-sinusoidal load, $q(x,y) = q_0 \sin(\pi x / a)\cos(\pi y / b)$, is distributed on the upper surface of the plate. Note that the x- and y-axes coincide with two joined edges of the plate. Again, only a quarter of the plate is discretized due to symmetries. Various discretizations and span-to-thickness ratios are considered. Each layer has only one element along the thickness direction in all discretizations. Table 9.1 lists the central deflection w, normal

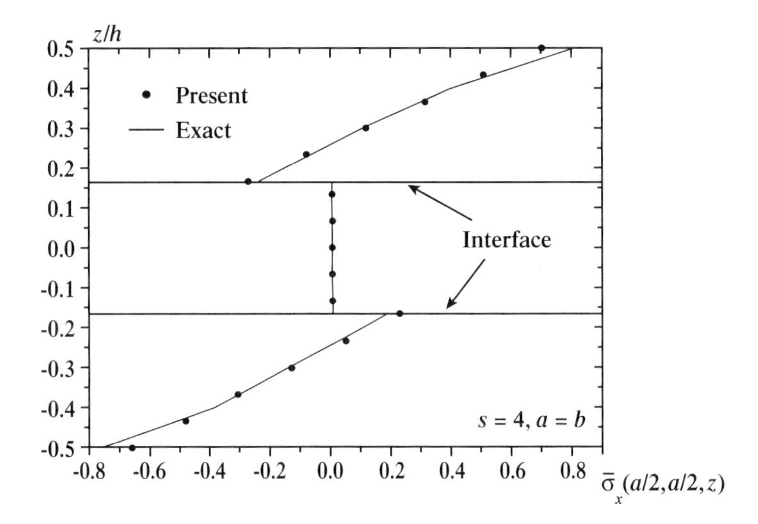

FIGURE 9.7
Inplane normal stress distribution across the thickness at the center of the simply supported [0°/90°/0°] square laminate ($x = y = a/2$, $s = 4$).

TABLE 9.1

Deflection and Stresses in the Rectangular Laminate ($b = 3a$)

a/h	Grid	$\bar{w}$ $(a/2, b/2, 0)$	$\bar{\sigma}_x$ $(a/2, b/2, h/2)$	$\bar{\sigma}_y$ $(a/2, b/2, h/6)^a$	$\bar{\tau}_{xz}$ $(0, b/2, 0)$
4	4 × 4 × 3	2.8295	0.9792	0.1029	0.3440
	8 × 8 × 3	2.8301	1.0027	0.1054	0.3490
	12 × 12 × 3	2.8302	1.0071	0.1059	0.3499
	16 × 16 × 3	2.8303	1.0086	0.1060	0.3502
	Exact [23]	2.82	1.14	0.109	0.351
10	4 × 4 × 3	0.8951	0.6669	0.03910	0.3986
	8 × 8 × 3	0.9128	0.6957	0.04072	0.4144
	12 × 12 × 3	0.9162	0.7012	0.04104	0.4175
	16 × 16 × 3	0.9174	0.7032	0.04115	0.4186
	Exact [23]	0.919	0.726	0.0418	0.420
20	4 × 4 × 3	0.5517	0.5683	0.02579	0.3656
	8 × 8 × 3	0.5937	0.6244	0.02813	0.4154
	12 × 12 × 3	0.6023	0.6358	0.02883	0.4258
	16 × 16 × 3	0.6053	0.6399	0.02902	0.4295
	Exact [23]	0.610	0.650	0.0294	0.434
50	4 × 4 × 3	0.3203	0.3762	0.01542	0.1594
	8 × 8 × 3	0.4494	0.5383	0.02207	0.3382
	12 × 12 × 3	0.4862	0.5844	0.02396	0.3901
	16 × 16 × 3	0.5005	0.6024	0.02470	0.4105
	Exact [23]	0.520	0.628	0.0259	0.439

[a]The values indicated for $\bar{\sigma}_y$ occur in the central layer.

stresses (σ_x, σ_y), and the transverse shear stress τ_{xz} at the plate edge, which are normalized as in Equation 9.21. The displacement and stresses converge stably and quickly for all the plates with different thickness. An error less than 4% can be reached for all the values when using 16×16×3 discrete grids. It is also demonstrated that the element is capable of thin laminated plate analysis without shear locking.

9.1.6 Conclusion

In this section, a new partially hybrid element for laminates analysis is presented. The assumed displacements $(u\,v\,w)$ and the transverse stresses $(\tau_{xz}\ \tau_{yz}\ \sigma_z)$ are chosen independently based on the state energy functional. By simply adopting the piecewise transverse stress distributions along the thickness direction and setting the stress parameters related to the interfaces or surfaces correctly, the interlaminar continuity requirements for the interface stresses and the surface traction-free conditions can be easily imposed and exactly satisfied. This model also ensures the continuous inplane strains and permits the discontinuity of the inplane stresses between layers. Numerical results show that this hybrid model is capable of thick and thin laminate analysis with good numerical accuracy and rapid convergence. It also can accurately capture the nonlinear stress distribution across the thickness direction.

9.2 Bimaterial Interface Hybrid Element for Piezoelectric Laminated Analysis

9.2.1 Introduction

By combining composite laminated structures with piezoelectric layers, the structures' mechanical behavior can be controlled by electric sensors and actuators. The analysis for this kind of smart structure has recently been the topic of numerous investigations. Using the finite element method, many workers have analyzed the dynamic response of these structures. Their attention has been focused on displacement control and vibration suppression. In contrast, only limited work has been done on the stress and strength analysis of piezoelectric laminated structures, which often have a relatively low transverse stiffness and are apt to delaminate. Moreover, the electromechanical coupling effect possibly makes the transverse stresses increase significantly and causes delamination, such that the piezoelectric system ends in failure.

It is difficult for conventional displacement-based elements to model composite systems, as they cannot take account of the traction-free condition on the top and bottom surfaces and the continuity requirement of the interlaminar transverse stresses. In order to overcome this problem, Spilker [6, 7] and Pian and Li

[9] studied laminated plates using the hybrid stress element method. To reduce the computational costs, Jing and Liao [11] used a partial hybrid element with assumed transverse shear stresses only. Spilker et al. [5] introduced a penalty method to satisfy the continuity of interlayer transverse stresses. Another penalty hybrid method employing a mixed variational principle based on the state space was suggested by Huang, Liu, and Wu [24]. However, all the above methods result in the order of the elemental stiffness matrix being dependent on the number of layers. For piezoelectric laminates, the transverse stresses induced by the electric-elastic coupling and anisotropy are very complex and highly nonlinear, which may affect the strength and safety of structures. In addition, the thickness of the piezoelectric layers is possibly far less than that of the substrates. Numerical difficulties may be encountered when conventional finite elements are employed. The current research in this field is, however, mostly based on analytical methods such as the displacement method and the state space approach [25–29]. The aim of this section is to find an effective hybrid stress element method to stimulate the piezoelectric laminated structures.

In this section, a mixed functional used for 3-D electromechanical analysis is suggested, in which the displacements (u, v, w), the transverse stresses $(\tau_{yz}, \tau_{xz}, \sigma_z)$, and the electric potential φ are assumed independently, and then a set of 3-D hybrid stress models are presented. We create two types of 3-D discrete model, namely the interface element and the surface element, to impose the surface traction-free condition and the continuity of the interlaminar transverse stresses.

9.2.2 Electro-Elastic Variational Formulation and the Governing Equations

Reissner [17] suggested a mixed variational principle to analyze elastic laminates. Introducing the electrostatic energy and work terms into the functional, and neglecting the body force and the body charge, the electric-elastic energy functional can be acquired as follows:

$$\Pi(u_i, \sigma_{ij}, D_i, \varphi) = \int_v B(\sigma_{ij}, D_i)dv - \int_v (u_{i,j}\sigma_{ij} + \varphi_{,j}D_j)dv$$
$$+\int_{S_u} \sigma_{ij}n_j(u_i - \overline{u}_i)dS + \int_{S_\varphi} D_j n_j(\varphi - \overline{\varphi})dS + \int_{S_\sigma} u_i \overline{T}_i dS + \int_{S_\omega} \varphi\overline{\omega}dS \tag{9.22}$$

in which $u_i; \sigma_{ij}; D_i$ and φ are the displacements, the stresses, the electric displacements, and the electric potential, respectively; also, on the boundary, $\overline{T}_i$ is the prescribed traction on the surface is the prescribed surface charge on S_ω; $\overline{u}_i$ is the prescribed displacement on S_u; $\overline{\varphi}$ is the prescribed electric potential on surface S_u; $\overline{\varphi}$ is the component of the unit normal vector to the boundary surface. In Equation 9.22, the complementary energy function

$$B(\sigma_{ij}, D_i) = \frac{1}{2}S_{ijkl}\sigma_{ij}\sigma_{kl} - \frac{1}{2}\beta_{ij}D_iD_j - g_{ijk}\sigma_{ij}D_k \tag{9.23}$$

where the coefficients S_{ijkl}, g_{ijk}, and β_{ij} in Equation 9.23 represent the elastic constants, the piezoelectric constants, and the dielectric permittivities, respectively. From Equation 9.23, the following constitutive laws can be obtained:

$$\varepsilon_{ij} = \frac{\partial B}{\partial \sigma_{ij}}, \quad E_i = -\frac{\partial B}{\partial D_i} \tag{9.24}$$

where ε_{ij} is the strains and E_i the electric fields.

The first-order variation of the variational functional $\Pi(u_i, \sigma_{ij}, D_i, \varphi)$ with respect to u_i, σ_{ij}, D_i, and φ can be expressed as

$$\delta\Pi = \int_v \{[\varepsilon_{ij} - \frac{1}{2}(u_{i,j} + u_{j,i})]\delta\sigma_{ij} - (E_i + \varphi_{,i})\delta D_i + \sigma_{ij,j}\delta u_i + D_{i,j}\delta\varphi\} dv$$

$$- \int_{S_u} (\bar{u}_i - u_i) n_j \delta\sigma_{ij} dS - \int_{S_\sigma} (\sigma_{ij} n_j - \bar{T}_i)\delta u_i dS$$

$$- \int_{S_\varphi} (\bar{\varphi} - \varphi) n_i \delta D_i dS - \int_{S_\omega} (D_i n_i - \bar{\omega})\delta\varphi dS$$

In terms of the stationary condition for the functional, $\delta\Pi = 0$, the electric-mechanical governing equations in v can be obtained as follows:

$$\sigma_{ij,j} = 0$$

$$D_{j,j} = 0$$

$$\varepsilon_{ij} = \frac{1}{2}(u_{i,j} + u_{j,i})$$

$$E_i = -\varphi_{,i}$$

The boundary conditions are

$$S_\sigma : \quad \sigma_{ij} n_j = \bar{T}_i$$

$$S_u : \quad u_i = \bar{u}_i$$

$$S_\varphi : \quad \varphi = \bar{\varphi}$$

$$S_\omega : \quad \mathrm{D}_i n_i = \bar{\omega}$$

For common piezoelectric materials of 2 mm crystal symmetry, the constitutive relations can be written as [30]

$$\boldsymbol{\varepsilon} = \mathbf{S}\boldsymbol{\sigma} + \mathbf{d}^{\mathrm{T}}\mathbf{E} \tag{9.25}$$

$$\mathbf{D} = \mathbf{e}\varepsilon + \in \mathbf{E} \tag{9.26}$$

with

$$\sigma = \{\sigma_{11}\ \sigma_{22}\ \sigma_{33}\ \tau_{23}\ \tau_{31}\ \tau_{12}\}^T$$

$$\mathbf{D} = \{D_1\ D_2\ D_3\}^T$$

$$\varepsilon = \{\varepsilon_{11}\ \varepsilon_{22}\ \varepsilon_{33}\ \gamma_{23}\ \gamma_{31}\ \gamma_{12}\}^T$$

$$\mathbf{E} = \{E_1\ E_2\ E_3\}^T$$

S, e, and ε are the compliance, piezoelectric stress, and dielectric matrices, respectively. $d = eS$ is the piezoelectric strain matrix. If

$$\sigma = \{\sigma_p^T\ \sigma_t^T\}^T$$

where

$$\sigma_p = \{\sigma_x\ \sigma_y\ \tau_{xy}\}^T, \qquad \sigma_t = \{\tau_{yz}\ \tau_{xz}\ \sigma_z\}^T$$

$$\varepsilon = \{\varepsilon_p^T\ \varepsilon_t^T\}^T$$

where

$$\varepsilon_p = \{\varepsilon_x\ \varepsilon_y\ \gamma_{xy}\}^T, \qquad \varepsilon_t = \{\gamma_{yz}\ \gamma_{xz}\ \varepsilon_z\}^T$$

and

$$\mathbf{S} = \begin{bmatrix} \mathbf{S_p} & \mathbf{S_{pt}} \\ \mathbf{S_{pt}^T} & \mathbf{S_t} \end{bmatrix}, \qquad \mathbf{d} = \begin{bmatrix} \mathbf{d_p} & \mathbf{d_t} \end{bmatrix}$$

then the mixed constitutive equation can be established from Equation 9.25:

$$\begin{Bmatrix} \sigma_p \\ \varepsilon_t \end{Bmatrix} = \begin{bmatrix} \mathbf{C}_p & \mathbf{C}_{pt} & \mathbf{C}_{p\varphi} \\ -\mathbf{C}_{pt}^T & \mathbf{S}_t^* & \mathbf{C}_{t\varphi} \end{bmatrix} \begin{Bmatrix} \varepsilon_p \\ \sigma_t \\ -\mathbf{E} \end{Bmatrix} \tag{9.27}$$

where

$$\mathbf{C}_p = \mathbf{S}_p^{-1}, \quad \mathbf{C}_{pt} = -\mathbf{C}_p\mathbf{S}_{pt}, \quad \mathbf{C}_{p\varphi}\mathbf{C}_p\mathbf{d}_p^T$$

$$\mathbf{S}_t^* = \mathbf{S}_t - \mathbf{S}_{pt}^T\mathbf{C}_p\mathbf{S}_{pt}, \qquad \mathbf{C}_{t\varphi} = \mathbf{S}_{pt}^T\mathbf{C}_p\mathbf{d}_p^T - \mathbf{d}_t^T$$

Substituting Equation 9.26 and Equation 9.27 into Equation 9.22 while eliminating D and σ_p from Π, the electromechanical mixed functional can be formulated as (ignoring the body force and the body charge and meeting the boundary conditions on S_u and S_φ)

$$\begin{aligned}\Pi(\mathbf{u},\varphi,\sigma_t) = &-\int_v \frac{1}{2}\varepsilon_p^{*T}\mathbf{C}_p^*\varepsilon_p^* dv - \int_v \sigma_t^T\mathbf{C}_{tp}^*\varepsilon_p^* dv \\ &+\int_v \frac{1}{2}\sigma_t^T\mathbf{S}_p^*\sigma_t dv - \int_v \sigma_t^T\varepsilon_t dv \\ &+\int_{S_\sigma} \overline{\mathbf{T}}^T\mathbf{u}dS + \int_{S_\omega} \overline{\omega}\varphi dS\end{aligned} \tag{9.28}$$

where

$$\mathbf{C}_p^* = \begin{bmatrix} \mathbf{C}_p & \mathbf{C}_p\mathbf{d}_p^T \\ \mathbf{d}_p\mathbf{C}_p^T & -\in + \mathbf{d}_p\mathbf{C}_p\mathbf{d}_p^T \end{bmatrix}, \quad \mathbf{C}_{tp}^* = \begin{bmatrix} \mathbf{C}_{tp} & -\mathbf{C}_{t\varphi} \end{bmatrix}$$

$$\varepsilon_\mathrm{p}^* = \begin{Bmatrix} \varepsilon_p \\ -\mathbf{E} \end{Bmatrix}$$

After elimination of the inplane stresses, the independent variables in the functional (Equation 9.28) are the transverse stressed displacements and electric potential. Based on the above mixed variational principle, a 3-D finite element model can be developed by independently assuming the independent variables in the ith element to be

$$\sigma_\mathrm{t}^i = \begin{Bmatrix} \tau_{yz}^i \\ \tau_{xz}^i \\ \sigma_z^i \end{Bmatrix} = \Psi^i\beta^i \tag{9.29}$$

$$\mathbf{u}^{*i} = \begin{bmatrix} u^i & v^i & \mathrm{w}^i & \varphi^i \end{bmatrix}^\mathrm{T} = \begin{Bmatrix} \mathbf{u}^i \\ \varphi^i \end{Bmatrix} = \mathbf{N}^i\mathbf{q}^i \tag{9.30}$$

in which ψ^i is the shape function of transverse stresses, and N^i is the shape function of displacements and electric potential; β^i and q^i are the stress parameters and the nodal displacement electric potential vector, respectively. In the next section, the details of the interpolation functions will be shown.

In terms of the strain-displacement and the electric field–electric potential relations, the inplane strains, the electric fields, and the transverse strains can be formulated as

$$\varepsilon_p^{*i} = \begin{Bmatrix} \varepsilon_p^i \\ -\mathbf{E}^i \end{Bmatrix} = \mathbf{D}_p \mathbf{u}^{*i} = \mathbf{D}_p \mathbf{N}^i \mathbf{q}^i = \mathbf{B}_p^i \mathbf{q}^i \tag{9.31}$$

and

$$\varepsilon_t^i = \mathbf{D}_t \mathbf{u}^{*i} = \mathbf{D}_t \mathbf{N}^i \mathbf{q}^i = \mathbf{B}_t^i \mathbf{q}^i \tag{9.32}$$

where

$$\mathbf{D}_p = \begin{bmatrix} \frac{\partial}{\partial x} & 0 & \frac{\partial}{\partial y} & 0 & 0 & 0 \\ 0 & \frac{\partial}{\partial y} & \frac{\partial}{\partial x} & 0 & 0 & 0 \\ 0 & 0 & 0 & 0 & 0 & 0 \\ 0 & 0 & 0 & \frac{\partial}{\partial x} & \frac{\partial}{\partial y} & \frac{\partial}{\partial z} \end{bmatrix}^{\mathrm{T}}$$

$$\mathbf{D}_t = \begin{bmatrix} \frac{\partial}{\partial x} & 0 & \frac{\partial}{\partial x} & 0 \\ 0 & \frac{\partial}{\partial z} & \frac{\partial}{\partial y} & 0 \\ 0 & 0 & \frac{\partial}{\partial z} & 0 \end{bmatrix}$$

Substituting Equation 9.29 through Equation 9.32 into Equation 9.28, the functional takes the form of

$$\Pi = \sum_i \left(\frac{1}{2}\mathbf{q}^{i^{\mathrm{T}}} \mathbf{K}_p^i \mathbf{q}^i + \beta^{i^{\mathrm{T}}} \mathbf{G}^i \mathbf{q}^i - \frac{1}{2}\beta^{i^{\mathrm{T}}} \mathbf{H}^i \beta\right) - \sum_j \mathbf{q}^{j^{\mathrm{T}}} \mathbf{F}^j \tag{9.33}$$

where

$$\mathbf{K}_p^i = \int_{v^i} \mathbf{B}_p^{i^T} \mathbf{C}_p^{*i} \mathbf{B}_p^i dv$$

$$\mathbf{G}^i = \int_{v^i} (\psi^{i^T} \mathbf{C}_{tp}^{*i} \mathbf{B}_p^i + \psi^{i^T} \mathbf{B}_t^i) dv$$

$$\mathbf{H}^i = \int_{v^i} \psi^{i^T} \mathbf{S}_t^{*i} \psi^i dv$$

$$\mathbf{F}^j = \int_{S_\sigma^j + S_\omega^j} -\mathbf{N}^{j^T} \begin{Bmatrix} \overline{\mathbf{T}} \\ \overline{\omega} \end{Bmatrix} dS$$

Here the superscript j denotes the element with prescribed force or charge. Then, the stationary condition $\partial \Pi / \partial \beta^i = 0$ lead to

$$\beta^i = (\mathbf{H}^i)^{-1} \mathbf{G}^i \mathbf{q}^i \tag{9.34}$$

Substituting Equation 9.34 into Equation 9.33 yields

$$\Pi = \sum_i \frac{1}{2} \mathbf{q}^{i^T} \mathbf{K}^i \mathbf{q}^i - \sum_j \mathbf{q}^{j^T} \mathbf{F}^j = \frac{1}{2} \mathbf{q}^T \mathbf{K} \mathbf{q} - \mathbf{q}^T \mathbf{F} \tag{9.35}$$

where $\mathbf{K}^i$ is the elemental stiffness matrix and can be expressed as

$$\mathbf{K}^i = \mathbf{K}_p^i + \mathbf{G}^{i^T} (\mathbf{H}^i)^{-1} \mathbf{G}^i \tag{9.36}$$

Once the nodal displacements and electric potentials are calculated by solving the global stiffness equation, the transverse stresses of each element can be obtained from Equations 9.34 and Equation 9.29, and the in-plane stresses can be obtained from Equation 9.27.

9.2.3 Element Formulations for the 3-D Hybrid Model

For the 3-D hybrid stress element as shown in Figure 9.8a, the displacement shape functions, for the ith element, are assumed to be

$$\begin{aligned} u^i = \sum_{j=1}^{4} N_j(\xi, \eta)[&-\frac{9}{16}(1-\zeta)(\frac{1}{9}-\zeta^2)u_j^i + \frac{27}{16}(1-\zeta^2)(\frac{1}{3}-\zeta)u_{j+4}^i \\ &\frac{27}{16}(1-\zeta^2)(\frac{1}{3}+\zeta)u_{j+8}^i - \frac{9}{16}(1+\zeta)(\frac{1}{9}-\zeta^2)u_{j+12}^i] \end{aligned} \tag{9.37}$$

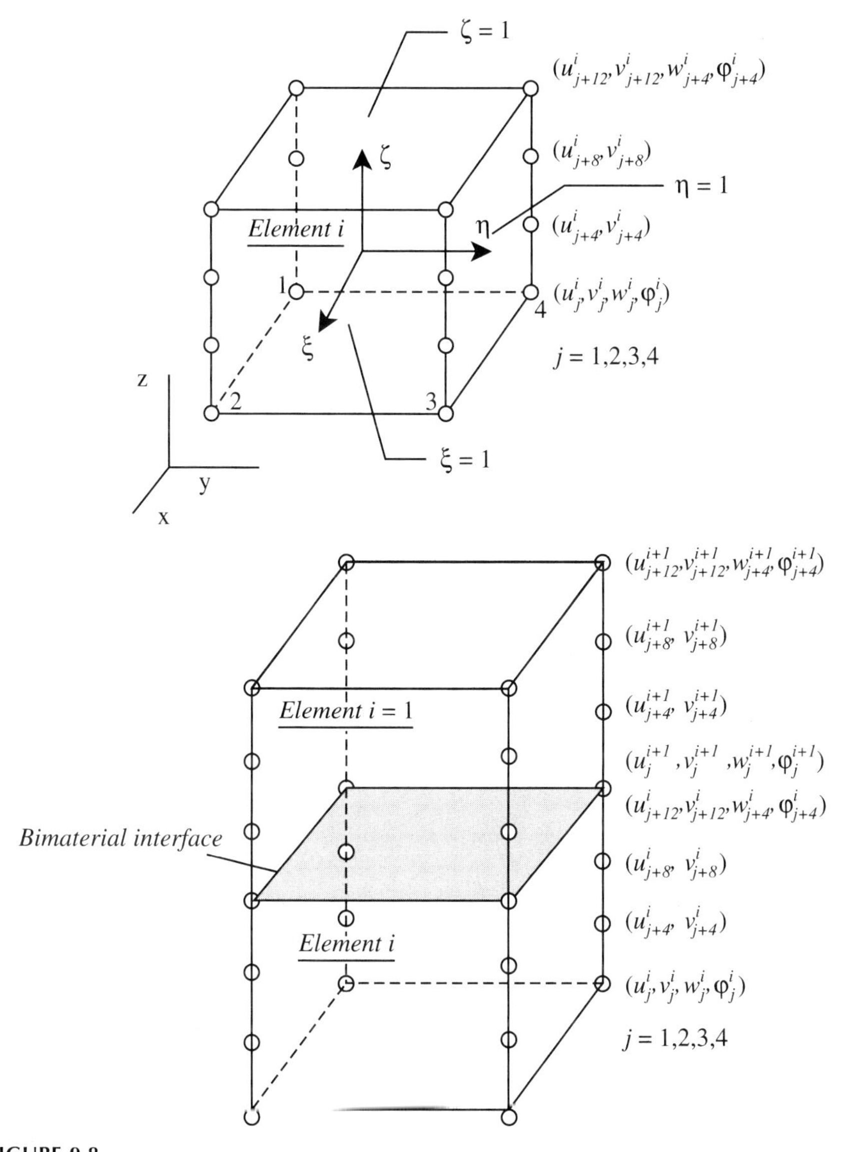

FIGURE 9.8
Element configuration.

$$
\begin{aligned}
v^i = \sum_{j=1}^{4} N_j(\xi,\eta)[&-\frac{9}{16}(1-\zeta)(\frac{1}{9}-\zeta^2)v^i_j + \frac{27}{16}(1-\zeta^2)(\frac{1}{3}-\zeta)v^i_{j+4} \\
&\frac{27}{16}(1-\zeta^2)(\frac{1}{3}+\zeta)v^i_{j+8} - \frac{9}{16}(1+\zeta)(\frac{1}{9}-\zeta^2)v^i_{j+12}]
\end{aligned}
\tag{9.38}
$$

$$w^i = \sum_{j=1}^{4} N_j(\xi,\eta)\left[\frac{1}{2}(1-\zeta)w_j^i + \frac{1}{2}(1+\zeta)w_{j+4}^i\right] \tag{9.39}$$

and the electric potential shape function is

$$\varphi^i = \sum_{j=1}^{4} N_j(\xi,\eta)\left[\frac{1}{2}(1-\zeta)\varphi_j^i + \frac{1}{2}(1+\zeta)\varphi_{j+4}^i\right] \tag{9.40}$$

In Equation 9.37 to Equation 9.40, $N_j(\xi,\eta)$ is always taken to be a bilinear interpolation function in the $\xi\eta$ plane. Some analyses and numerical tests [31] show that it is very helpful to improve the transverse stress distribution in the ζ direction when the element displacements u^i and v^i are taken to be a third-order function of ζ. Therefore the four-node cubic interpolations are introduced into the trial functions for both u^i in Equation 9.37 and v^i in Equation 9.38.

In the transverse direction (ζ-direction) w^i and φ^i are linear, while u^i and v^i are taken as a third-order function so as to obtain more accurate transverse stress distribution along the thickness direction (see Reference 31 for details).

In accordance with the theory of 3-D laminated plates, the transverse shear stresses τ^i_{yz}, $\tau^i_{xz} \in P_2(\zeta)$ are a quadratic function of ζ, while the transverse normal stress $\sigma^i_z \in P_3(\zeta)$ is a cubic function. Therefore, the element interpolation functions for the transverse stresses can be assumed to be

$$\begin{aligned}\tau^i_{yz} &= L_1(\zeta)(\beta_1^i + \beta_2^i\xi + \beta_3^i\eta) + L_2(\zeta)(\beta_4^i + \beta_5^i\xi + \beta_6^i\eta) \\ &\quad + L_3(\zeta)(\beta_7^i + \beta_8^i\xi + \beta_9^i\eta)\end{aligned} \tag{9.41}$$

$$\begin{aligned}\tau^i_{xz} &= L_1(\zeta)(\beta_{10}^i + \beta_{11}^i\xi + \beta_{12}^i\eta) + L_2(\zeta)(\beta_{13}^i + \beta_{14}^i\xi + \beta_{15}^i\eta) \\ &\quad + L_3(\zeta)(\beta_{16}^i + \beta_{17}^i\xi + \beta_{18}^i\eta)\end{aligned} \tag{9.42}$$

$$\sigma^i_z = M_1(\zeta)\beta_{19}^i + M_2(\zeta)\beta_{20}^i + M_3(\zeta)\beta_{21}^i + M_4(\zeta)\beta_{22}^i \tag{9.43}$$

where

$$L_3(\zeta) = \frac{1}{2}\zeta(1+\zeta)$$

$$M_1(\zeta) = -\frac{1}{16}(1-3\zeta)(1+3\zeta)(1-\zeta)$$

$$M_2(\zeta) = \frac{9}{16}(1+\zeta)(1-3\zeta)(1-\zeta)$$

$$M_3(\zeta) = \frac{9}{16}(1+\zeta)(1+3\zeta)(1-\zeta)$$

$$M_4(\zeta) = -\frac{1}{16}(1+\zeta)(1-3\zeta)(1+3\zeta)$$

9.2.4 Interface and Surface Elements

Let us consider the upper and lower surfaces of an element in Figure 9.8a. Only the β parameters relating to $L_3(\xi)$ or $M_4(\xi)$ contribute to the stresses on the upper surface, while the β parameters relating to $L_1(\xi)$ or $M_1(\xi)$ contribute to the lower surface, so that the interlayer traction-continuity and surface traction–free conditions can be easily satisfied when the transverse stresses (Equation 9.42) and (Equation 9.43) are adopted.

Observing the transverse interface of two adjacent 3-D elements in Figure 9.8b, the continuity requirement of the interlaminar transverse stress can be expressed as

$$\sigma_z^i \mid_{\zeta=1} = \sigma_z^{i+1} \mid_{\zeta=-1} \tag{9.44}$$

where i denotes the lower element and $i + 1$ denotes the upper element. The condition (Equation 9.44) can also be expressed by β parameters as

$$j = (1,2,3,10,11,12) \tag{9.45}$$

On enforcing the conditions (Equation 9.25), the transverse stresses of the elements i and $i + 1$ can be assembled together. The shape functions of displacements and the electric potential must also be united. This results in a pair of 3-D elements with a common interface element, which exactly meets all the interface continuity conditions for the transverse stresses, the displacements, and the electric potentials.

Let i_U and i_V indicate elements located on the upper and the lower surfaces, respectively, of a laminate. Thus the traction-free condition can be expressed as

$$\begin{aligned} &\tau_{yz}^{i_L} \mid_{\zeta=-1} = \tau_{xz}^{i_L} \mid_{\zeta=-1} = \sigma_z^{i_L} \mid_{\zeta=-1} = 0 \\ &\tau_{yz}^{i_U} \mid_{\zeta=1} = \tau_{xz}^{i_U} \mid_{\zeta=1} = 0 \end{aligned} \tag{9.46}$$

or equivalently expressed as

$$\begin{aligned} &\beta_j^{i_L} = 0, \qquad j = 1,2,3,10,11,12,19 \\ &\beta_j^{i_U} = 0, \qquad j = 7,8,9,16,17,18 \end{aligned} \tag{9.47}$$

If the surface constraint (Equation 9.47) is imposed on the element parameter β, the above 3-D element will become a surface element that exactly satisfies the traction-free condition. An interface element can also be a surface element. In this case, there is no additional difficulty in the theory and programming.

The interface element discussed here is easily implemented to model a given laminated system, and only the following requirement is necessary: One interface should be correspondingly modeled by one layer of the interface elements.

9.2.5 Numerical Example

Two intelligent laminates with piezoelectric layers bonded to the upper and lower surfaces are considered here to investigate the numerical method. The first is a simply supported three-layer laminate of infinite length under cylindrical bending (Figure 9.9) and the second is a simply supported square laminated plate with applied double-sinusoidal surface potential. The exact solutions of the above problems have been provided by Ray et al. [29] and Heyliger [25], respectively.

In the first numerical example, the thickness of the substrate, h, is 1 mm and the thickness of the piezoelectric layer, $h1$, is 40 μm. The total thickness of the plate is $H = h + 2h_1$. The span-thickness ratio $s = L/H = 4$. The material constants of the substrate and the piezoelectric layer are listed in Table 9.2 and Table 9.3, respectively.

The simply supported boundary of the laminate is idealized as such a constraint, by which the tangential and transverse displacements are prevented, while the normal displacement in the laminate plane is still free. The electrical potential is fixed at zero at the simply supported edge. Thus the following boundary conditions exist along the axial edge $x = 0, L$:

$$w = 0, \qquad v = 0, \qquad \varphi = 0$$

The prescribed boundary electric potentials on the lateral surfaces of the plate take the form of

$$\varphi\left(x, y, \frac{H}{2}\right) = V_1 \sin\left(\frac{\pi x}{L}\right)$$

$$\varphi\left(x, y, \frac{H}{2}\right) = \varphi\left(x, y, -\frac{h}{2}\right) = 0$$

The bottom surface of the plate is charge free:

$$\bar{\varpi}\left(x, y, -\frac{H}{2}\right) = 0$$

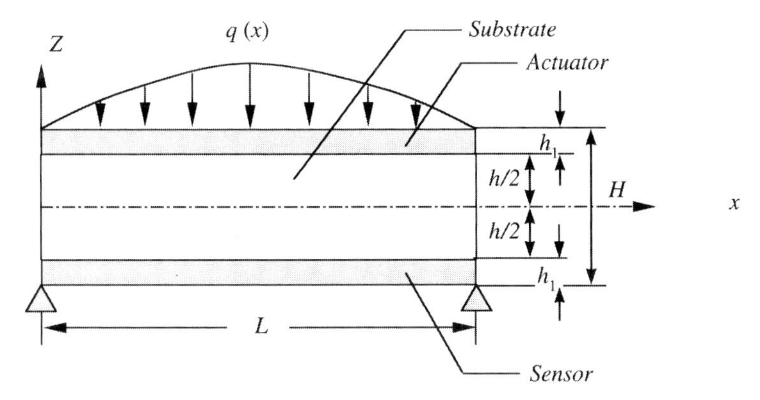

FIGURE 9.9
Geometry of the plate under cylindrical bending.

TABLE 9.2
Material Properties of Poly (Vinylidene Fluoride) [28]

Property	Value	Units
e_{31}	0.046	c/m^2
e_{32}	0.046	c/m^2
e_{33}	0.000	c/m^2
e_{15}	0.000	c/m^2
e_{24}	0.000	c/m^2
ε_{11}	0.1062×10^{-9}	F/m
ε_{22}	0.1062×10^{-9}	F/m
ε_{33}	0.1062×10^{-9}	F/m
Poisson's ratio	0.29	
Elastic modulus	2.0×10^9	N/m^2

TABLE 9.3
Material Properties of the Graphite-Epoxy Composite [28]

Property[a]	Value	Units
Longitudinal elastic modulus, E_L	172.5×10^9	N/m^2
Transverse elastic modulus, E_T	6.9×10^9	N/m^2
Longitudinal shear modulus, G_{LT}	3.45×10^9	N/m^2
Major Poisson's ratio, υ_{LT}	0.25	
Transverse Poisson's ratio, υ_{TT}	0.25	

[a] L and T represent the directions parallel and perpendicular to the fiber, respectively.

and the transverse distributed loading is given by

$$q\left(x, y, \frac{H}{2}\right) = q_0 \sin\left(\frac{\pi x}{L}\right)$$

For symmetry, only half of the plate strip needs to be calculated. The half-strip of unit width is divided into 20×3 elements in the xz plane. In the *z* direction, besides the one layer of the general 3-D elements that is located in the middle of the substrate, two layers of the interface elements are used to model the two interfaces between the substrate and the sensor-actuator.

Figure 9.10 to Figure 9.13 show the results for two cases: $V_1 = 0, q_0 = 1$ and $V_1 = 100, q_0 = 1$; all these data are nondimensionalized as follows [11]:

$$\bar{\tau}_{xz} = \frac{\tau_{xz}}{q_0}$$

These results are in agreement with the analytical solutions provided by Ray et al. [29]. Although the thickness ratio of the elements in the substrate

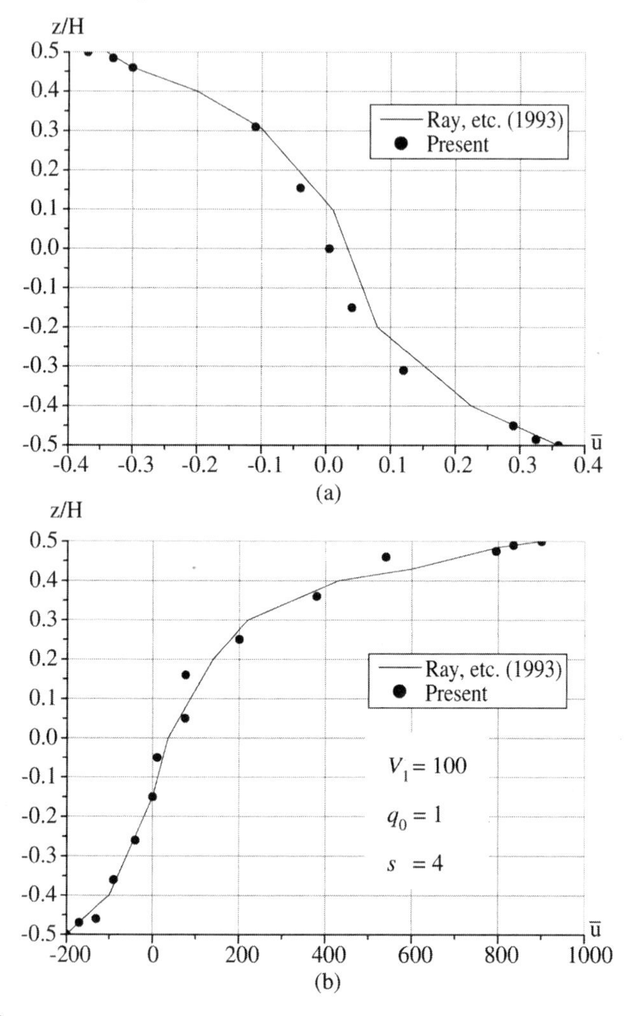

FIGURE 9.10
Distribution of the axial displacement across the thickness at the edge of the plate (x = 0).

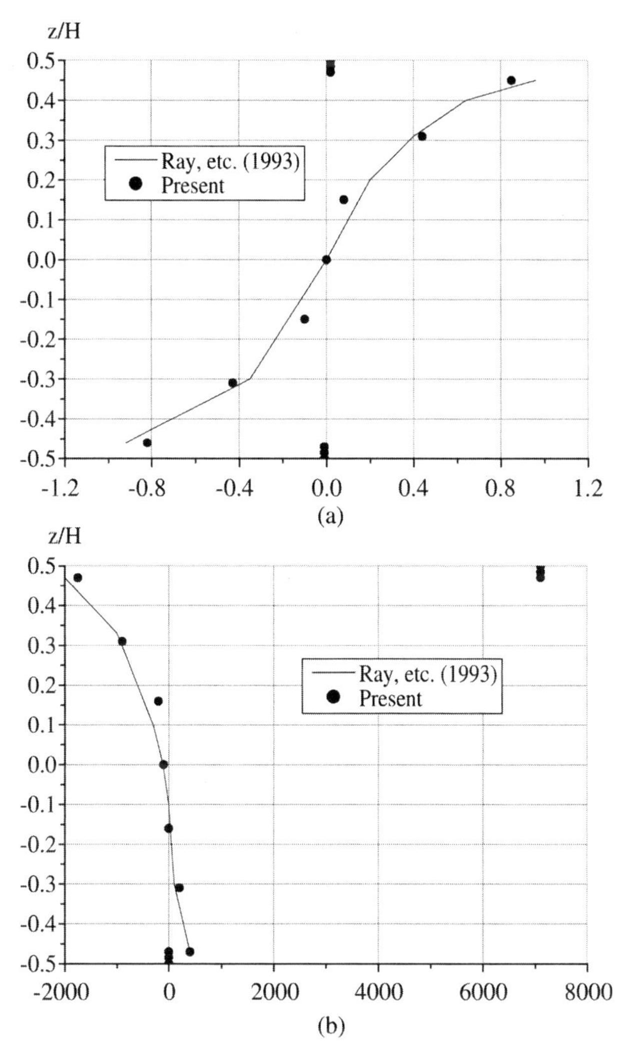

FIGURE 9.11
Distribution of the inplane normal stress across the thickness at the center of the plate (x = L/2).

to those in the sensor-actuator is nearly a hundred, there is no shear locking in the results. The results also show that the prescribed mechanical loading and surface electric potential produce more complex and highly nonlinear transverse stress distributions than mechanical loading alone. The successful numerical simulation exhibits the high performance of the present 3-D hybrid elements.

The second example is chosen because the first example is a 2-D problem. The square plate had a width and a total thickness $H = 1\ m$. The substrate is made of a [0/90] cross-ply laminate and has a thickness 0.4 H, with the piezoelectric layer thickness 0.1 H. The material properties of the substrate

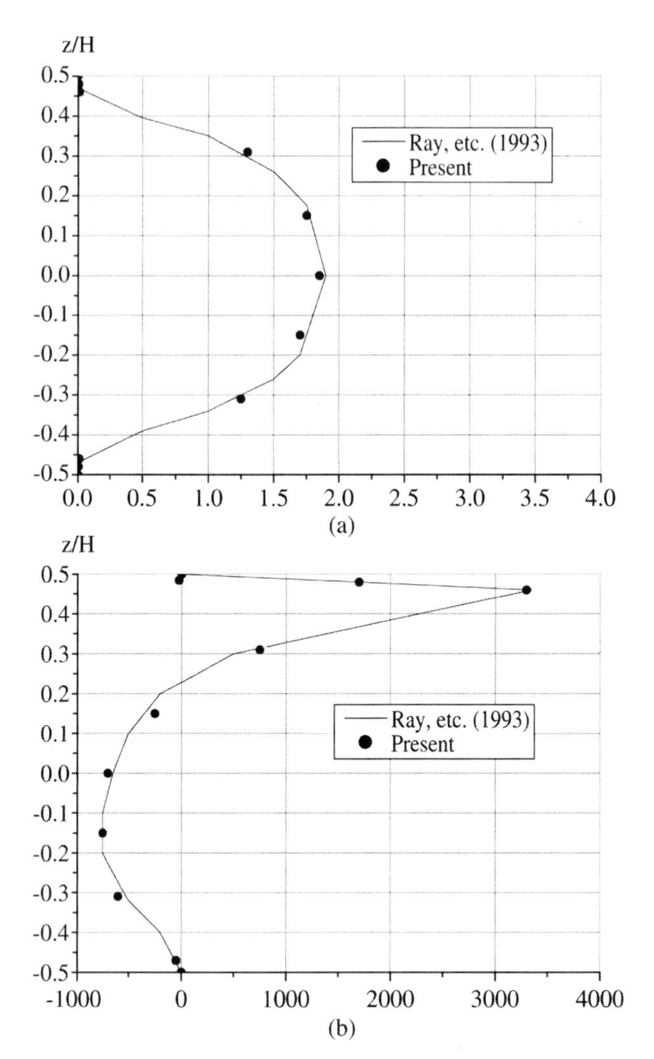

FIGURE 9.12
Distribution of the transverse shear stress across the thickness at the edge of the plate (x = 0).

and piezoelectric layers can be found in Reference 25. The applied surface electric potential takes the form of $\overline{\varphi} = \sin(\pi x / L)\sin(\pi y / L)$.

Considering the symmetry, only an 8×8 mesh in the xy plane of a quarter of a plate is used. In the thickness direction, the [0/90] cross-ply laminate is divided into [0/0/90/90] to ensure that the interface elements can be implemented on all physical interfaces. No other divisions of piezoelectric layers along the thickness are needed.

The distributions of the maximum inplane displacement, electric potential, and transverse shear stress across the thickness are shown in Figure 9.14 to Figure 9.16. The numerical results are compared with the exact solution obtained by Heyliger [25] and match very well except for some difference

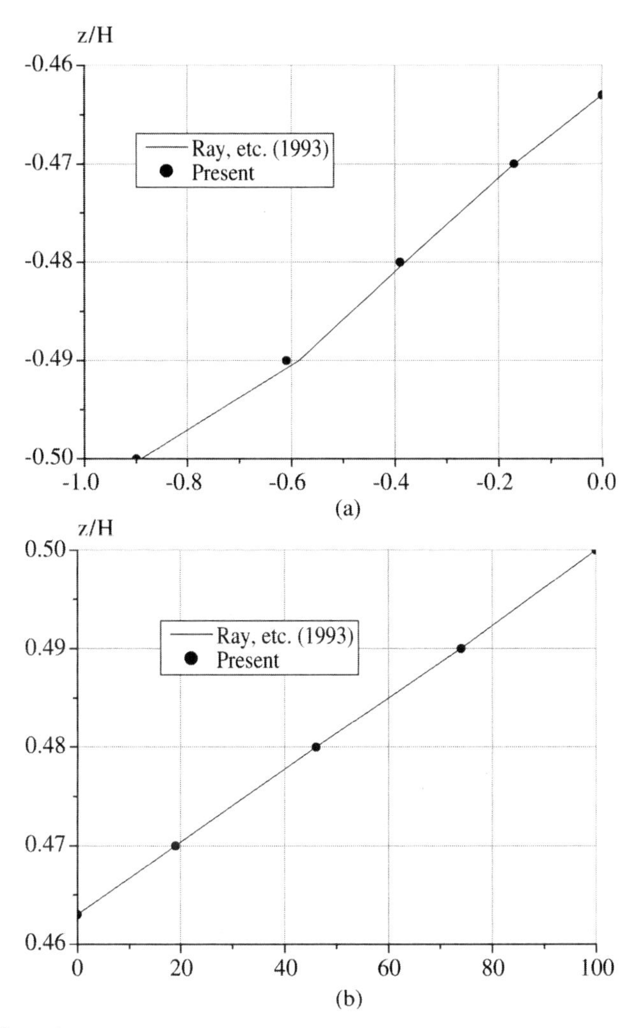

FIGURE 9.13
Distribution of the electric potential at the center of the plate ($x = L/2$) (a) across the thickness of the sensor and (b) across the thickness of the actuator.

in the transverse stress magnitude on the interface between the substrate and the piezoelectric layer.

9.2.6 Conclusion

This section has presented a variational formulation for a laminated system with a piezoelectric medium, in which the displacements, the transverse stresses, and the electric potential are taken to be individual variables. Two 3-D hybrid stress models, namely the interface element and the surface element, are well constructed, such that the interlaminar traction-continuity

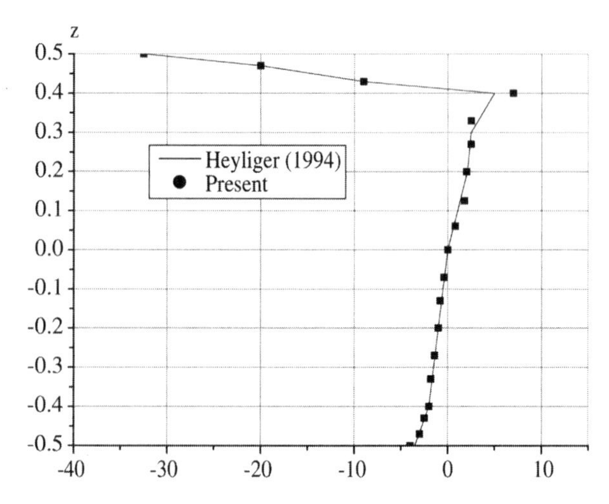

FIGURE 9.14
Distribution of the inplane displacement across the thickness at the midedge of the plane.

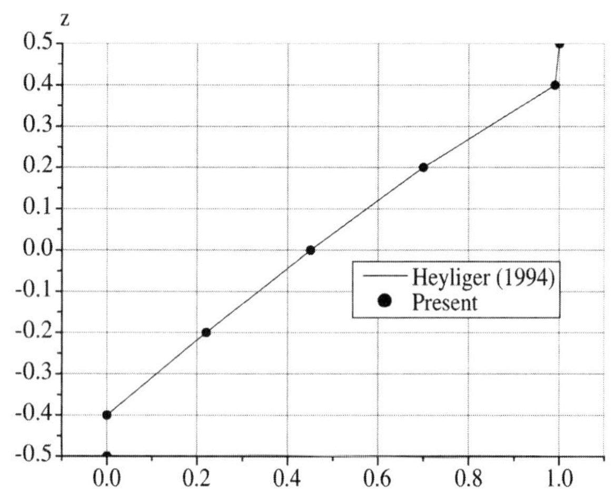

FIGURE 9.15
Distribution of the electric potential across the thickness at the center of the plate.

and surface traction-free conditions can easily be satisfied. In addition, the order of the resulting elemental stiffness matrix is independent of the number of layers.

Two numerical examples have been presented to inspect the validity of the suggested method. The numerical results show that the method presented here provides an efficient way to make an electromechanical analysis of thick laminated structures. According to the numerical results, the electric-elastic coupling may produce significant effects on the transverse stresses. This results in highly nonlinear stress distributions on the cross section of piezoelectric laminates.

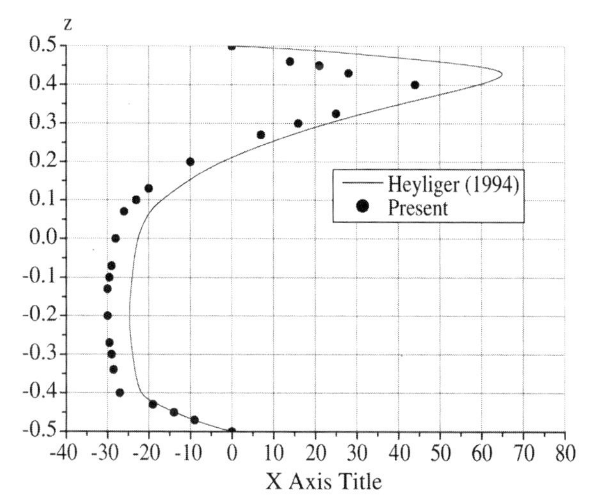

FIGURE 9.16
Distribution of the transverse shear stress across the thickness at the midedge of the plate.

9.3 Numerical Solutions on Fractures of Piezoelectric Materials

9.3.1 Introduction

Piezoelectric ceramics have been indispensable for electromechanical transducers, sensors, actuators, and adaptive structures as well as various resonators. Piezoelectric ceramics are brittle in nature. When subjected to loading, these materials can fail prematurely due to the propagation of flaws or defects induced during the manufacturing process and by the in-service electromechanical loading. Hence, it is important to understand and be able to analyze the fracture characteristics of piezoelectric materials so that reliable service life predictions of the pertinent devices can be made.

Among the theoretical studies of cracked piezoelectric bodies, permeable [42] and impermeable [32] conditions at the crack face are mostly adopted. Under Parton's permeable condition, the electric potential and the normal component of the electric displacement are assumed to be continuous across the crack. This is considered to be acceptable for slender cracks in which the separation between the faces is negligible. Deeg's impermeable condition is based on the observation that the permittivity of air and vacuum is typically three orders of magnitude lower than that of the piezoelectric materials. The crack cavity induces an insulating condition in which the normal component of the electric displacement vanishes. In other words, the crack face is charge free. However, the exact or most realistic interfacial conditions at the crack face (between material and air and vacuum) should be the continuities of the normal component of the electric displacement and the tangential component of the electric field [35].

Using Deeg's impermeable condition, Pak [39] derived the closed-form solutions for the model III crack by the semiinverse method and proposed a path-independent integral for computing the energy release rate by Eshelby's method. Later, Pak [40] used the method of distributed dislocations and electric dipoles to calculate the electro-elastic fields for antiplane and plane strain cracks in infinite piezoelectric bodies subjected to far-field electromechanical loads. Using Stroh's formalism, Park and Sun [41] presented the full-field closed-form solutions for all three modes of fracture for an insulated crack embedded in an infinite piezoelectric medium. The formulae for the energy release rates of PZT-5 and PZT-4 piezoelectric ceramics were also worked out respectively by Pak [40] and Park and Sun [41]. The electric loading always resists the crack from propagation when the energy release rate is the governing fracture criteria. This contradicts the experimental evidence that the electric field can both impede and promote crack propagation. Based on the conjecture that fracture is purely a mechanical phenomenon, Park and Sun [41] proposed to adopt the strain energy release rate as the fracture criterion. Sosa [44] again employed Deeg's impermeable condition to derive the asymptotic expressions for the electromechanical fields near the crack tip in an infinite plane strain body using complex potentials. The characteristic singularity at the crack tip for all variables was found to be $1/\sqrt{r}$, and the angular distributions of stress and electric displacement at the crack-tip zone depend not only on the geometry and loading, but also on the material properties. The results of Pak [40], Park and Sun [41], and Sosa [44] all indicate that the maximum circumferential or crack opening stress does not occur at the angular position in line with the crack when the electric loading is sufficiently large. This agrees with the experimental observation of the crack deviating from the original crack line.

Dunn [33] considered both Parton's permeable and Deeg's impermeable conditions and noted significant differences in the energy release rate. Using the exact interfacial conditions at the elliptic cavity face, Sosa and Khutoryansky [45] reexamined the closed-form expressions for the elastic and electric variables induced inside and outside the cavity in a plane-strain infinite body. They found that the electric fields at the crack tip are large but not singular. They also pointed out that the impermeable condition would lead to significant error for slender ellipses or sharp cracks.

Compared to the amount of theoretical works, numerical studies of cracked piezoelectric bodies are rare. Kumar and Singh [36] studied the double-edge cracked PZT-5 panel using the eight-node element. Although the computed $\sigma_r\sqrt{r}$, $\sigma_\theta\sqrt{r}$, $\tau_{r\theta}\sqrt{r}1$, $D_r\sqrt{r}$, and $D_\theta\sqrt{r}$ at the crack-tip zone always change with respect to the distance from the crack tip, and the analytical $1/\sqrt{r}$ -singularity cannot be confirmed, the numerical results at a particular radius are close to the theoretical predictions of Pak [40]. Later, Kumar and Singh [37] also studied a centrally cracked panel. They examined the angular distribution of the circumference stress and energy release rates at the crack-tip zone under electromechanical loading. A substantial difference was noted in the predicted directions of crack propagation when stress- and energy-based fracture criteria were used.

In this section, a four-node plane piezoelectric hybrid finite element mode that is markedly more accurate than the isoparametric one will be derived using a multifield functional that involves displacement, electric potential, stress, and electric displacement as the independent variables.

With the new finite element model, the effect of Parton's permeable and Deeg's impermeable conditions on the crack-tip solution is studied. The path-independent J integral is also calculated to examine the influence of the electric loading on the energy release rate.

9.3.2 Constitutive Relations and Variational Functional

In most publications [36, 37, 39–41], constitutive constants for transverse isotropic piezoelectric materials like PZT-4 and PZT-5 are given with respect to the following relations:

$$\begin{Bmatrix} \sigma_{11} \\ \sigma_{22} \\ \sigma_{33} \\ \sigma_{32} \\ \sigma_{31} \\ \sigma_{12} \end{Bmatrix} = \begin{bmatrix} C_{11} & C_{12} & C_{13} & 0 & 0 & 0 \\ C_{12} & C_{11} & C_{13} & 0 & 0 & 0 \\ C_{13} & C_{13} & C_{33} & 0 & 0 & 0 \\ 0 & 0 & 0 & C_{33} & 0 & 0 \\ 0 & 0 & 0 & 0 & C_{44} & 0 \\ 0 & 0 & 0 & 0 & 0 & (C_{11}-C_{12})/2 \end{bmatrix} \begin{Bmatrix} \varepsilon_{11} \\ \varepsilon_{22} \\ \varepsilon_{33} \\ 2\varepsilon_{32} \\ 2\varepsilon_{31} \\ 2\varepsilon_{12} \end{Bmatrix} - \begin{bmatrix} 0 & 0 & e_{31} \\ 0 & 0 & e_{31} \\ 0 & 0 & e_{33} \\ 0 & e_{15} & 0 \\ e_{15} & 0 & 0 \\ 0 & 0 & 0 \end{bmatrix} \begin{Bmatrix} E_1 \\ E_2 \\ E_3 \end{Bmatrix} \tag{9.48}$$

$$\begin{Bmatrix} D_1 \\ D_2 \\ D_3 \end{Bmatrix} = \begin{bmatrix} 0 & 0 & 0 & 0 & e_{15} & 0 \\ 0 & 0 & 0 & e_{15} & 0 & 0 \\ e_{31} & e_{31} & e_{33} & 0 & 0 & 0 \end{bmatrix} \begin{Bmatrix} \varepsilon_{11} \\ \varepsilon_{22} \\ \varepsilon_{33} \\ 2\varepsilon_{32} \\ 2\varepsilon_{31} \\ 2\varepsilon_{12} \end{Bmatrix} + \begin{bmatrix} \in_{11} & 0 & 0 \\ 0 & \in_{11} & 0 \\ 0 & 0 & \in_{33} \end{bmatrix} \begin{Bmatrix} E_1 \\ E_2 \\ E_3 \end{Bmatrix} \tag{9.49}$$

in which the 1-2 plane is the plane symmetry and 3 is the poling direction, σ_{ij} and ε_{ij} are the stress and the strain tensors, E_i and D_i are the electric field strength and electric displacement vectors, C_{ij} is the material elasticity constants measured at constant electric field, $\in_{ij}$ is the dielectric constants measured at constant strain, and e_{ij} is the piezoelectric constants. After incorporating the plane strain assumptions and taking the y-axes as the poling direction, the constitutive relations can be degenerated and expressed in the following forms:

$$\begin{Bmatrix} \varepsilon_x \\ \varepsilon_y \\ \gamma_{xy} \end{Bmatrix} = \begin{bmatrix} S_{11} & S_{12} & 0 \\ S_{11} & S_{22} & 0 \\ 0 & 0 & S_{33} \end{bmatrix} \begin{Bmatrix} \sigma_x \\ \sigma_y \\ \tau_{xy} \end{Bmatrix} + \begin{bmatrix} 0 & g_{21} \\ 0 & g_{22} \\ g_{13} & 0 \end{bmatrix} \begin{Bmatrix} D_x \\ D_y \end{Bmatrix} \quad \text{or } \varepsilon = \mathbf{S}\sigma + \mathbf{g}^T\mathbf{D} \tag{9.50}$$

$$\begin{Bmatrix} E_x \\ E_y \end{Bmatrix} = -\begin{bmatrix} 0 & 0 & g_{13} \\ g_{21} & g_{22} & 0 \end{bmatrix} \begin{Bmatrix} \sigma_x \\ \sigma_y \\ \tau_{xy} \end{Bmatrix} + \begin{bmatrix} f_1 & 0 \\ 0 & f_2 \end{bmatrix} \begin{Bmatrix} D_x \\ D_y \end{Bmatrix} \quad \text{or } \mathbf{E} = -\mathbf{g}\sigma + \mathbf{f}\mathbf{D} \tag{9.51}$$

in which all symbols are self-defined. Furthermore, the strain-displacement relation and electric field–electric potential relation are

$$\begin{Bmatrix} \varepsilon_x \\ \varepsilon_y \\ \gamma_{xy} \end{Bmatrix} = \begin{bmatrix} \partial/\partial x & 0 \\ 0 & \partial/\partial y \\ \partial/\partial y & \partial/\partial x \end{bmatrix} \begin{Bmatrix} u \\ \upsilon \end{Bmatrix} \text{or } \varepsilon = \mathbf{D}_m\mathbf{u} \tag{9.52}$$

$$\begin{Bmatrix} E_x \\ E_y \end{Bmatrix} = -\begin{bmatrix} \partial/\partial x \\ \partial/\partial y \end{bmatrix} \varphi \text{ or } \mathbf{E} = -\mathbf{D}_e\varphi \tag{9.53}$$

where φ is the electric potential and all other symbols are self-defined. The following multifield variational functional for plane piezoelectricity is considered [34]:

$$\begin{aligned} &\Pi(\mathbf{u},\varphi,\sigma,\mathbf{D}) \\ &= \int_V (h_m + \sigma^T\mathbf{D}_m\mathbf{u} + \mathbf{D}^T\mathbf{D}_e\varphi)d\upsilon - \int_{S_\sigma} \overline{\mathbf{T}}^T\mathbf{u}ds - \int_{S_{\overline{\omega}}} \overline{\omega}^T\varphi ds \end{aligned} \tag{9.54}$$

where V denotes the domain of the 2-D body, $h_m = -(1/2)\sigma^T\mathbf{S}\sigma - \sigma^T\mathbf{g}^T\mathbf{D} +$ $\mathbf{S}\sigma - \sigma^T\mathbf{g}^T\mathbf{D} + (1/2)\mathbf{D}^T\mathbf{f}\mathbf{D}$ is the mechanical enthalpy, S_σ denotes the boundary portion that the prescribed traction $\overline{\mathbf{T}}$ is acting on, and S_ω denotes the boundary portion that the prescribed surface charge density $\overline{\omega}$ is acting on.

Euler's equations for the functional are:

$$\text{Stress equilibrium condition: } \mathbf{D}_m^T\sigma = \mathbf{0} \quad \text{in } V \tag{9.55}$$

$$\text{Charge conservation condition: } \mathbf{D}_e^T\mathbf{D} = \mathbf{0} \quad in\ V \tag{9.56}$$

$$\text{Constitutive relations: } \mathbf{D}_m\mathbf{u} = \mathbf{S}\sigma + \mathbf{g}^T\mathbf{D}\ ,\ \mathbf{D}_e\varphi = -\mathbf{g}\sigma + \mathbf{f}\mathbf{D} \quad in\ V \tag{9.57}$$

$$\text{Mechanical natural boundary condition: } \mathbf{n}_m\sigma = \overline{\mathbf{T}} \quad \text{at } S_\sigma \tag{9.58}$$

$$\text{Electric natural boundary condition: } \mathbf{n}_m\mathbf{D} = -\overline{\omega} \quad \text{at } S_\varphi \tag{9.59}$$

where

$$\mathbf{n}_m = \begin{bmatrix} n_x & 0 & n_y \\ 0 & n_y & n_x \end{bmatrix} \text{ and } \mathbf{n}_e = \begin{bmatrix} n_x & n_y \end{bmatrix} \tag{9.60}$$

in which n_x and n_y constitute the outward normal of the domain boundary. The subsidiary conditions of the functional are the strain-displacement relation, the electric field–electric potential relation, the mechanical essential-boundary condition (displacement prescribed), and the electric essential-boundary condition (electric potential prescribed).

9.3.3 Piezoelectric Hybrid Element: PZT-Q4

Four-node elements are very popular in finite element analysis due to their good balance of accuracy, computational cost, and small bandwidth and frontwidth. However, the standard isoparametric element is rather poor. In this section, a four-node plane piezoelectric hybrid element will be developed based on the functional given in Equation 9.54 as shown in Figure 9.17 [38, 79].

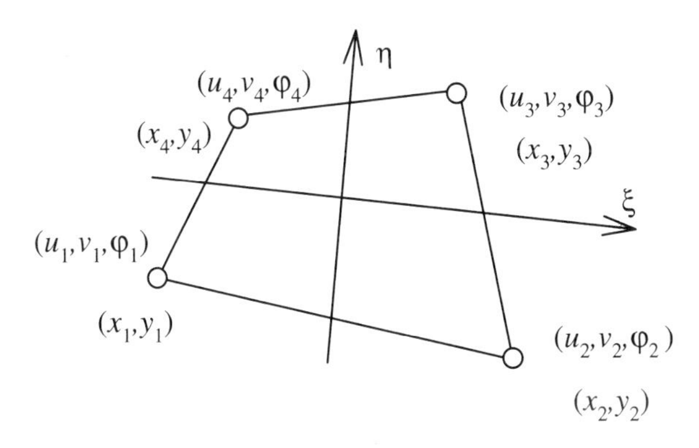

FIGURE 9.17
Description of the four-node piezoelectric plane element PZT-Q4.

The element nodal DOFs include the two displacement components and the electric potential. These and the coordinates are interpolated, that is,

$$\begin{Bmatrix} u \\ \upsilon \\ \varphi \end{Bmatrix} = \frac{1}{4}\sum_{i=1}^{4}(1+\xi_i\xi)(1+\eta_i\eta)\begin{Bmatrix} u_i \\ \upsilon_i \\ \varphi_i \end{Bmatrix} = \mathbf{Nq}$$

and

$$\begin{Bmatrix} x \\ y \end{Bmatrix} = \frac{1}{4}\sum_{i=1}^{4}(1+\xi_i\xi)(1+\eta_i\eta)\begin{Bmatrix} x_i \\ y_i \end{Bmatrix} \tag{9.61}$$

where the variables with subscripts denote their nodal values, and ξ and η bounded by -1 and $+1$ are the natural coordinates. Moreover,

$$\mathbf{N} = \frac{1}{4}[(1-\xi)(1-\eta)\mathbf{I}_3 \quad (1+\xi)(1-\eta)\mathbf{I}_3 \quad (1+\xi)(1+\eta)\mathbf{I}_3 \quad (1-\xi)(1+\eta)\mathbf{I}_3] \tag{9.62}$$

$$\mathbf{q} = \{u_1, \upsilon_1, \varphi_1, \cdots, u_4, \upsilon_4, \varphi_4\}^T \tag{9.63}$$

The stress field of the Pian–Sumihara element is employed [43]:

$$\begin{Bmatrix} \sigma_x \\ \sigma_y \\ \tau_{xy} \end{Bmatrix} = \begin{bmatrix} 1 & 0 & 0 & a_3^2\xi & a_1^2\eta \\ 0 & 1 & 0 & b_3^2\xi & b_1^2\eta \\ 0 & 1 & 0 & a_3b_3\xi & a_1b_1\eta \end{bmatrix}\begin{Bmatrix} \beta_1 \\ \vdots \\ \beta_5 \end{Bmatrix} = \mathbf{P}_m\beta_m \tag{9.64}$$

in which the geometric parameters are

$$\begin{bmatrix} a_1 & b_1 \\ a_2 & b_2 \\ a_3 & b_3 \end{bmatrix} = \frac{1}{4}\begin{bmatrix} -1 & 1 & 1 & -1 \\ 1 & -1 & 1 & -1 \\ -1 & -1 & 1 & 1 \end{bmatrix}\begin{bmatrix} x_1 & y_1 \\ x_2 & y_2 \\ x_3 & y_3 \\ x_4 & y_4 \end{bmatrix} \tag{9.65}$$

The element stress (Equation 9.64) can be degenerated from uncoupled complete linear expansions by the mechanical energy consistency condition [14, 47].

The element electric displacement, it is initially assumed as a linear function:

$$\mathbf{D} = \begin{Bmatrix} D_x \\ D_y \end{Bmatrix} = \begin{bmatrix} 1 & 0 & \xi & \eta & 0 & 0 \\ 0 & 1 & 0 & 0 & \xi & \eta \end{bmatrix} \begin{Bmatrix} \beta_1 \\ \vdots \\ \beta_6 \end{Bmatrix} = \mathbf{D}_c + \mathbf{D}_h \tag{9.66}$$

where $\mathbf{D}_c$ and $\mathbf{D}_h$ are the constant mode and the high-order mode, respectively. To improve the numerical performance of the piezoelectric hybrid element, the electric energy consistency condition [38, 79] will be introduced into $\mathbf{D}_h$ such that two β-parameters will be eliminated from Equation 9.66, and the desirable electric displacement can be formulated as

$$\begin{Bmatrix} D_x \\ D_y \end{Bmatrix} = \begin{bmatrix} 1 & 0 & a_3\xi & a_1\eta \\ 0 & 1 & b_3\xi & b_1\eta \end{bmatrix} \begin{Bmatrix} \beta_6 \\ \vdots \\ \beta_9 \end{Bmatrix} = \mathbf{P}_e \beta_e \tag{9.67}$$

For finite element formulation, the functional given in Equation 9.54 is rewritten as

$$\Pi(\mathbf{u}, \varphi, \sigma, \mathbf{D}) = \sum_{\text{elements}} \Pi^e(\mathbf{u}^e, \varphi^e, \sigma^e, \mathbf{D}^e) \tag{9.68}$$

where

$$\Pi^e(\mathbf{u}^e, \varphi^e, \sigma^e, \mathbf{D}^e) = \int_{V^e} (h_m^e + (\sigma^e)^T D_m \mathbf{u}^e + (\mathbf{D}^e)^T D_e \varphi^e)\, dv - \int_{S_\sigma^e} \overline{\mathbf{T}}^T \mathbf{u}^e ds - \int_{S_\omega^e} \overline{\omega}^T \varphi^e ds \tag{9.69}$$

is the element-wise functional. The superscript e denotes an individual element. After invoking Equation 9.51, Equation 9.64, and Equation 9.67, the element-wise functional becomes

$$\Pi^e = -\frac{1}{2} \begin{Bmatrix} \beta_m \\ \beta_e \end{Bmatrix}^T \mathbf{H} \begin{Bmatrix} \beta_m \\ \beta_e \end{Bmatrix} + \begin{Bmatrix} \beta_m \\ \beta_e \end{Bmatrix}^T \mathbf{G}\mathbf{q} - \text{load terms} \tag{9.70}$$

where

$$\mathbf{H} = \int_{V^e} \begin{bmatrix} \mathbf{P}_m & \mathbf{0}_{3\times 4} \\ \mathbf{0}_{2\times 5} & \mathbf{P}_m \end{bmatrix}^T \begin{bmatrix} \mathbf{S} & \mathbf{g}^T \\ \mathbf{g} & -\mathbf{f} \end{bmatrix} \begin{bmatrix} \mathbf{P}_m & \mathbf{0}_{3\times 4} \\ \mathbf{0}_{2\times 5} & \mathbf{P}_m \end{bmatrix} dv$$

$$\mathbf{G} = \int_{V^e} \begin{bmatrix} \mathbf{P}_m & \mathbf{0}_{3\times 4} \\ \mathbf{0}_{2\times 5} & \mathbf{P}_m \end{bmatrix}^T \left(\begin{bmatrix} D_m & \mathbf{0}_{3\times 1} \\ \mathbf{0}_{2\times 2} & D_m \end{bmatrix} \mathbf{N} \right) dv$$

As the stationary condition of Π we have

$$\mathbf{H} \begin{Bmatrix} \beta_m \\ \beta_e \end{Bmatrix} = \mathbf{Gq} \quad \text{and thus } \Pi^e = \frac{1}{2}\mathbf{q}^T(\mathbf{G}^T\mathbf{H}^{-1}\mathbf{G})\mathbf{q} - \text{load terms} \tag{9.71}$$

Hence, the generalized element matrix is $\mathbf{G}^T\mathbf{H}^{-1}\mathbf{G}$. The element stress and electric displacement can be retrieved by using Equation 9.64, Equation 9.67, and Equation 9.71 after solving $\mathbf{q}$.

To demonstrate the relative accuracy of the hybrid element PZT-Q4 and the standard four-node (Q4) element, the PZT-4 plane-strain cantilever depicted in Figure 9.18 is considered. The cantilever is modeled by 10 elements and is subjected to an end moment. Electric potential along the bottom face is set to zero. While the analytic bending stress is a linear function of y, both the analytic deflection and electric potential are quadratic functions of y. Table 9.4 lists the computed deflection and electric potential at C as well as the computed bending stresses at A and B. The PZT-Q4 element model is much more accurate than the standard one.

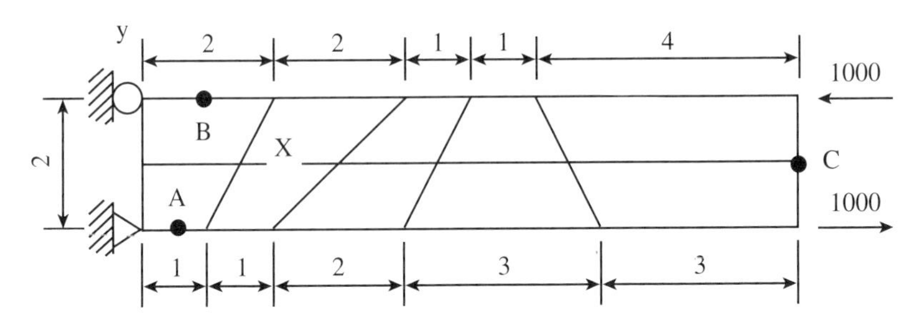

FIGURE 9.18
Unit thickness PZT-4 plane-strain cantilever beam under pure bending.

TABLE 9.4

Predictions of Stress, Deflection, and Electric Potential for the Cantilever Beam Shown in Figure 9.18

	v^C	φ^C	σ_x^A	σ_x^B	τ_{xy}^A	τ_{xy}^B
Q4	0.755×10^{-6}	−13.84	2843	−2359	−97.65	197.7
PZT-Q4	1.141×10^{-6}	−27.12	3012	−3021	6.960	−29.62
Theory	1.184×10^{-6}	−26.67	3000	−3000	0	0

The element computational cost can be reduced by replacing ξ and η in Equation 9.64 and Equation 9.67 with

$$\xi' = \xi - \frac{a_1 b_2 - a_2 b_1}{3(a_1 b_3 - a_3 b_1)} \quad \text{and} \quad \eta' = \eta - \frac{a_2 b_3 - a_3 b_2}{3(a_1 b_3 - a_3 b_1)} \tag{9.72}$$

respectively. The volume integrals of the above two terms are equal to zero. This induces large sparsity in the H-matrix and thus saves a substantial portion of algebraic operations [46].

9.3.4 Numerical Analyses: Crack-Tip Field Simulation

In this section, the effect of permeable and impermeable conditions on the crack-tip solution is studied with the previously derived finite element model. From here onward, for simplicity, permeable and impermeable conditions abbreviate, respectively, Parton's permeable and Deeg's impermeable conditions. Pure mechanical, pure electric, and mixed loading are considered. The path-independent J integral is also calculated to examine the influence of the far-field electromechanical loading on the energy release rate. The results are compared with the previously reported theoretical solutions.

To mimic a cracked infinite plate subjected to far-field electromechanical loading using the finite element method, we consider the centrally cracked plane strain panel depicted in Figure 9.19. The panel dimension is $2w \times 2w$, the half crack length is a, and $w/a = 8$. The y-direction far-field stress r1 and electric displacement 1 are considered. The poling direction is also aligned with the y-axis. It has been confirmed that the crack-tip finite element solution remains virtually unchanged for larger w/a. Owing to symmetry, a quarter of the panel is modeled. Figure 9.20 and Figure 9.21 give the details of the finite element mesh. The mesh involves 702 elements and 760 nodes. To capture the strong gradients expected at the crack-tip zone, high mesh density is employed as portrayed in Figure 9.21. The side length of the elements at the crack tip is around $10^{-5}a$. Figure 9.20 also shows the three different paths along which J integrals are computed. The boundary conditions at the crack faces for the permeable and impermeable conditions are

$$\text{Impermeable: } \sigma_y = \tau_{xy} = 0 \quad \text{and} \quad D_y = 0$$

$$\text{Permeable: } \sigma_y = \tau_{xy} = 0\,,\ \varphi\,|_{y=0^-} = \varphi\,|_{y=0^+} \quad \text{and} \quad D_y\,|_{y=0^-} = D_y\,|_{y=0^+}$$

9.3.4.1 *Pure Mechanical Loading*

In this case, only the far field stress r1 is acting. Figure 9.22 and Figure 9.23 show the angular distributions of $\sigma_r\sqrt{r}$, $\sigma_\theta\sqrt{r}$, and $\tau_{r\theta}\sqrt{r}$ for PZT-4 and PZT-5

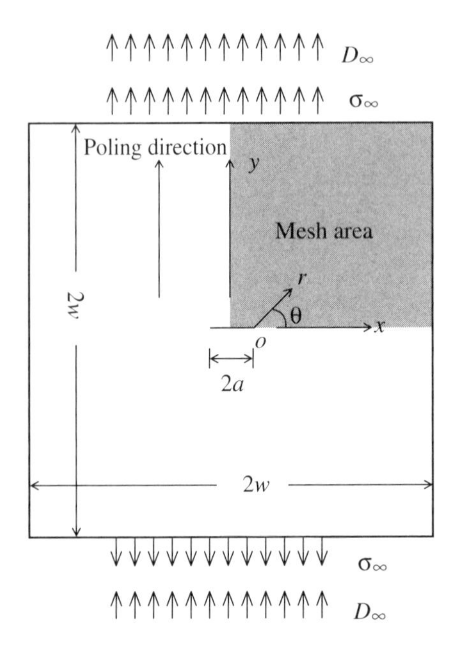

FIGURE 9.19
Centrally cracked piezoelectric tensile plane strain panel of dimension $w \times w$, half-crack length a and $w/a = 8$.

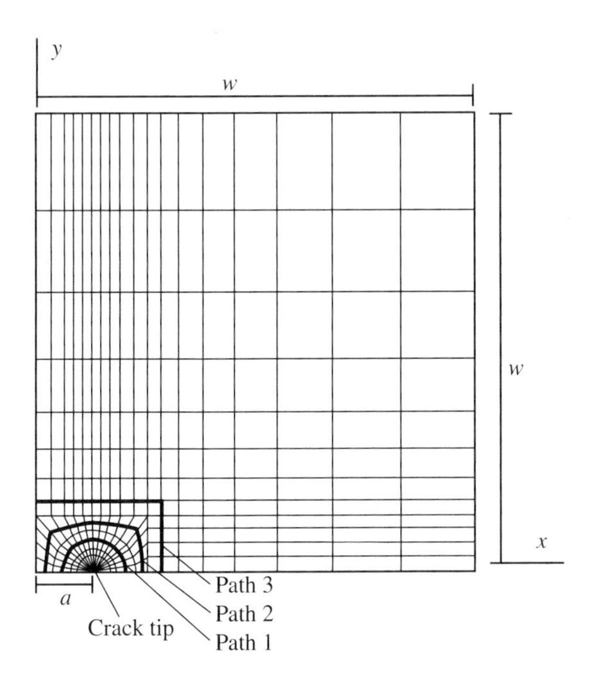

FIGURE 9.20
Finite element mesh for the upper right-hand quarter of the panel shown in Figure 9.18. J integrals are computed along the three highlighted paths.

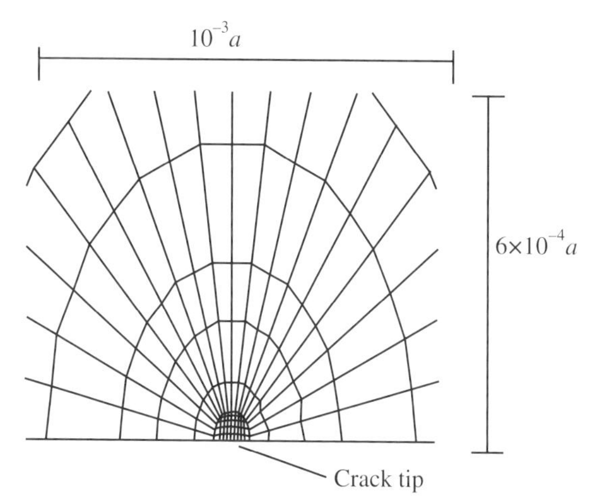

FIGURE 9.21
Enlarged view of the finite element mesh at the crack tip.

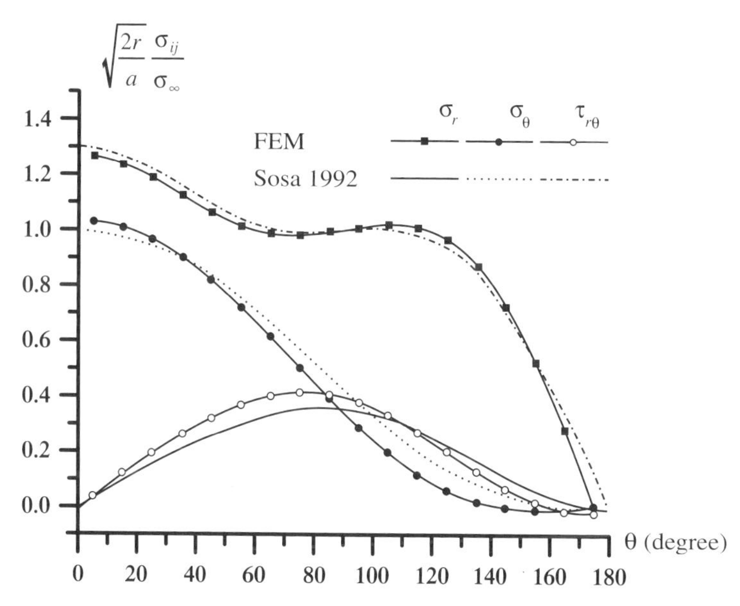

FIGURE 9.22
Angular distributions of stresses for PZT-4 under mechanical loading and impermeable conditions.

at the crack-tip region using the impermeable condition. Material constants for PZT-4 and PZT-5 can be found in Table 9.5. Figure 9.24 and Figure 9.25 give the angular distributions of $D_r\sqrt{r}$ and $D_\theta\sqrt{r}$. The theoretical results of Pak [40] and Sosa [44] are also plotted in the figures for comparison. The results of Park and Sun [41] are not shown, as their normalizing factors were not specified. All of our data are taken at the element origins closest

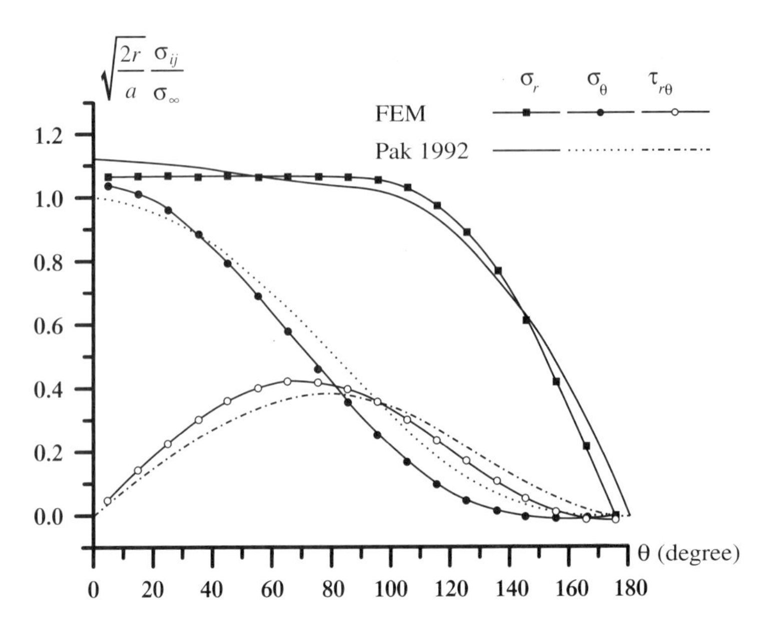

FIGURE 9.23
Angular distributions of stresses for PZT-5 under pure mechanical loading and impermeable conditions.

TABLE 9.5
Material Constants for PZT-4 [41] and PZT-5 [40], where C_{ij} s Are in 10 GN/m², e_{ij} s Are in C/m², and $\in_{ij}$ s Are in μ C/V/m

	C_{11}	C_{12}	C_{13}	C_{33}	C_{44}	e_{31}	e_{33}	e_{15}	ε_{11}	ε_{33}
PZT-4	13.9	7.78	7.43	11.3	2.56	–6.98	13.84	13.44	6.00	5.47
PZT-5	12.6	5.5	5.3	11.7	3.53	–6.5	23.3	17.0	15.1	13.0

to $r/a = 3.86\times10^{-3}$. For $r/a = 10^{-4}$ to 10^{-2}, the predicted $D_r\sqrt{r}$ and $D_\theta\sqrt{r}$ remain practically unchanged. In other words, the $1/\sqrt{r}$ -singularities of D_r and D_θ are clearly verified. However, slight changes in $\sigma_r\sqrt{r}$, $\sigma_\theta\sqrt{r}$, and $\tau_{r\theta}\sqrt{r}$ are observed for $r/a = 10^{-4} - 10^{-2}$. Nevertheless, the $1/\sqrt{r}$ -singularities of σ_r, σ_θ, and $\tau_{r\theta}$ can still be confirmed. These present results are close to the analytical solutions of Pak [40] and Sosa [44]. However, the $\sigma_\theta\sqrt{r}$ -distribution is quite different from that presented by Park and Sun (see Reference 41). The latter is similar to the $\sigma_r\sqrt{r}$ -distribution. The invariancy of $D_r\sqrt{r}$, $D_\theta\sqrt{r}$, $\sigma_r\sqrt{r}$, $\sigma_\theta\sqrt{r}$, and $\tau_{r\theta}\sqrt{r}$ within the above range of r is in contrast with the numerical results of Kumar and Singh [36].

Under the permeable condition, singularities are not noted in D_r and D_h. The stress predictions are very close to the one computed under the impermeable condition. The difference is of the order of 10^3 and therefore not repeated.

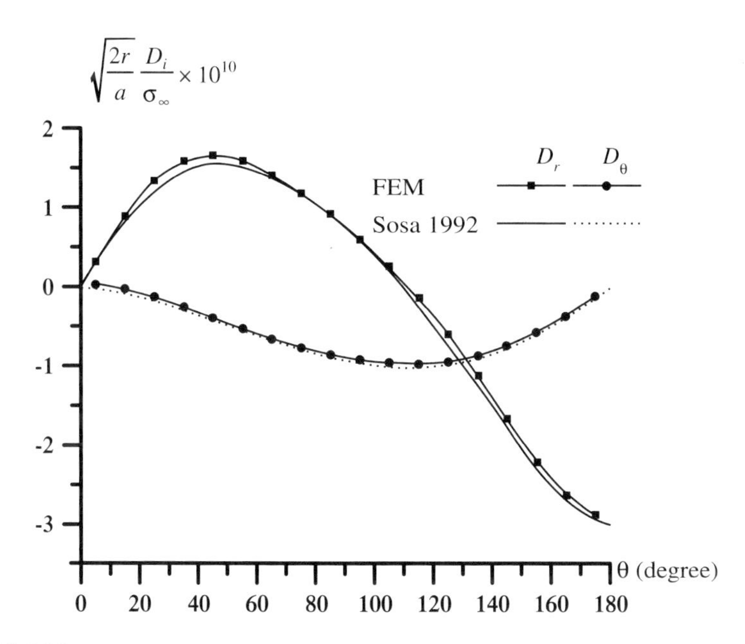

FIGURE 9.24
Angular distribution of electric displacements for PZT-4 under pure mechanical loading and impermeable conditions.

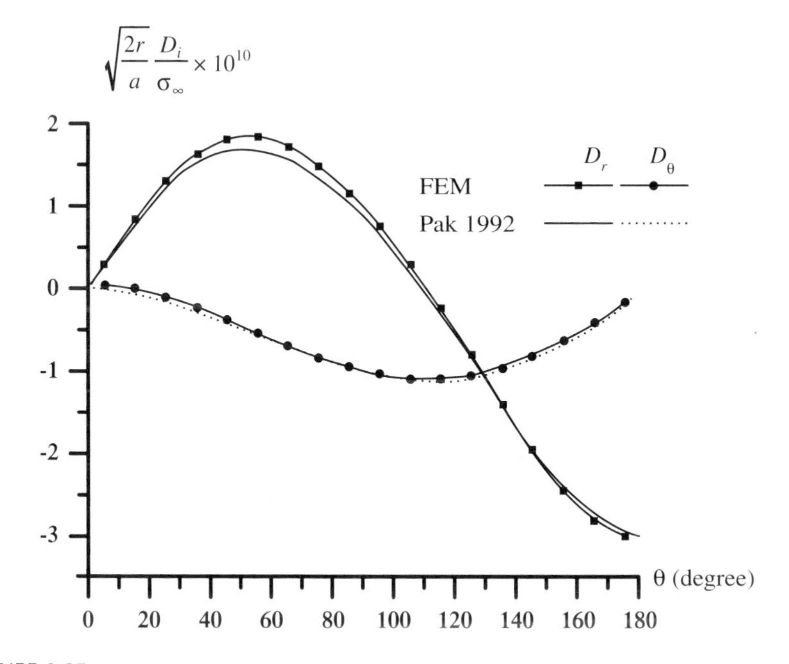

FIGURE 9.25
Angular distribution of electric displacements for PZT-5 under pure mechanical loading and impermeable conditions.

To examine the energy release rate, the following J integral derived by Pak [39] is computed for the three paths shown in Figure 9.20:

$$J = \int_{\Gamma} (h_e n_x - \sigma_{ij} n_j u_{i,1} + D_i n_i E_1) ds \tag{9.73}$$

where h_e is the electric enthalpy, n_x and n_y constitute the unit outward normal to path C, and repeated subscripts denote the summation over x and y. For linear material, h_e has the same value as the mechanical enthalpy h_m defined in Equation 9.54. The J integral results are listed in Table 9.6. The impermeable and permeable conditions do not affect the path independence of the integral. However, the permeable condition leads to a higher J value. If the energy release rate is the governing fracture criterion, the permeable crack is more vulnerable than the impermeable crack under pure mechanical loading.

9.3.4.2 *Pure Electric Loading*

In pure electric loading, only the far-field electric displacement D_∞ is acting. Figure 9.26 and Figure 9.27 show the angular distributions of $\sigma_r\sqrt{r}$, $\sigma_\theta\sqrt{r}$, and $\tau_{r\theta}\sqrt{r}$ for PZT-4 and PZT-5 at the crack-tip region under the impermeable condition. Figure 9.28 and Figure 9.29 give the angular distributions of $D_r\sqrt{r}$ and $D_\theta\sqrt{r}$. Most of the predicted stresses are close to the analytical solutions of Pak [40] and Sosa [44], both in distribution and magnitude, except that there appear to be magnitude differences in $\sigma_r\sqrt{r}$ for PZT-4 and in $\sigma_\theta\sqrt{r}$ for PZT-5. As under the pure mechanical loading, the predicted $\sigma_\theta\sqrt{r}$ is quite different from that of Park and Sun (Figure 9.19 in Reference 41). As θ increases, Park and Sun's circumferential stress changes from positive to negative, but one is basically negative for the entire range of θ. Table 9.7 lists the J integrals that are negative and agree with the results of Pak [39, 40] and Park and Sun [41]. In other words, the electric loading always impedes crack propagation if energy release rate is the governing fracture criterion.

Under the permeable condition, the computed electric displacements are uniform over the finite element mesh, while the stress is negligible compared to that of the impermeable model (the magnitude ratio $\approx 10^{-9}$). Crack-tip

TABLE 9.6

Computed J Integrals (in $10^{-11}a$ N/m) for Pure Mechanical Loading ($\sigma_\infty = 1$ N/m^2) (see Figure 9.19 and Figure 9.20)

Crack face idealization	PZT-4			PZT-5		
	Path 1	Path 2	Path 3	Path 1	Path 2	Path 3
Impermeable	2.869	2.869	2.871	2.646	2.644	2.655
Permeable	3.768	3.758	3.737	3.216	3.209	3.209

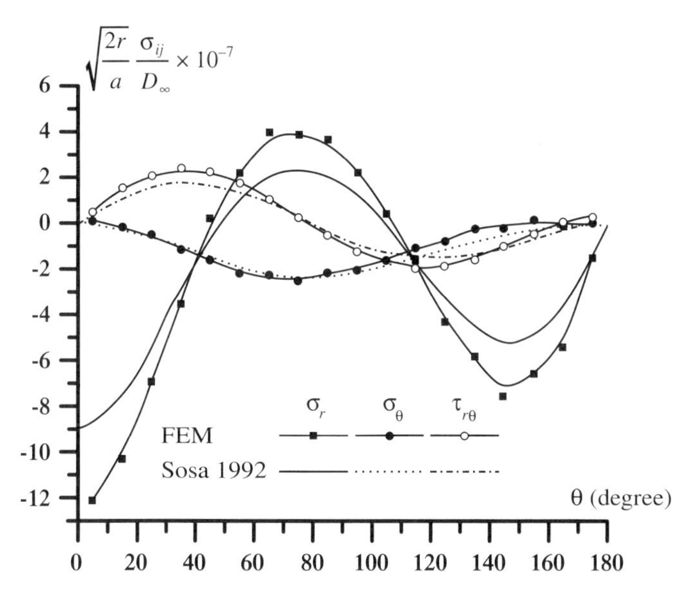

FIGURE 9.26
Angular distribution of stresses of PZT-4 under pure electric loading and impermeable conditions.

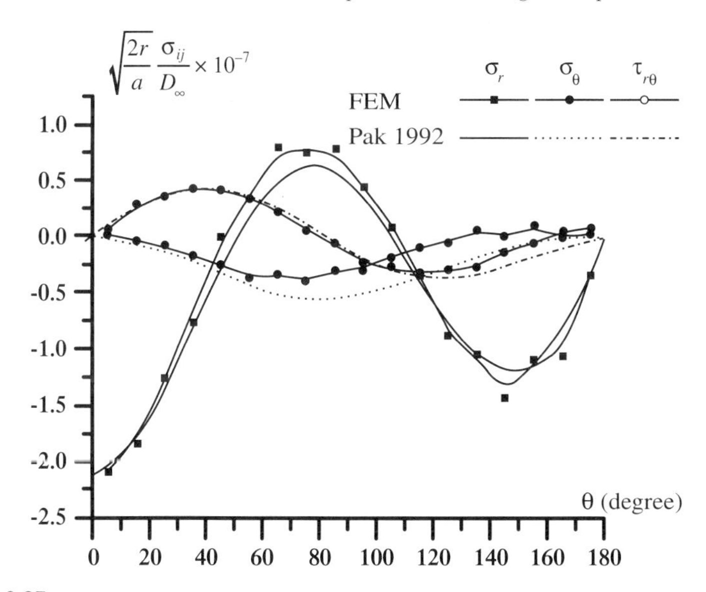

FIGURE 9.27
Angular distribution of stresses for PZT-5 under pure electric loading and impermeable conditions.

singularity is not noted, and thus, the computed J integrals are practically zero. These agree with the conclusions of Dunn [33] and Sosa and Khutoryansky [45] in the sense that the electric loading has no effect on crack propagation under the permeable condition regardless of whether a stress- or energy-based fracture criterion is adopted.

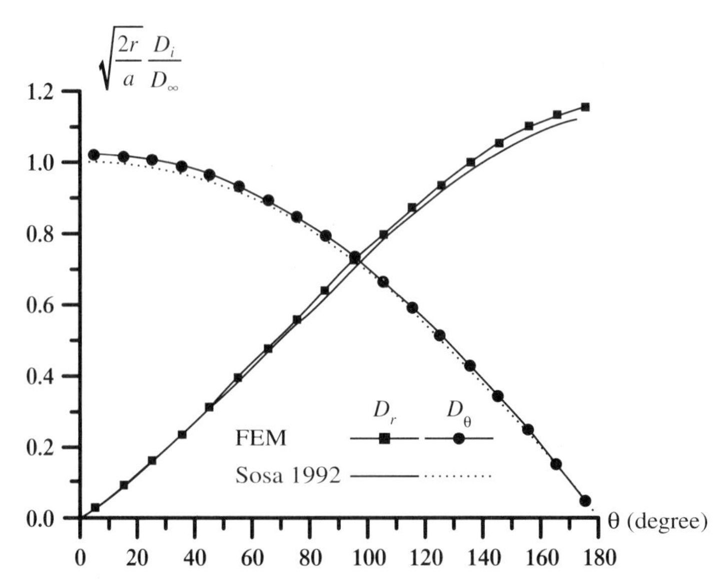

FIGURE 9.28
Angular distribution of electric displacements for PZT-4 under pure electric loading and impermeable conditions.

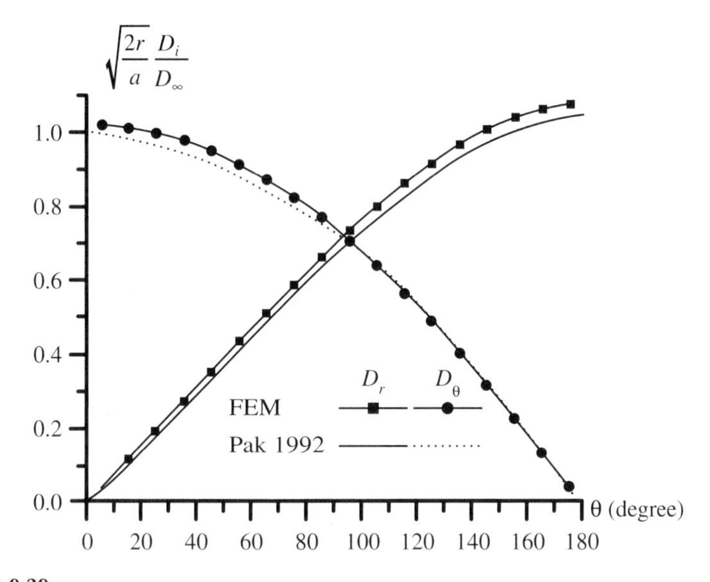

FIGURE 9.29
Angular distribution of electric displacements for PZT-5 under pure electric loading and impermeable conditions.

9.3.4.3 Mixed Loading

In mixed loading, the ratio of the far-field electric displacement D1 to the far field stress r1 is varied. r1 is always kept positive to avoid crack closure.

TABLE 9.7

Computed J integrals (in $10^{-11}a$ N/m) for Pure Electric Loading ($D_\infty = 1\mu$ C/m²) (see Figure 9.19 and Figure 9.20)

Crack face idealization	PZT-4 Path 1	PZT-4 Path 2	PZT-4 Path 3	PZT-5 Path 1	PZT-5 Path 2	PZT-5 Path 3
Impermeable	−13.81	−13.77	−13.87	−7.229	−7.208	−7.258
Permeable	10^{-10}	10^{-10}	10^{-10}	10^{-10}	10^{-10}	10^{-10}

The stress induced by the electric loading is negligible under the permeable condition, and therefore, only the impermeable condition is considered here. To compare the result with that of Sosa [44], the following D_∞ / σ_∞ is considered for PZT-4:

$$\text{PZT-4: } D_\infty / \sigma_\infty = \pm 10^{-9}, \pm 10^{-8}, \pm 10^{-8}$$

For linear materials, the results for mixed loading can be obtained by linear superposition of the predictions obtained from pure mechanical loading and electric loading. As the crack opening or circumferential stress is mostly concerned from the fracture point view, it is plotted for examining whether the crack will deviate from its initial direction. Figure 9.30 and Figure 9.31 show the stress in PZT-4 for positive and negative D_∞ s, respectively. All solutions indicate that the maximum circumference stress will shift away from $\theta = 0$ when a large enough negative D_∞ is applied.

Energy release rate is another concern in fracture mechanics. Pak [40] and Park and Sun [41] gave the following formulae for the energy release rates of PZT-5 and PZT-4:

$$G_{PZT-5} = \frac{a}{8}(2.0189 \times 10^{-10} \sigma_\infty^2 + 0.3258 D_\infty / \sigma_\infty - 5.7528 \times 10^8 D_\infty^2) N/m$$

$$G_{PZT-4} = \frac{\pi a}{2}(1.8270 \times 10^{-11} \sigma_\infty^2 + 0.04510 D_\infty / \sigma_\infty - 8.56 \times 10^7 D_\infty^2) N/m$$

where the half-crack length a is in meters. The above two relations are plotted in Figure 9.32 for D_∞ / σ_∞, ranging from -10^{-9} to $+10^{-9}$. Within the same range, 11 J integrals are computed here for each material as shown in Figure 9.32. Based on these results, the following energy release rates are obtained by fitting a quadratic curve to our acquired data:

$$G_{PZT-5} = \frac{a}{8}(2.1188 \times 10^{-10} \sigma_\infty^2 + 0.3258 D_\infty / \sigma_\infty - 5.7856 \times 10^8 D_\infty^2) N/m$$

$$G_{PZT-4} = \frac{\pi a}{2}(1.8270 \times 10^{-11} \sigma_\infty^2 + 0.04510 D_\infty / \sigma_\infty - 8.7949 \times 10^7 D_\infty^2) N/m$$

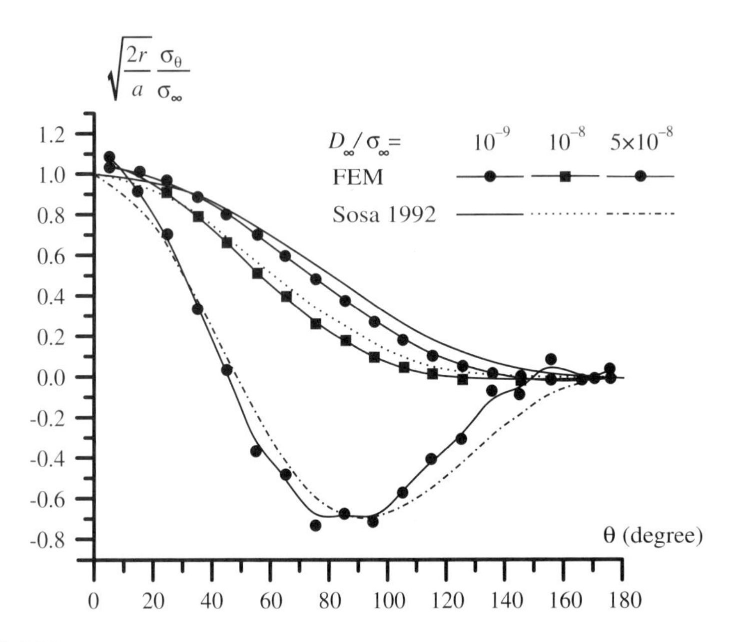

FIGURE 9.30
Effect of positive D_∞ on the angular distribution of circumferential stress in PZT-4 under mixed loading and impermeable conditions.

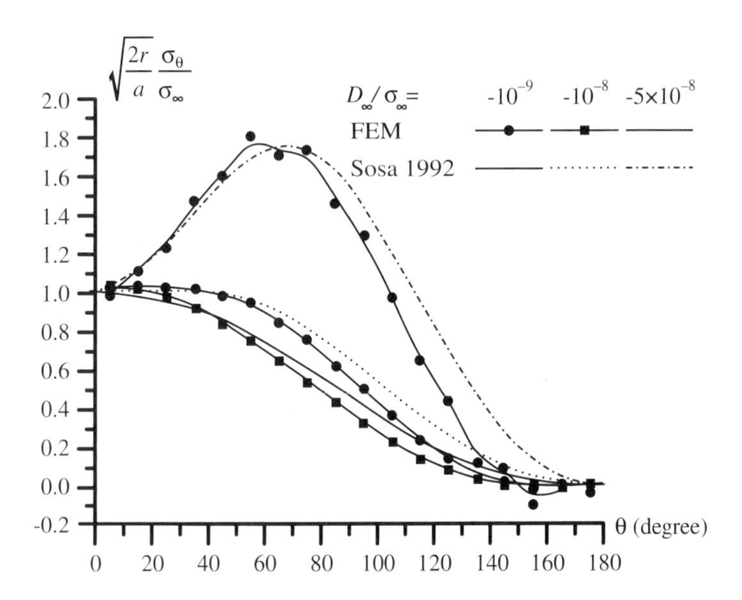

FIGURE 9.31
Effect of negative D_∞ on the angular distribution of circumferential stress in PZT-4 under mixed loading and impermeable conditions.

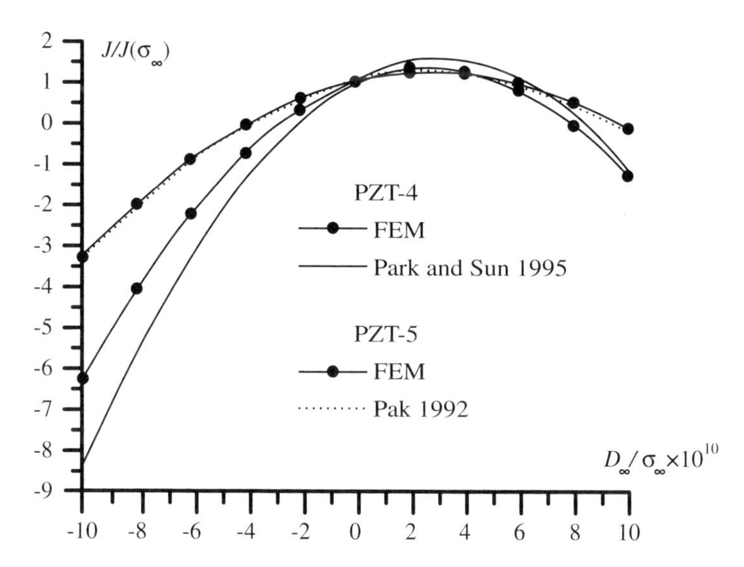

FIGURE 9.32
J integrals vs. D_∞/σ_∞.

The two expressions for PZT-5 are very close, but the ones for PZT-4 are quite different.

9.3.5 Conclusion

In this section, a piezoelectric hybrid element is formulated for computing the electromechanical coupling problem. In contrast with the conventional displacement-based element, this hybrid model can efficiently simulate the singularity fields near the crack tip. The stress σ and the electric displacement $\mathbf{D}$ are taken as independent variables in the element formulation.

Under the impermeable crack condition, the finite element solutions formulated here reproduce the $1/\sqrt{r}$ -singularity of the angular distributions for σ and $\mathbf{D}$ at the crack-tip zone. No matter which kind of loading (mechanical, electrical, or mixed) is considered, some theoretical solutions of concern are verified numerically.

The computed results indicate that under the impermeable condition, the mechanical loading (σ_∞) will induce crack propagation in its initial direction. On the contrary, the negative electric loading ($-D_\infty$) may cause the crack to deviate from its initial direction when the electric displacement becomes a governing loading.

The numerical solutions under the permeable condition show that in the case of pure mechanical loading, the stresses σ still hold the $1/\sqrt{r}$ -singularity at the crack-tip zone, but the singularity of the electric displacements $\mathbf{D}$ will disappear. However, in the case of pure electric loading, the computed **D** takes a uniform distribution, while r and the J integral and the energy release rates are practically zero.

Pak et al.'s energy release rate formulas are well examined by means of the calculations of the J integral. The suggested numerical fitting approach can also be developed to seek some approximate energy formulas for various piezoelectric crack problems with the typical $\sigma_\infty \sim D_\infty$ coupling solutions.

9.4 Homogenization-Based Hybrid Element for Torsion of Composite Shafts

9.4.1 Introduction

In the mechanics of composite materials, it is important to determine the effective properties of the composite from the distribution and basic properties of constituents and the detailed distribution of fields on the scale of microconstituents [48–51]. The hybrid stress element introduced by the authors [52] is a useful tool in the analysis of composite shafts because it gives accurate results for the warping displacement, the angle of twist per unit length, as well as the shear stress by using a relatively coarse mesh. However, if the volume fraction of reinforcement is very large, it is not realistic to obtain the microfields by this method because too many DOFs are needed to model the entire macrodomain with a grid size comparable to that of the microscale features. In this case, the mathematical homogenization method, which has received considerable attention in recent years [53–63], seems to be the most suitable for problems with boundary layers [64] that exist at regions near the interfaces of different phases in a heterogeneous medium. With the help of multiple scale expansion, mathematical homogenization gives not only the effective properties of the composite, but also detailed distribution of fields on the scale of microconstituents with acceptable cost. In contrast to the most widely used methods for determining the macro properties (i.e., the Eshelby method, the self-consistent method, the Mori–Tanaka method, the differential scheme, and the bound theories [48–51]), the homogenization method naturally takes into account the interaction between phases and avoids assumptions other than the assumption of the periodic distribution of constituents. However, it accounts for microstructural effects on the macroscopic response without explicitly representing the details of the microstructure in the global analysis. This computational model at the lower scales is only needed if and when there is a necessity. In recent years this model has been employed for the solution of complex problems in conjunction with the finite element method [55–57, 59–61]. Since the accuracy of the widely used isoparametric element method is not satisfactory, high-performance multivariable elements must be introduced in the homogenization method to improve the accuracy [61].

In this study, we will employ the homogenization method for the solution of pure torsion of a composite shaft with fibers aligned along its axis and study the application of high-performance multivariable elements. In accordance with the mathematical homogenization method, control differential equations for the representative unit cell (RUC) are obtained in Section 9.4.2. The corresponding variational principles are then deduced as the basis of the finite element method in Section 9.4.3. The formulation and practical application of incompatible and enhanced-strain elements to the analysis of RUC are discussed in detail in Sections 9.4.4 and 9.4.5. A four-node incompatible element is also introduced. In Section 9.4.6, a penalty function method is discussed to enforce the periodicity boundary condition of the RUC. In Section 9.4.7, composite shafts of square as well as rectangular cross sections reinforced with circular and elliptic fibers are analyzed as examples. For a square shaft containing 16 fibers, the problem is also solved directly with the hybrid stress element. The shear modulus from the homogenization method is compared with that obtained with the hybrid stress element and the Voigt–Reuss theory [51]. A comparison of the computed results shows common features of the local fields. Conclusions and discussion follow in Section 9.4.8.

9.4.2 Mathematical Homogenization

Consider a uniform composite shaft of arbitrary cross section twisted by couples applied at the ends. Without a loss of generality, the origin of coordinates is taken at the left-end cross section, with the x_1- and x_2-axes as the principal axes of inertia, and along the x_3-axis of the shaft and pointing to its other end. According to St. Venant's theory of torsion, the displacement components are [65]

$$\begin{aligned} u_1 &= -\theta x_3 x_2 \\ u_2 &= \theta x_3 x_1 \\ u_3 &= w(x_1, x_2) \end{aligned} \tag{9.74}$$

where θ represents the angle of twist per unit length (clockwise about the x_3-axis).

Assume the microstructure of the cross section Ω^{ε} to be locally periodic with a period defined by a statistically homogeneous volume element, denoted by the representative volume element or unit cell Y, as shown in Figure 9.33. In other words, the composite material is formed by a spatial repetition of the unit cell. The shaft has two length scales: a global length scale D that is of the section size and a local length scale d that is of the order of the microstructure to the wavelength of the variation of the microstructure. The size of the unit cell is much larger than that of the constituents but much smaller than that of the section. The relation between the global

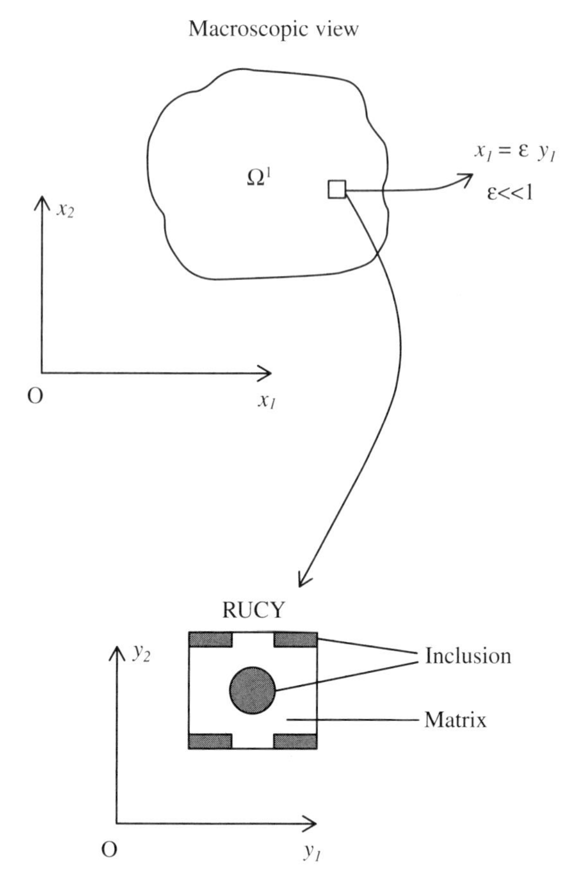

FIGURE 9.33
Illustration of a problem with two length scales.

coordinate system x_i for the section and the local system y_i for the minimum repeated unit cell can then be written as

$$i = 1,2 \tag{9.75}$$

where ε is a very small positive number representing the scale factor between the two length scales. The local coordinate vector y_i is regarded as a stretched coordinate vector in the microscopic domain. For an actual heterogeneous body subjected to external forces, field quantities such as displacements, strains, and stresses are assumed to have slow variations from point to point with macroscopic (global) coordinate x as well as fast variations with local microscopic coordinate y within a small neighborhood of size ε of a given point x. That is, displacements, strains, and stresses have two explicit dependencies: one on the macroscopic level with coordinates x_i, and the other on the level of microconstituents with coordinates y_i

$$u_3^\varepsilon = u_3^\varepsilon(x,y)$$
$$\gamma_{3j}^\varepsilon = \gamma_{3j}^\varepsilon(x,y) \tag{9.76}$$
$$\tau_{3j}^\varepsilon = \tau_{3j}^\varepsilon(x,y)$$

where $j = 1,2$. Due to the periodicity of the microstructure, functions u_3^ε, γ_{3j}^ε, and τ_{3j}^ε are assumed to be Y-periodic, that is, $u_3^\varepsilon(x,y) = u_3^\varepsilon(x,y+kY)$, $\gamma_{3j}^\varepsilon(x,y) = \gamma_{3j}^\varepsilon(x,y+kY)$, and $\tau_{3j}^\varepsilon(x,y) = \gamma_{3j}^\varepsilon(x,y+kY)$, where $Y(y_i)$ is the size of the unit cell, or the basic period of the stretched coordinate system y and k is a nonzero integer.

The unknown displacement u_3^ε, the angle of twist per unit length θ, and the nonzero strain γ_{3j}^ε and stress τ_{3j}^ε can be solved from the following equations:

$$\text{Equilibrium: } \frac{\partial \tau_{3j}^\varepsilon}{\partial x_j} = 0 \text{ in } \Omega^\varepsilon \tag{9.77}$$

$$\text{Kinematical: } \begin{Bmatrix} \gamma_{31}^\varepsilon \\ \gamma_{32}^\varepsilon \end{Bmatrix} = \begin{bmatrix} \dfrac{\partial}{\partial x_1} & -x_2 \\ \dfrac{\partial}{\partial x_2} & x_1 \end{bmatrix} \begin{Bmatrix} u_3^\varepsilon \\ \theta \end{Bmatrix} \text{ in } \Omega^\varepsilon \tag{9.78}$$

$$\text{Constitutive: } \tau_{3i}^\varepsilon = C_{ij}^\varepsilon \gamma_{3j}^\varepsilon \text{ in } \Omega^\varepsilon \tag{9.79}$$

and

$$M_t = \int_{\Omega^\varepsilon} (\tau_{32}x_1 - \tau_{31}x_2)\, d\Omega \tag{9.80}$$

together with the traction-free condition on the surface of the shaft and the traction and displacement conditions at the interfaces between the microconstituents. For the sake of simplicity and clarity, we assume that the fields are continuous across the interfaces. The material property tensor C_{ij}^ε is symmetric with respect to indices (i,j). M_1 is the torque applied at the ends. The superscript ε denotes the Y-periodic of the corresponding function. The convection of summation over the repeated indices is used.

The displacement $u_3^\varepsilon(x,y)$ is expanded in powers of the small number ε [53–61]:

$$u_3^\varepsilon(x,y) = u_3^{(0)}(x,y) + \varepsilon u_3^{(1)}(x,y) + \varepsilon^2 u_3^{(2)}(x,y) + \cdots \tag{9.81}$$

where $u_3^{(0)}, u_3^{(1)}, u_3^{(2)}, \ldots,$ are Y-periodic functions with respect to y. Substituting Equation 9.81 into Equation 9.78 gives the expansion of the strain $\gamma_{3j}^{\varepsilon}$:

$$\gamma_{3j}^{\varepsilon}(x,y) = \varepsilon^{-1}\gamma_{3j}^{(-1)} + \gamma_{3j}^{(0)} + \varepsilon\gamma_{3j}^{(1)} + \cdots \tag{9.82}$$

where

$$\begin{aligned}
\gamma_{3j}^{(-1)} &= \frac{\partial u_3^{(0)}}{\partial y_j} \\
\gamma_{3j}^{(0)} &= \gamma_{x3j}^{(0)} + \gamma_{y3j}^{(0)} \\
\gamma_{x31}^{(0)} &= \frac{\partial u_3^{(0)}}{\partial x_1} - \theta x_2 \\
\gamma_{x32}^{(0)} &= \frac{\partial u_3^{(0)}}{\partial x_2} + \theta x_1 \\
\gamma_{y3j}^{(0)} &= \frac{\partial u_3^{(1)}}{\partial y_j} \\
\gamma_{3j}^{(1)} &= \frac{\partial u_3^{(1)}}{\partial x_j} + \frac{\partial u_3^{(2)}}{\partial y_j}
\end{aligned} \tag{9.83}$$

Substituting Equation 9.82 into the constitutive relation (Equation 9.79) gives the expansion of the stress τ_{3j}^{ε}:

$$\tau_{3j}^{\varepsilon}(x,y) = \varepsilon^{-1}\tau_{3j}^{(-1)} + \tau_{3j}^{(0)} + \varepsilon\tau_{3j}^{(1)} + \cdots \tag{9.84}$$

where

$$\tau_{3i}^{(-1)} = C_{ij}\gamma_{3j}^{(-1)} \tag{9.85}$$

$$\tau_{3i}^{(0)} = C_{ij}\gamma_{3j}^{(0)} \tag{9.86}$$

$$\tau_{3i}^{(1)} = C_{ij}\gamma_{3j}^{(1)} \tag{9.87}$$

Inserting the asymptotic expansion for the stress field (Equation 9.84) into the equilibrium equation (Equation 9.77) and collecting the terms of like powers in ε gives

$$O(\varepsilon^{-2}) \quad : \frac{\partial \tau_{3j}^{(-1)}}{\partial y_j} = 0 \tag{9.88}$$

$$O(\varepsilon^{-1}) \quad : \frac{\partial \tau_{3j}^{(-1)}}{\partial x_j} + \frac{\partial \tau_{3j}^{(0)}}{\partial y_j} = 0 \tag{9.89}$$

$$O(\varepsilon) \quad : \frac{\partial \tau_{3j}^{(0)}}{\partial x_j} + \frac{\partial \tau_{3j}^{(1)}}{\partial y_j} = 0 \tag{9.90}$$

We first consider the $o(\varepsilon^{-2})$ equilibrium equation (Equation 9.88) in Y. Premultiplying it by $u_3^{(0)}$ and integrating over Y, followed by integration by parts, yields

$$\int_Y u_3^{(0)} \frac{\partial \tau_{3j}^{(-1)}}{\partial y_j} dY = \oint_{\partial Y} u_3^{(0)} \tau_{3j}^{(-1)} n_j d\Gamma - \int_Y \frac{\partial u_3^{(0)}}{\partial y_j} C_{ji} \frac{\partial u_3^{(0)}}{\partial y_i} dY = 0 \tag{9.91}$$

where ∂Y denotes the boundary of Y. The boundary integral term in Equation 9.91 vanishes due to the periodicity of the boundary conditions in Y, because $u_3^{(0)}$ and τ_{3j}^{-1} are identical on the opposite sides of the unit cell, while the corresponding normals n_j are in opposite directions.

Taking into account the positive definiteness of the symmetric constitutive tensor C_{ij}, we have

$$\frac{\partial u_3^{(0)}}{\partial y_j} = 0 \Rightarrow u_3^{(0)} = u_3^{(0)}(x) \tag{9.92}$$

and

$$\tau_{3j}^{(-1)}(x, y) = 0 \tag{9.93}$$

Next, we proceed to the $o(\varepsilon^{-1})$ equilibrium equation (Equation 9.89). From Equation 9.83 and Equation 9.86 and taking into account Equation 9.93, it follows that

$$\frac{\partial}{\partial y_j}(C_{ji}\gamma_{y3i}^{(0)}(u_3^{(1)})) = -\frac{\partial C_{ji}}{\partial y_j}\gamma_{x3i}^{(0)}(u_3^{(0)}) \tag{9.94}$$

Based on the form of the right-hand side of Equation 9.94, which permits a separation of variables, $u_3^{(1)}$ may be expressed as

$$u_3^{(1)}(x, y) = \chi_3^{3j}(y)\gamma_{x3j}^{(0)}(u_3^{(0)}) \tag{9.95}$$

where $\gamma_3^{3j}(y)$ is a Y-periodic function defined in the unit cell Y. Substituting Equation 9.95 into Equation 9.94 and taking into account the arbitrariness of the macroscopic strain field, $\gamma_{x3j}(u_3^{(0)})$, within a unit cell, we have

$$\frac{\partial}{\partial y_j}(C_{ji}\gamma_{y3i}^{(0)}(\chi_3^{3k}(y))) = -\frac{\partial C_{jk}}{\partial y_j} \tag{9.96}$$

We now consider the $o(\varepsilon)$ equilibrium equation (Equation 9.90). Substituting Equation 9.95 into Equation 9.83, followed by the latter into Equation 9.86, and substituting the result into Equation 9.90 yields

$$\frac{\partial}{\partial x_j}\left[C_{ji}(\delta_{ik} + \chi_{3,y_i}^{3k})\gamma_{x3k}^{(0)}(u_3^{(0)})\right] + \frac{\partial \tau_{3j}^{(1)}}{\partial y_j} = 0 \tag{9.97}$$

where δ_{ik} is the Kronecker Delta. Integrating Equation 9.97 over the unit cell domain Y and taking into account the periodicity of $\tau_{3j}^{(1)}$ yields

$$\frac{\partial}{\partial x_j}\left[C_{jk}^H\gamma_{x3k}^{(0)}(u_3^{(0)})\right] = 0 \tag{9.98}$$

This is an equilibrium equation for a homogeneous medium (cf. Equation 9.77) with constant material properties C_{jk}^H, which are usually termed as the homogenized or effective material properties and are given by

$$C_{jk}^H = \frac{1}{Y}\int_Y C_{ji}(\delta_{ik} + \chi_{3,y_i}^{3k})\, dY \tag{9.99}$$

where Y is the area of the unit cell.

9.4.3 Variational Principles and Finite Elements

To solve the torsion of composite shafts by the homogenization method, together with numerical methods (e.g., the finite element method adopted here), we will first solve for $\chi_3^{3j}(y)$ from Equation 9.96, assuming it to be a Y-periodic function defined in Y. The effective material properties C_{jk}^H are given by Equation 9.99. We then solve the homogeneous St. Venant torsion problem by the hybrid stress element [52] and obtain the macroscopic fields: warping displacement $u_3^{(0)}$, the angel of twist per unit length θ, strains $\gamma_{x3j}^{(0)}$, and stresses (given by $C_{ji}^H\gamma_{x3i}^{(0)}$). If the distribution of the microscopic fields in the neighborhood of point x is of interest, we use Equation 9.95 to calculate the higher-order displacement term, and then Equation 9.83 and Equation 9.86 to calculate the higher-order strain and stress terms.

The key problem here is to develop powerful finite element methods to solve Equation 9.96. Corresponding to the equilibrium equation (Equation 9.96), the virtual work principle states that

$$\int_Y \delta\chi_3^{3k} \frac{\partial}{\partial y_j}\left(C_{ji}\frac{\partial \chi_3^{3k}}{\partial y_i}\right)dY + \int_Y \delta\chi_3^{3k}\frac{\partial C_{jk}}{\partial y_j}dY = 0 \tag{9.100a}$$

where $\delta\chi_3^{3k}$ are arbitrary Y-periodic functions defined in the unit cell Y. Integration of Equation 9.100a by parts yields

$$\oint_{\partial Y} \delta\chi_3^{3k} C_{ji}\frac{\partial \chi_3^{3k}}{\partial y_i} n_j ds + \oint_{\partial Y} \delta\chi_3^{3k} C_{jk} n_j ds - \int_Y \frac{\partial \delta\chi_3^{3k}}{\partial y_j} C_{ji} \times \frac{\partial \chi_3^{3k}}{\partial y_i} dY - \int_Y \frac{\partial \delta\chi_3^{3k}}{\partial y_j} C_{jk} dY = 0$$

The boundary integral terms in the above equation vanish due to the Y-periodicity of χ_3^{3k} and $\delta\chi_3^{3k}$. Thus, we have

$$\int_Y \frac{\partial \delta\chi_3^{3k}}{\partial y_j} C_{ji}\frac{\partial \chi_3^{3k}}{\partial y_i} dY + \int_Y \frac{\partial \delta\chi_3^{3k}}{\partial y_j} C_{jk} dY = 0 \tag{9.100b}$$

Based on Equation 9.100b, displacement elements can be established in a standard manner.

It is easy to prove that Equation 9.100a and Equation 9.100b are the first-order variation of the following potential functional

$$\Pi_P(\chi_3^{3k}) = \int_Y \frac{1}{2}\frac{\partial \chi_3^{3k}}{\partial y_j} C_{ji}\frac{\partial \chi_3^{3k}}{\partial y_i} dY + \int_Y \frac{\partial \chi_3^{3k}}{\partial y_j} C_{jk} dY \tag{9.101}$$

If we define the strain

$$\tilde{\gamma}_{3i}^k = \frac{\partial \chi_3^{3k}}{\partial y_i}$$

and the stress

$$\tilde{\tau}_{3j}^k = C_{ji}\tilde{\gamma}_{3i}^k$$

so that

$$\tilde{\gamma}_{3i}^k = C_{ij}^{-1}\tilde{\tau}_{3j}^k$$

which are Y-periodic functions in the unit cell, we have a two-field Hellinger–Reissner functional:

$$\Pi_{HR}(\chi_3^{3k}, \tilde{\tau}_{3j}^k) = \int_Y \left[-\frac{1}{2} \tilde{\tau}_{3i}^k C_{ij}^{-1} \tilde{\tau}_{3j}^k + \tilde{\tau}_{3i}^k \frac{\partial \chi_3^{3k}}{\partial y_i} - \frac{\partial C_{jk}}{\partial y_j} \chi_3^{3k} \right] dY \quad (9.102a)$$

or equivalently

$$\Pi_{HR}(\chi_3^{3k}, \tilde{\tau}_{3j}^k) = \int_Y \left[-\frac{1}{2} \tilde{\tau}_{3i}^k C_{ij}^{-1} \tilde{\tau}_{3j}^k + \tilde{\tau}_{3i}^k \frac{\partial \chi_3^{3k}}{\partial y_i} + C_{jk} \frac{\partial \chi_3^{3k}}{\partial y_j} \right] dY \quad (9.102b)$$

By making use of the Lagrange multiplier method and relaxing the compatibility condition in the potential principle (Equation 9.101), or by employing the Legendre transformation on the Hellinger–Reissner principle (Equation 9.102b), one arrives at the three-field Hu–Washizu functional

$$\Pi_{HW}(\chi_3^{3k}, \tilde{\gamma}_{3j}^k, \tilde{\tau}_{3j}^k) = \int_Y \left[\frac{1}{2} \tilde{\gamma}_{3i}^k C_{ij} \tilde{\gamma}_{3j}^k - \tilde{\tau}_{3i}^k \left(\tilde{\gamma}_{3i}^k - \frac{\partial \chi_3^{3k}}{\partial y_i} \right) + C_{jk} \frac{\partial \chi_3^{3k}}{\partial y_j} \right] dY \quad (9.103)$$

Based on the functionals (Equation 9.102 and Equation 9.103), multivariable finite elements can be established.

Although hybrid elements based on the Hellinger–Reissner principle or the Hu–Washizu principle can in general improve the accuracy of the approximate displacement and stress solutions, they will not be used here, as it is difficult to meet the Y-periodicity condition of the stress on the boundary of the unit cell. The general isoparametric elements are also not satisfactory because of the gradients of χ_3^{3k} that appear in Equation 9.99 in the evaluation of the homogenized material properties. For these reasons, in this section we will introduce displacement-incompatible elements based on the potential (Equation 9.101) and enhanced-strain elements based on Equation 9.103.

9.4.4 Displacement-Incompatible Elements

Subdivide the unit cell domain Y into finite element subdomains Y_e, such that $\cup Y_e = Y, Y_a \cap Y_b = \phi$ and $\partial Y_a \cap \partial Y_b = S_{ab}$ (a, b are arbitrary elements).

In each element, $\chi_3^{3k}\phi$ is divided into a compatible part χ_{3q}^{3k} and an incompatible part $\chi_{3\lambda}^{3k}$, so that the functional (Equation 9.101) can be rewritten as

$$\Pi_P(\chi_3^{3k} = \chi_{3q}^{3k} + \chi_{3\lambda}^{3k}) = \sum_e \int_{Y_e} \frac{1}{2} \frac{\partial \chi_3^{3k}}{\partial y_j} C_{ji} \frac{\partial \chi_3^{3k}}{\partial y_i} dY + \int_{Y_e} \frac{\partial \chi_{3q}^{3k}}{\partial y_j} C_{jk} dY \quad (9.104)$$

Taking the variation of the above functional, integrating by parts, and making use of the periodicity condition on the outer boundary of the unit cell yields

$$\delta\Pi_P(\chi_3^{3k}) = \sum_e -\int_{Y_e} \delta\chi_{3q}^{3k}\left[\frac{\partial}{\partial y_j}\left(C_{ji}\frac{\partial \chi_3^{3k}}{\partial y_i}\right) + \frac{\partial C_{jk}}{\partial y_j}\right]dY$$

$$+\sum_{a,b}\int_{S_{ab}} \delta\chi_{3q}^{3k}\left\{\left[\left(C_{ji}\frac{\partial \chi_3^{3k}}{\partial y_i} + C_{jk}\right)n_j\right]^{(a)} + \left[\left(C_{ji}\frac{\partial \chi_3^{3k}}{\partial y_i} + C_{jk}\right)n_j\right]^{(b)}\right\}ds$$

$$+\sum_e \int_{Y_e} \frac{\partial \delta\chi_{3\lambda}^{3k}}{\partial y_j} C_{ji} \frac{\partial \chi_3^{3k}}{\partial y_i} dY$$

The stationary condition of the functional (Equation 9.104) gives the equilibrium equation (Equation 9.96) and the equilibrium of traction between the elements if the following condition is met *a priori*

$$\sum_e \int_{Y_e} \frac{\partial \delta\chi_{3\lambda}^{3k}}{\partial y_j} C_{ji} \frac{\partial \chi_3^{3k}}{\partial y_i} dY = 0$$

A convenient way to meet this condition is to satisfy the following strong form (i.e., the sufficient but not the necessary condition) in each element:

$$\int_{Y_e} \frac{\partial \delta\chi_{3\lambda}^{3k}}{\partial y_j} C_{ji} \frac{\partial \chi_3^{3k}}{\partial y_i} dY = 0$$

Since a constant stress state is recovered in each element as its size is reduced to zero and since $\delta\chi_{3\lambda}^{3k}$ is arbitrary, the above constraint reduced to the general patch test condition (PTC) [15, 66] is

$$\int_{Y_e} \frac{\partial \chi_{3\lambda}^{3k}}{\partial y_j} dY = 0 \quad \text{or equivalently} \oint_{\partial Y_e} \chi_{3\lambda}^{3k} n_j ds = 0 \qquad (9.105)$$

The incompatible functions meeting the PTC can now be easily formulated.

If we refer to the four-node isoparametric element shown in Figure 9.34, the compatible displacement χ_{3q}^{3k} is related to the nodal values ${}^e q^k$ via the bilinear interpolation functions

$$\chi_{3q}^{3k} = N\,{}^e q^k \qquad (9.106)$$

where

$$N = \begin{bmatrix} N_1 & N_2 & N_3 & N_4 \end{bmatrix}$$

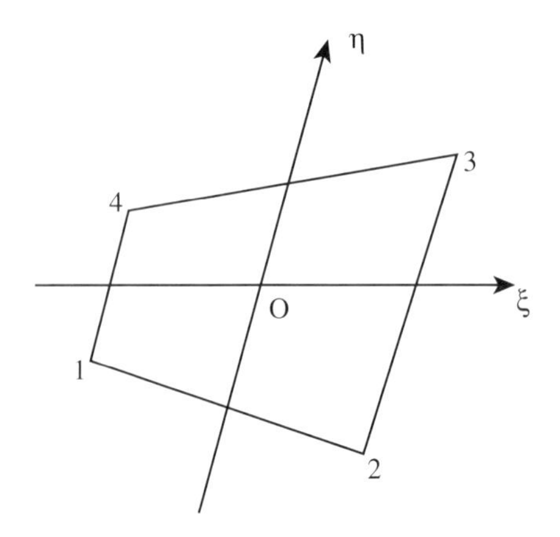

FIGURE 9.34
A four-node element.

and

$$N_i = \frac{1}{4}(1+\xi_i\xi)(1+\eta_i\eta)$$

(ξ, η) represents the isoparametric coordinates, and (ξ_i, η_i) is the isoparametric coordinates of point i with the global coordinates $(x_i, y_i), i = 1,2,3,4.$

The incompatible term $\chi_{3\lambda}^{3k}$ is related to the element inner parameters ${}^e\lambda^k$ via the shape functions N_λ:

$$\chi_{3\lambda}^{3k} = N_\lambda\, {}^e\lambda^k \tag{9.107}$$

Here, two incompatible terms are employed in each element as derived in References 15 and 66:

$$N_{\lambda 2} = \eta^2 + \Delta \tag{9.108}$$

where J_0, J_1, and J_2 are employed in each element Jacobian as follows:

$$|J| = J_0 + J_1\xi + J_2\eta = (a_1b_3 - a_3b_1) + (a_1b_2 - a_2b_1)\xi + (a_2b_3 - a_3b_2)\eta \tag{9.109}$$

and coefficients a_i and b_i $(i = 1,2,3)$ are dependent on the element nodal coordinates

$$\begin{bmatrix} a_1 & b_1 \\ a_2 & b_2 \\ a_3 & b_3 \end{bmatrix} = \frac{1}{4}\begin{bmatrix} -1 & 1 & 1 & -1 \\ 1 & -1 & 1 & -1 \\ -1 & -1 & 1 & 1 \end{bmatrix}\begin{bmatrix} x_1 & y_1 \\ x_2 & y_2 \\ x_3 & y_3 \\ x_4 & y_4 \end{bmatrix} \tag{9.110}$$

With the above assumed displacement (Equation 9.106 and Equation 9.107), we have

$$\begin{Bmatrix} \dfrac{\partial \chi_3^{3k}}{\partial y_1} \\ \dfrac{\partial \chi_3^{3k}}{\partial y_2} \end{Bmatrix} = \begin{Bmatrix} \dfrac{\partial}{\partial y_1} \\ \dfrac{\partial}{\partial y_2} \end{Bmatrix} \begin{bmatrix} N & N_\lambda \end{bmatrix} \begin{Bmatrix} {}^e q^k \\ {}^e \lambda^k \end{Bmatrix} = \begin{bmatrix} B & B_\lambda \end{bmatrix} \begin{Bmatrix} {}^e q^k \\ {}^e \lambda^k \end{Bmatrix} \tag{9.111}$$

where

$$B_\lambda = \begin{Bmatrix} \dfrac{\partial}{\partial y_1} \\ \dfrac{\partial}{\partial y_2} \end{Bmatrix} N_\lambda$$

Substituting Equation 9.111 into Equation 9.104 and making use of the stationary condition yields

$${}^e\tilde{K}\, {}^e q^k = {}^e f^k \tag{9.112}$$

where the element stiffness matrix and nodal vector are

$${}^e f^k = -\int_{Y_e} B^T \begin{Bmatrix} C_{1k} \\ C_{2k} \end{Bmatrix} dY \tag{9.113}$$

in which

$$\begin{bmatrix} K_{qq} & K_{q\lambda} \\ K_{q\lambda}{}^T & K_{\lambda\lambda} \end{bmatrix} = \int_{Y_e} \begin{bmatrix} B^T \\ B_\lambda^T \end{bmatrix} C \begin{bmatrix} B & B_\lambda \end{bmatrix} dY$$

The element inner parameters ${}^e\lambda^k$ are recovered as follows:

$${}^e\lambda^k = -K_{\lambda\lambda}{}^{-1} K_{q\lambda}{}^T\, {}^e q^k$$

This element is designated NQ6, and 2×2 Gauss quadrature is employed for the element formulation.

9.4.5 Enhanced-Strain Element Based on the Hu–Washizu Principle

Only the compatible displacement field is used here, and the strain field is enhanced by appending an incompatible field to the strain corresponding to the compatible displacement [67, 68]:

$$\tilde{\gamma}_{3i}^{k} = \frac{\partial \chi_3^{3k}}{\partial y_i} + \tilde{\gamma}_{\lambda 3i}^{k}$$

The Hu–Washizu functional (Equation 9.103) can now be rewritten as

$$\Pi_{\mathrm{HW}}(\chi_3^{3k}, \tilde{\gamma}_{\lambda 3j}^{k}, \tilde{\tau}_{3j}^{k}) = \sum_e \int_{Y_e} \left[\frac{1}{2} \tilde{\gamma}_{3i}^{k} C_{ij} \tilde{\gamma}_{3j}^{k} - \tilde{\tau}_{3i}^{k} \tilde{\gamma}_{\lambda 3i}^{k} + C_{jk} \frac{\partial \chi_3^{3k}}{\partial y_j} \right] dY \tag{9.114}$$

Taking the variation of the above functional, integrating by parts, and making use of the periodicity condition on the outer boundary of the unit cell yields

$$\begin{aligned}
\delta \Pi_{\mathrm{HW}}(\chi_3^{3k}, \tilde{\gamma}_{\lambda 3j}^{k}, \tilde{\tau}_{3j}^{k}) = {} & \sum_e \int_{Y_e} \left[-\delta \chi_3^{3k} \frac{\partial}{\partial y_j} (C_{ji} \tilde{\gamma}_{3i}^{k} + C_{jk}) + \delta \tilde{\gamma}_{\lambda 3i}^{k} (C_{ij} \tilde{\gamma}_{3j}^{k} - \tilde{\tau}_{3i}^{k}) \right] dY \\
& + \sum_{a,b} \oint_{S_{ab}} \delta \chi_3^{3k} \{ [(C_{ji} \tilde{\gamma}_{3i}^{k} + C_{jk}) n_j]^{(a)} + [(C_{ji} \tilde{\gamma}_{3i}^{k} + C_{jk}) n_j]^{(b)} \} ds \\
& - \sum_e \int_{Y_e} \delta \tilde{\tau}_{3i}^{k} \tilde{\gamma}_{\lambda 3i}^{k} dY
\end{aligned}$$

The stationary condition of the functional (Equation 9.114) gives the equilibrium equation (Equation 9.96), the stress–strain relations, and the equilibrium of traction between the elements if the following condition is met *a priori*:

$$\sum_e \int_{Y_e} \delta \tilde{\tau}_{3i}^{k} \tilde{\gamma}_{\lambda 3i}^{k} dY = 0$$

Following the procedure employed in Section 9.4.4, the above constraint can be simplified to the PTC [15, 69]:

$$\int_{Y_e} \tilde{\gamma}_{\lambda 3i}^{k} dY = 0 \tag{9.115}$$

It is evident that Equation 9.115 is an alternative formulation of the PTC (Equation 9.105), if the enhanced-strain $\tilde{\gamma}_{\lambda 3i}^{k}$ corresponds to the incompatible displacement $\chi_{3\lambda}^{3k}$.

The finite element based on the stationary condition of the functional (Equation 9.114) requires an independent approximation of three fields: χ_3^{3k}, $\tilde{\gamma}_{\lambda 3j}^{k}$, and $\tilde{\tau}_{3j}^{k}$. In the enhanced strain element, however, the independent stress field is eliminated by selecting it to be orthogonal to the enhanced strain field, that is,

$$\int_{Y_e} \tilde{\tau}_{3i}^k \, \tilde{\gamma}_{\lambda 3i}^k \, dY = 0 \tag{9.116}$$

Thus, the two independent fields for the enhanced-strain formulation are the displacement χ_3^{3k} and the enhanced assumed strain $\tilde{\gamma}_{\lambda 3j}^k$. The formulation here is the same as that in Section 9.4.4 above, provided that $\tilde{\gamma}_{\lambda 3j}^k$ is interpolated from the element inner parameters as follows:

$$\begin{Bmatrix} \tilde{\gamma}_{\lambda 31}^k \\ \tilde{\gamma}_{\lambda 32}^k \end{Bmatrix} = B_\lambda \lambda^k \tag{9.117}$$

Moreover, if the assumed strain $\tilde{\gamma}_{\lambda 3j}^k$ in Equation 9.117 corresponds to the incompatible displacement $\chi_{3\lambda}^{3k}$ in Equation 9.107, the enhanced-strain formulation will be equivalent to the incompatible-displacement formulation discussed in Section 9.4.4. Therefore, only NQ6, introduced in Section 9.4.4, will be employed in the enhanced-strain formulation and can recover with the help of the orthogonalization condition (Equation 9.116), as suggested in Reference 67.

9.4.6 Enforcing the Periodicity Boundary Condition in the Analysis of the RUC

Assembling the discretized equations of equilibrium of all elements yields the following system of equilibrium equations:

$$Kq^k = f^k \tag{9.118}$$

Two different loading cases need to be analyzed in order to determine the characteristic deformations of the unit cell. The periodicity of the boundary displacement can conveniently be enforced by a penalty function technique [70]. Equation 9.118 is the Euler–Lagrange equation of the following functional:

$$\Pi(q^k) = \frac{1}{2} q^{k^T} K q^k - q^{k^T} f^k \tag{9.119}$$

The periodicity condition yields the following constraint: $Rq^k = 0$.

If two nodes, i and j, on the boundary have the same displacement because of the periodicity condition, that is,

$$q_i^k = q_j^k$$

then the above condition is equivalent to

$$R(i, l \neq i, j) = 0$$

In order to satisfy the above periodicity constraint with a penalty function technique, the functional (Equation 9.119) is modified as

$$\tilde{\Pi}(q^k) = \frac{1}{2} q^{k^T} K q^k - q^{k^T} f^k + \frac{\alpha}{2} q^{k^T} R^T R q^k \tag{9.120}$$

where α is a large positive number taken to be 10^4 in our computations. Thus, instead of Equation 9.118, we will solve the following equation:

$$(K + \alpha R^T R) q^k = f^k \tag{9.121}$$

9.4.7 Numerical Examples

As the performance of the element using incompatible functions defined in Equation 9.108 and of the hybrid stress element to be used here for the torsion of shafts has been extensively studied in References 52, 66, and 15, no standard performance tests for these elements are included in this section. However, to illustrate the method described above, we solve the torsion of a composite shaft with a square cross section (length of side = 80 mm), as shown in Figure 9.35a. Assume that the microstructure of the cross section is locally periodic with a period defined by an RUC shown in Figure 9.35b, that is, it consists of an isotropic circular fiber of diameter $2a$ embedded in an isotropic square matrix with side $4a$, a = 5 mm is adapted in this study. The problem is solved in two stages. First, we solve the RUC by using the incompatible element NQ6 introduced in Section 9.4.4, with the periodicity boundary condition enforced by the penalty function approach discussed in Section 9.4.6. We obtain the field χ_3^{3k} and its derivatives $(\partial / \partial y_j)\chi_3^{3k}$ and calculate the homogenized moduli from Equation 9.99. Second, we solve the torsion of the square shaft shown in Figure 9.35a with the homogenized moduli obtained in the first step above, by using the hybrid stress element introduced in Reference 52. In this way, we calculate the warping displacement, torsional rigidity, and angle of twist per unit length, as well as the shear stresses and strains. With the results, we can calculate the first-order warping displacement from Equation 9.95 and the local strain and stress fields from Equation 9.83 and Equation 9.86, respectively. For our example, we chose $\varepsilon = 0.25$. The complete shaft section from which the RUC has been extracted is shown in Figure 9.35c. In the following figures, filled triangles represent computed data. In all the figures that illustrate the stress distribution, a line segment represents the distribution within an element. In Figure 9.37 and Figure 9.38b,c, the solid lines represent the polynomial fit of the corresponding computed data that is not satisfactorily smooth.

The RUC shown in Figure 9.35b is discretized into 896 quadrilateral elements and 929 nodes, as shown in Figure 9.36a. According to the definition of the RUC, its size should be enlarged four times as $\varepsilon = 0.25$. However, numerical results show that the results are unaffected by whether or not the

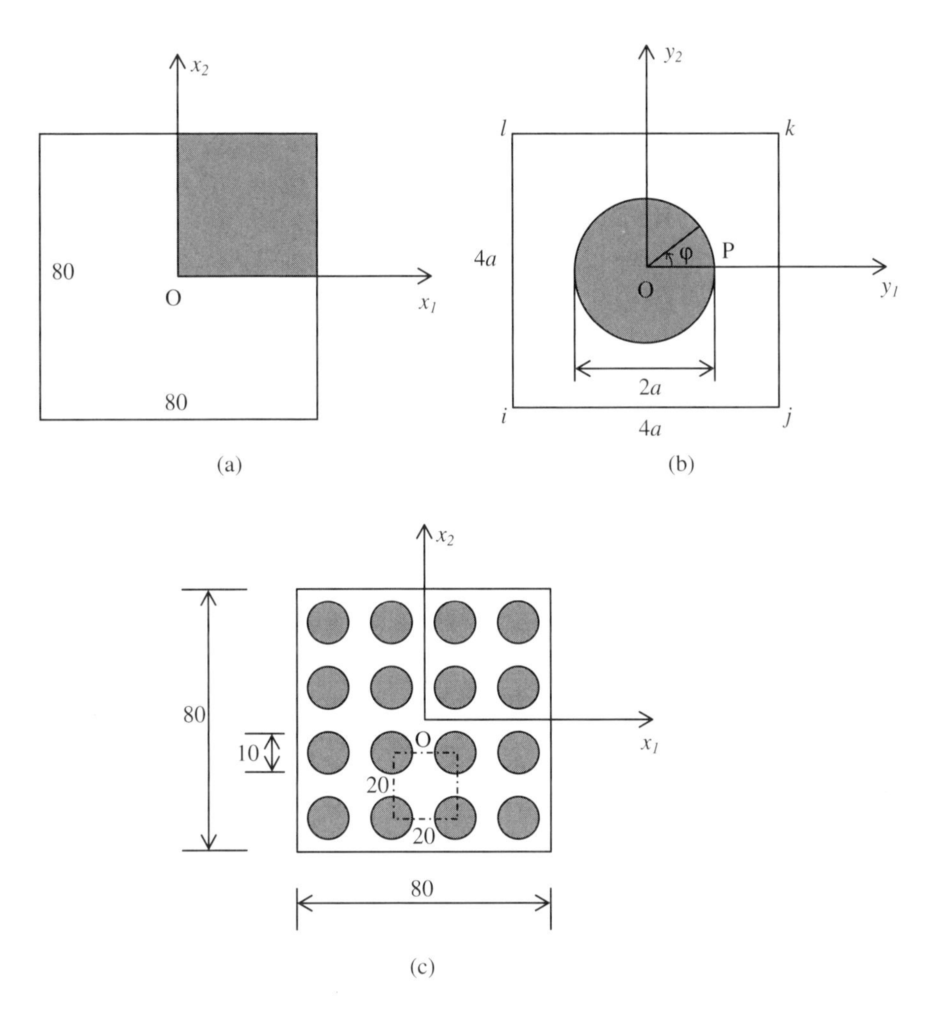

FIGURE 9.35
Geometry of a composite shaft of square profile. (a) Square profile; (b) RUC; (c) square shaft with 16 fibers.

RUC size is enlarged, allowing us to use the original RUC size. Care must be taken in enforcing the periodicity boundary condition at corner nodes. For the four corner nodes, i, j, k, and l in Figure 9.35b, the periodicity condition yields

$$q_i^k = q_j^k = q_k^k = q_l^k$$

The above condition can be rewritten as

$$q_i^k = q_j^k$$

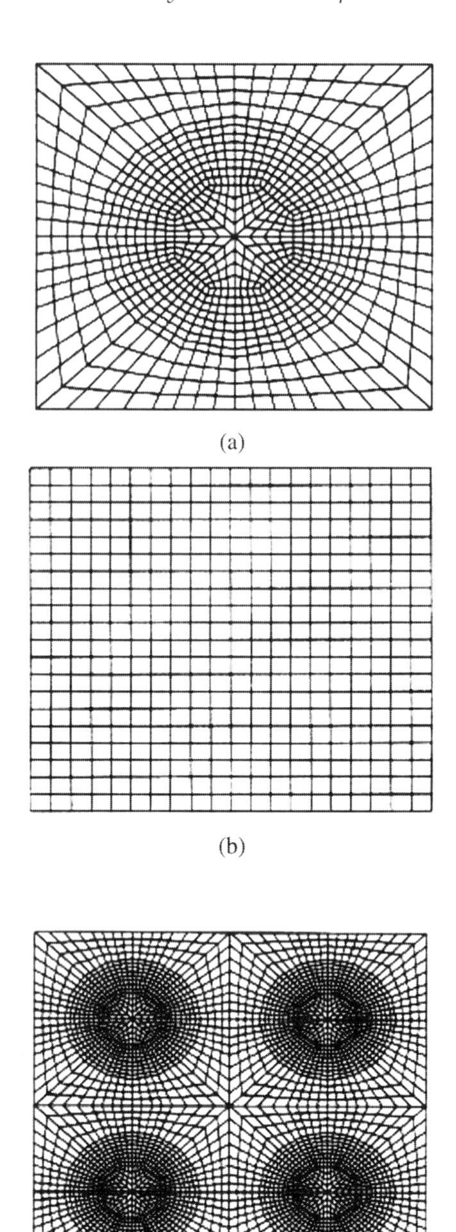

FIGURE 9.36
Discretized meshes used in the computation. (a) Mesh of the RUC shown in Figure 9.35b; (b) mesh of a quarter of the cross section shown in Figure 9.35a; (c) mesh of a quarter of the cross section shown in Figure 9.35c.

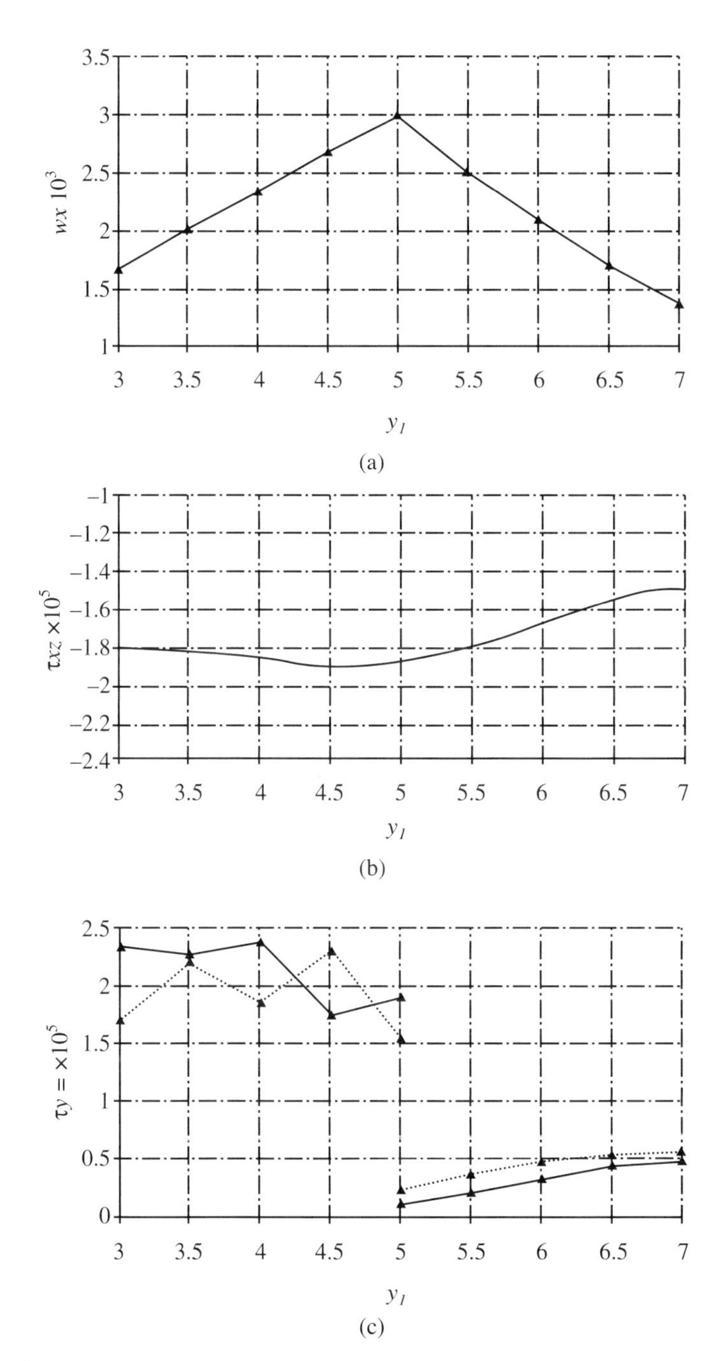

FIGURE 9.37
Numerical results on the line $3 \leq y_1 \leq 7$, $y_2 = 0$, from the homogenization method. (a) distribution of warping displacement; (b) distribution of τ_{xz}; (c) distribution of τ_y.

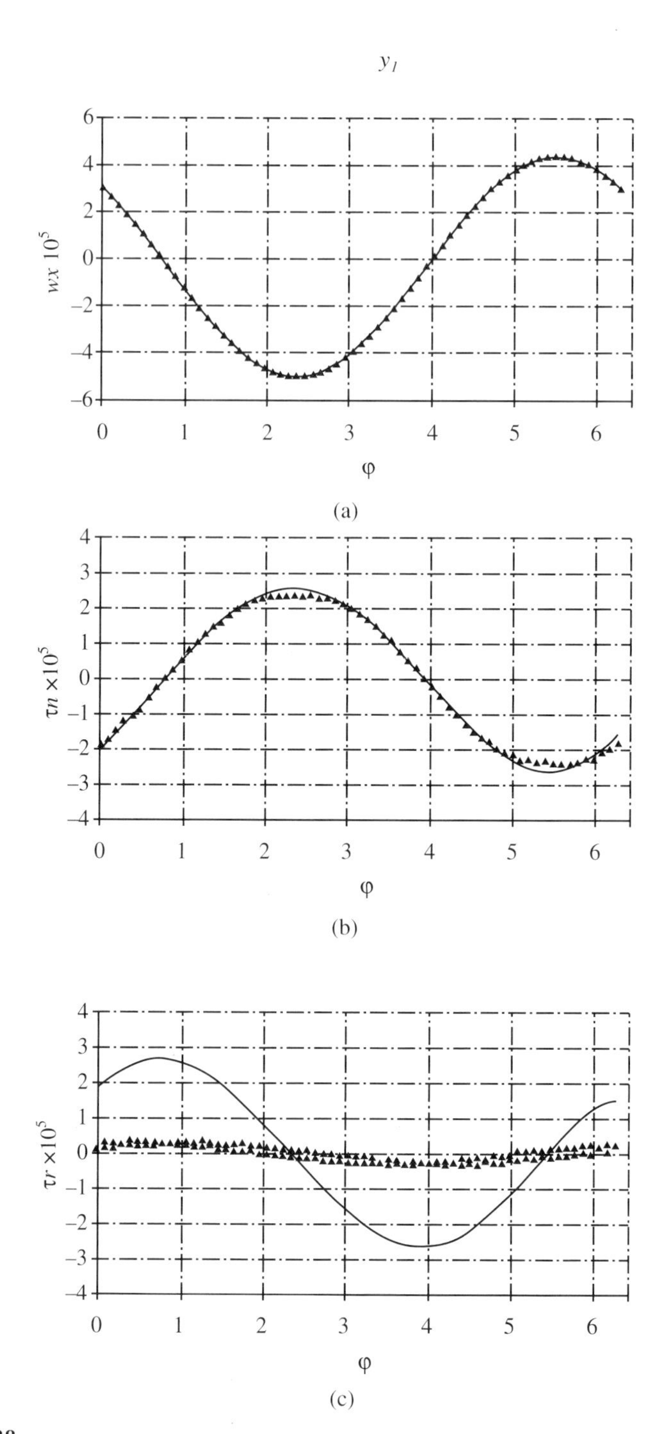

FIGURE 9.38
Numerical results along the interface from the homogenization method. (a) Distribution of warping displacement; (b) distribution of the normal shear stress τ_n; (c) distribution of the tangential shear stress τ_r.

$$q_j^k = q_k^k$$

$$q_k^k = q_l^k$$

and treated conveniently by the procedure discussed in Section 9.4.5. The fiber and the matrix are considered to be isotropic with the shear moduli, $G_f = 10$ and $G_m = 1$, respectively. The computed homogenized shear moduli are

$$\begin{bmatrix} C_{11} & C_{12} \\ Sym & C_{22} \end{bmatrix} = \begin{bmatrix} 1.38271 & -0.00138 \\ Sym & 1.38467 \end{bmatrix} \tag{9.122}$$

Thus the macroscopic behavior of the composite shaft is also isotropic.

The isotropic shaft of the square cross section shown in Figure 9.35a is now analyzed with the homogenized shear moduli (Equation 9.122). Only a quarter of the cross section, the shaded part shown in Figure 9.35a, is discretized because of symmetry. The warping displacements are fixed on the axes of symmetry. The employed finite element mesh with 400 quadrilateral elements and 441 nodes is shown in Figure 9.36b. One unit of torque is applied on the quarter section, with its units being consistent with those of the shear moduli. The computed result for the torsional rigidity $4 \times 1.9927 \times 10^6$ is very close to the accurate value 7.9856×10^6 obtained from the formula [63]

$$\text{Torsional rigidity} = 0.141G(2b)^4 \tag{9.123}$$

where the shear modulus $G = 1.38271$ and the length of the side of the square cross section $2b = 80$. The numerical results for the local fields near or along the interface between the fiber and the matrix adjacent to the point with global coordinates $(x_1 = 30, x_2 = 30)$ are shown in Figure 9.37 and Figure 9.38. Figure 9.37a–c shows the results along the line $3 \le y_1 \le 7, y_2 = 0$ near point P in Figure 9.35b. Figure 9.37b shows the polynomial fitting of the computed shear stress τ_{xz}. On the scale of the figure, results given by the upper and lower elements adjacent to the line cannot be distinguished. Figure 9.37c shows the computed shear stress τ_{yz}, data linked by solid and broken lines, which represent, respectively, the results obtained from the upper and lower elements adjacent to the line in question. The results show that the gradient of the warping displacement changes rapidly across the interface $(y_1 = 5)$ and that the distribution of τ_{xz} but not of τ_{yz} is continuous across the interface. The distribution of warping displacement, and of normal and tangential shear stresses along the interface, are given by

$$\begin{aligned} \tau_n &= \tau_{xz}\cos\varphi + \tau_{yz}\sin\varphi \\ \tau_t &= -\tau_{xz}\sin\varphi + \tau_{yz}\cos\varphi \end{aligned} \tag{9.124}$$

where φ is the angle from the axis y_1 as shown in Figure 9.35b. These distributions are plotted in Figure 9.38a–c. In Figure 9.38b,c, data linked by broken lines represent the results obtained from the matrix side, and the continuous solid line represents the polynomial fit of the results obtained from the fiber side of the interface. These results show that the warping displacement and normal shear stress τ_n vary continuously across the interface, whereas the tangential shear stress τ_t has a significant discontinuity.

Although it will not be possible to compare the results with those obtained by the homogenization method used above, we will directly solve the torsion of the composite shaft shown in Figure 9.35c with the hybrid stress element introduced in Reference 52 to illustrate some typical features of local fields adjacent to the interface. Again, only a quarter of the cross section needs to be discretized because of symmetry. The warping displacements are fixed on the axes of symmetry. The finite element mesh with 3584 quadrilateral elements and 3649 nodes is shown in Figure 9.36c. One unit of torque is applied on the quarter section, with its units being consistent with those of the shear modulus. The computed result for torsional rigidity is $4\times 1.9456356\times 10^6$, which according to the formula (Equation 9.123) corresponds to an isotropic shaft with shear modulus 1.34754. The result is reasonably close to that obtained by the homogenization method (Equation 9.122). The latter predicts larger values of the moduli because the employment of the periodic boundary condition makes the system stiffer. The result given by the homogenization method is also within the lower bound of 1.215 and the upper bound of 2.767 as per the Voigt–Reuss theory [51]. Zhao and Weng [71] have derived the nine effective elastic constants of an orthotropic composite reinforced with monotonically aligned, uniformly dispersed elliptic cylinders using the Eshelby–Mori–Tanaka method. The problem studied above is the special case in which the reinforcements are fibers with circular cross sections. The two shear moduli relevant to torsion given by Zhao and Weng [71] are

$$\frac{C_{22}}{G_m} = 1 + \frac{c_f}{\dfrac{c_m}{1+\alpha} + \dfrac{G_m}{G_f - G_m}} \tag{9.125}$$

where c_f and c_m are volume fractions of fiber and matrix, respectively, and α is the cross-sectional aspect ratio of the reinforced fiber. In our case, $c_f = \pi/4, c_m = 1-\pi/4$ and $\alpha = b/a = 1$, and hence the effective shear moduli $C_{11} = 4.595947 = C_{22}$ given by Equation 9.125 are unreasonably higher than the results of the direct finite element analysis, as well as the results (Equation 9.122) from the homogenization method mentioned above. They are also above the upper bound of the Voigt–Reuss theory.

The Eshelby–Mori–Tanaka method cannot give good results, especially for a high volume fraction of reinforcements, because Eshelby's tensor is based on the inclusion of an infinite matrix, which takes into account of the interaction between reinforcements in a very weak sense. However, it is evident that the homogenization method has the advantage of taking the interaction between phases into account naturally and of not having to make assumptions such as isotropy of the material.

The distribution of warping displacement and shear stresses along the line corresponding to Figure 9.37 and the interface corresponding to Figure 9.38 are plotted in Figure 9.39 and Figure 9.40. Equation 9.124 has been used to obtain the normal and tangential shear stresses in Figure 9.40b,c. A comparison of Figure 9.37 and Figure 9.38 with Figure 9.39 and Figure 9.40, respectively, shows the obvious differences of the results obtained by the homogenization method and the direct hybrid stress element. The differences are to be expected in view of the limited number of fibers that can be handled by the hybrid stress element. The homogenization method is suitable for problems involving many periodically distributed reinforcements so that the RUC occupies only a point in the physical domain [59]. The computed stress fields from the hybrid stress element are smoother than those obtained from the homogenization method, and smoothing techniques are unnecessary for the former since differentiations are avoided in the computations. Notwithstanding these differences, the results of the two methods reveal the common features of the local fields: A significant discontinuity exists in the tangential shear stress, while other fields are continuous adjacent to the interface.

Having gained confidence in the accuracy of the incompatible element NQ6 [66] developed from the homogenization theory in predicting the effective shear moduli, we study below the effect of the cross-sectional shape of the reinforcing fibers. The RUC used is illustrated in Figure 9.41a, that is, an elliptic cylindrical fiber is embedded in the rectangular matrix; the pattern of discretization is similar to Figure 9.36b. The material properties of the fiber as well as the matrix and their volume fractions are selected as above. The variation of the computed C_{11} and C_{22} with the aspect ratio of the fiber is plotted in Figure 9.41b. Again, the results from Equation 9.125 (i.e., the Eshelby–Mori–Tanaka method) are unreasonably higher. However, the predicted trend is the same—within an increase in b/a, C_{11} decreases and C_{22} increases.

9.4.8 Conclusion and Discussion

The homogenization method is most suitable for problems involving a large number of periodically distributed reinforcements so that the RUC can be regarded as a point in the physical domain. It gives not only the equivalent material properties but also detailed information of local fields with much lower computational costs. Such detailed information about the fields on the

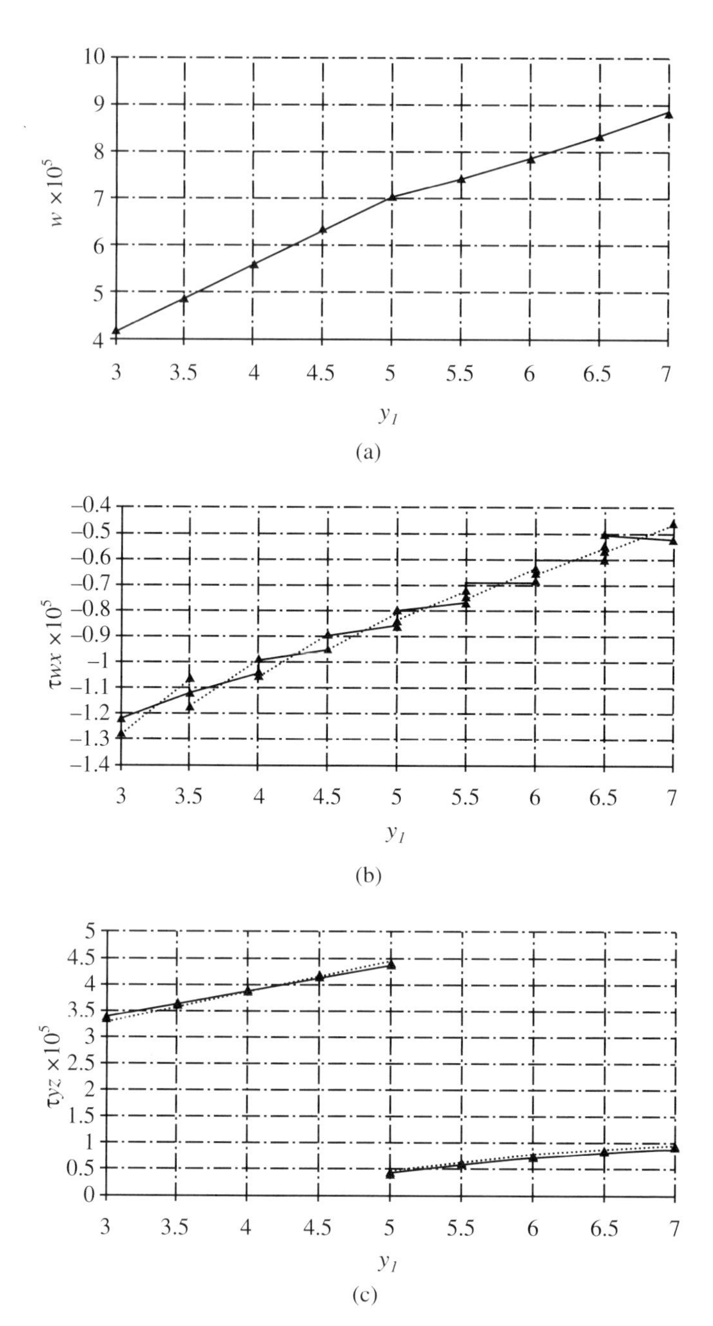

FIGURE 9.39
Numerical results on the line $3 \leq y_1 \leq 7$, $y_2 = 0$ from the hybrid stress element method. (a) Distribution of warping displacement; (b) distribution of τ_{wx}; (c) distribution of τ_{yz}.

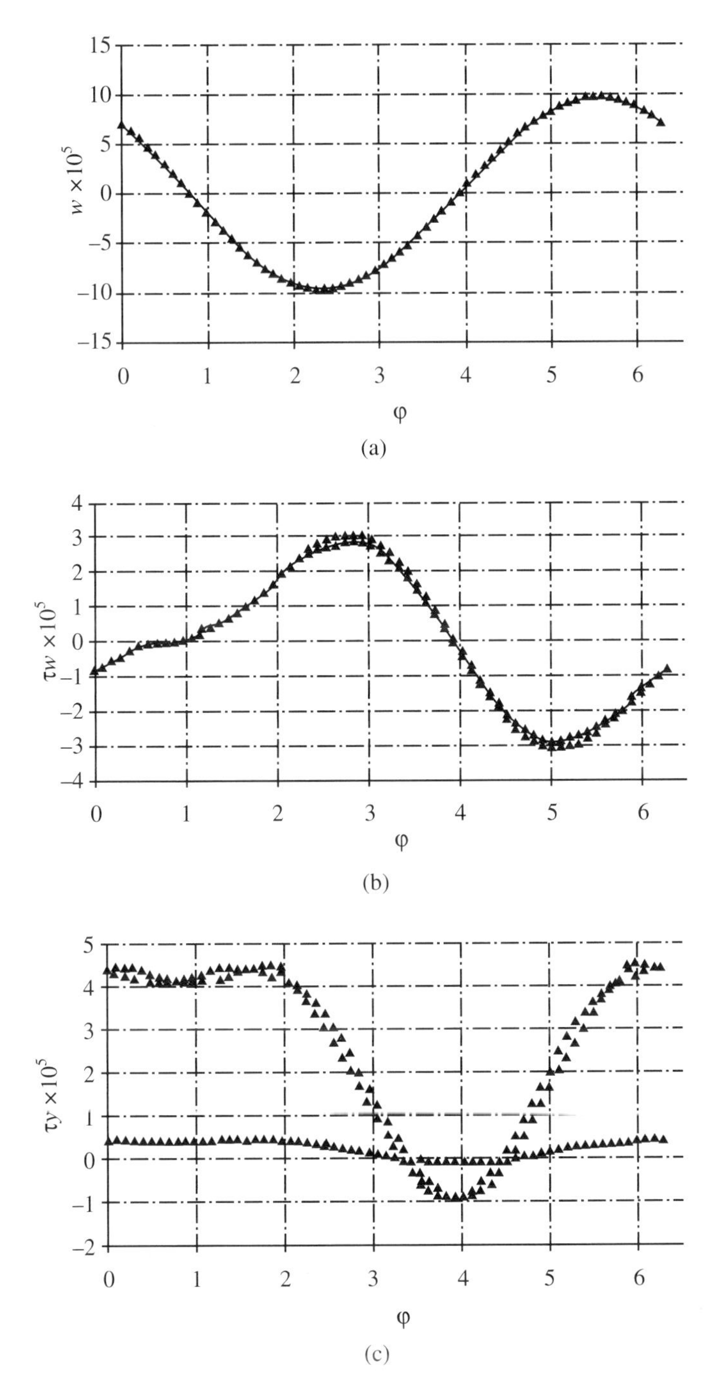

FIGURE 9.40
Numerical results along the interface from the hybrid stress element method. (a) Distribution of warping displacement; (b) distribution of the normal shear stress τ_n ; (c) distribution of the tangential shear stress τ_y.

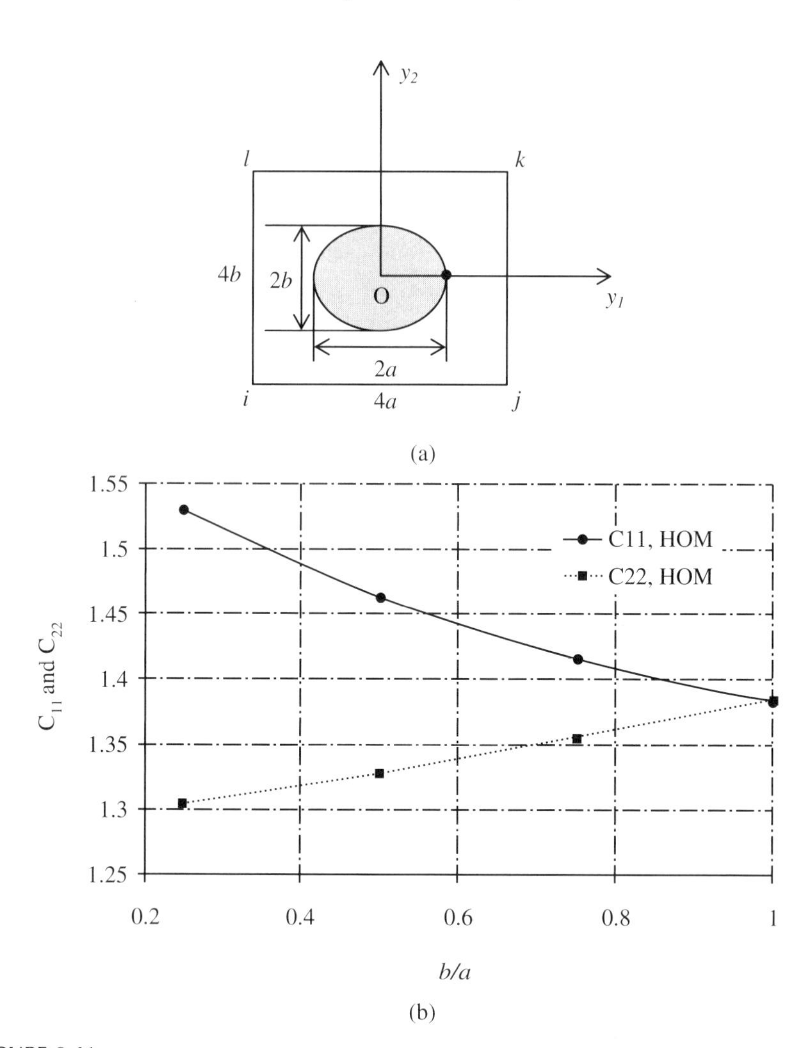

FIGURE 9.41
Effect of the cross-sectional shape of the fiber on the effective shear moduli. (a) Rectangular RUC with an embedded elliptic cylindrical fiber; (b) variation of C_{11} and C_{22} with the aspect ratio of the fiber.

scale of microconstituents is almost impossible to obtain by using the hybrid stress element because of the enormous degrees of freedom needed to model the entire macrodomain with a grid size comparable to that of the microscale features. When the number of the reinforcement is not very large, numerical results by the homogenization method without the terms of order higher than one are usually quantitatively different from those obtained by the direct hybrid stress element. The inclusion of higher terms may improve numerical accuracy, but it inevitably complicates the properties, and in the evaluation of the local fields, the derivatives of the characteristic displacement χ_3^{3k} are needed. Therefore, the widely used isoparametric elements are not suitable for analysis of the RUC. The hybrid stress element is also limited because it is difficult to enforce the periodicity condition on the assumed stresses. From a practical point of view, the incompatible element based on the modified potential principle and the enhanced-strain element based on the three-field Hu–Washizu principle are the most appropriate for analyzing the RUC, whereas the hybrid stress element is appropriate for the macrohomogenized problem.

For torsion of fiber-reinforced composite shafts, a significant discontinuity exists in the tangential shear stress, while other fields are continuous along the interface between the fiber and the matrix.

9.5 A Study of 3-D Braided Piezoceramic Composites

9.5.1 Introduction

Piezoelectric materials have been widely used in smart structures. Most of these materials are laminated. Traditional laminated materials have low intra- and interlaminar strength, lower failure-resistant properties, such as low through-thickness strength, low damage tolerance, and low transverse and shear properties. In recent years, textile structure composites have been developed for structural applications by textile forming techniques. Textile composites include weaving, knitting, stitching, and braiding processes with 2-D and 3-D techniques. 3-D braided composites have presented improved properties and can overcome the weaknesses of traditional laminar structures. 3-D braided composites compare favorably with other advanced textile composites. They present a good balance of in-plane and out-of-plane properties and are well suited for complex shapes and fully integrated structures. However, little research has been done on piezoceramic textile composites due to the difficulties in handling the forming techniques.

Safari et al. [72] have given the first experimental contribution to the research on the two-step braiding performing technique of 3-D piezoceramic composites. The theoretical analysis of piezoceramic fiber composites is still incipient. Ruan et al. [73] have studied the effective properties of the braided and axial yarns of the 3-D two-step connective unit cell and predicted the elastic, piezoelectric, and dielectric constants of the braided composites using an averaging technique.

In general, the object is regarded as a large-scale macroscopic structure. The common approach to modeling the macroscopic properties of 3-D braided composites is to create a representative volume element (RVE) or a unit cell that captures the major features of the underlying microstructure and braided perform. The mechanical and physical properties of the constituent material are always regarded as a small-scale microscopic structure.

The homogenization method was first developed in the early 1980s [74]. As a theoretical and numerical technique for solving macro- and microcoupling problems, homogenization is particularly well suited for understanding the macro- and micromechanics of composite materials [75–77].

Consequent upon the use of multiple-scale expansion, the homogenization method can provide the effective properties of the composite, as well as detailed distribution of fields at the scale of the microconstituents at an acceptable cost. In contrast to the most widely used methods in determining the macro properties, namely, the Eshelby method, the self-consistent method, the Mori–Tanaka method, the differential scheme, and the bound theories, the homogenization method takes into account the interaction between phases and avoids other assumptions except for a periodic distribution of constituents. However, it accounts for microstructure effects on the macroscopic response without explicitly representing the details of the microstructure in the global analysis. In recent years it has been employed in the solution of complex problems in conjunction with the finite element method.

Studies of 3-D four-step piezoceramic braided composites using theoretical and numerical analysis based on the homogenization and finite element methods cannot be found in the specialized literatures. In this section, we will employ the homogenization method for predicting the effective properties in this type of composite.

9.5.2 Basic Equations

Consider a heterogeneous elastomer with periodic structure, consisting of two components, as illustrated in Figure 9.42. This body has two length scales, a global length scale D that is of the order of the body size and a local length scale d that is proportional to the wavelength of the variation of the microstructure. The size of the unit cell is assumed to be much smaller than the size of the body. The relation between the global coordinate system x_i for the body and the local system y_i for the minimum repeated unit cell can be written as

$$y_i = \frac{x_i}{\varepsilon} \tag{9.126}$$

where, is the scaling factor between the two length scales. For the actual heterogeneous body, when it is subject to external forces, the field quantities such as displacements and stresses will vary with not only the global coordinate system x, but also the local system y. Because of the inhomogeneity of the microstructure, these quantities will vary rapidly within a short wavelength.

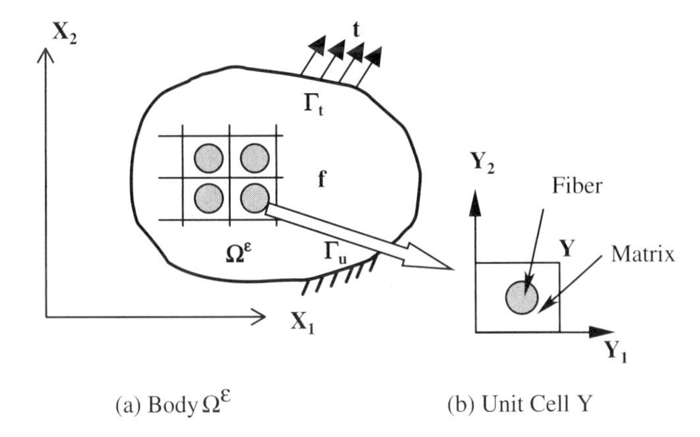

FIGURE 9.42
Two-component periodic structure body and its unit cell.

In the piezoelectric problem, the mechanical and electric field quantities are coupled, so it is more difficult to solve than the purely mechanical or electric problem. The unknown field quantities of the piezoelectric problem can be broken into two parts. The first is relative to mechanics and is described by following the equilibrium and compatibility equations. On the stress σ_{ij}, strain e_{ij}, and displacement u_i components:

$$\sigma^{\varepsilon}_{ij,j} + f_i = 0 \qquad \text{in} \quad \Omega^{\varepsilon} \tag{9.127}$$

$$e^{\varepsilon}_{ij} = \frac{1}{2}(u^{\varepsilon}_{i,j} + u^{\varepsilon}_{j,i}) \quad \text{in} \;\; \Omega^{\varepsilon} \tag{9.128}$$

For the others, such as the electric displacement D_i, the electric field vector E_i, and the electric potential φ,

$$D^{\varepsilon}_{i,i} + \rho_b = 0 \qquad \text{in} \quad \Omega^{\varepsilon} \tag{9.129}$$

$$E^{\varepsilon}_i = -\varphi^{\varepsilon}_{,i} \qquad \text{in} \quad \Omega^{\varepsilon} \tag{9.130}$$

The boundary conditions of the problem are

$$\sigma^{\varepsilon}_{ij} n_j = \overline{T_i} \qquad \text{on} \quad \Gamma^{\varepsilon}_T \tag{9.131}$$

$$D^{\varepsilon}_j n_j = \overline{\varpi} \qquad \text{on} \quad S^{\varepsilon}_{\varpi} \tag{9.132}$$

$$u^{\varepsilon}_i = \overline{u_i} \qquad \text{on} \quad \Gamma^{\varepsilon}_u \tag{9.133}$$

$$\varphi^{\varepsilon} = \overline{\varphi} \qquad \text{on} \quad S^{\varepsilon}_{\varphi} \tag{9.134}$$

f_i and ρ_b are the body force and the extrinsic bulk charge, respectively, in the body Ω^ε, and n_j is the outward normal vector of the boundary. $\overline{T}_i$ and $\overline{\omega}$ are the given external force on the boundary Γ_T^ε and the charge on the boundary S_ω^ε, respectively, and $\overline{u}_i$ and $\overline{\varphi}$ are the given displacement on the boundary Γ_u^ε and the given electric potential on the boundary S_φ^ε, respectively.

The constitutive relations of piezoelectric materials are coupled. Although these relations are nonlinear for an actual material, they can be treated as similarly linear to an extent. Linearization of the constitutive relations of a piezoelectric material is used here to simplify the analysis.

In the mechanical–electric coupled constitutive equations

$$\sigma_{ij}^\varepsilon = C_{ijkl}e_{kl}^\varepsilon - \breve{e}_{ijk}^T E_k^e \tag{9.135a}$$

$$D_i^\varepsilon = \breve{e}_{ijk}e_{jk} + \varepsilon_{ij}E_j^\varepsilon \tag{9.135b}$$

C_{ijkl}, $\hat{e}_{ijk}$, and ε_{ij} are the elastic, piezoelectric, and dielectric moduli of the material, respectively. The equation also can be written in matrix form:

$$\begin{Bmatrix} \sigma^\varepsilon \\ \boldsymbol{D} \end{Bmatrix} = \begin{bmatrix} \widehat{C} & -\hat{e}^T \\ \hat{e} & \hat{\varepsilon} \end{bmatrix} \begin{Bmatrix} \boldsymbol{e}^\varepsilon \\ \boldsymbol{E}^\varepsilon \end{Bmatrix} = C \begin{Bmatrix} \boldsymbol{e}^\varepsilon \\ \boldsymbol{E}^\varepsilon \end{Bmatrix} \tag{9.136}$$

where

$$\sigma^\varepsilon = \{\sigma_{11} \quad \sigma_{22} \quad \sigma_{33} \quad \tau_{23} \quad \tau_{13} \quad \tau_{12}\}^T,$$

$$\mathbf{e}^\varepsilon = \{e_{11} \quad e_{22} \quad e_{33} \quad \gamma_{23} \quad \gamma_{13} \quad \gamma_{12}\}^T,$$

$$\mathbf{D}^\varepsilon = \{D_1 \quad D_2 \quad D_3\}^T,$$

and

$$\mathbf{E}^\varepsilon = \{E_1 \quad E_2 \quad E_3\}^T$$

9.5.3 Asymptotic Expansion

We look for an asymptotic expansion of unknown displacement u_i^ε and electric potential φ^ε as follows:

$$u_i^\varepsilon(x) = u_i^{(0)}(x,y) + \varepsilon u_i^{(1)}(x,y) + \varepsilon^2 u_i^{(2)}(x,y) + \ldots \tag{9.137a}$$

$$\varphi^\varepsilon(x) = \varphi^{(0)}(x,y) + \varepsilon\varphi^{(1)}(x,y) + \varepsilon^2\varphi^{(2)}(x,y) + \ldots \tag{9.137b}$$

where the functions $u_i^{(0)}, u_i^{(1)}, u_i^{(2)}$ ... and $\varphi^{(0)}, \varphi^{(1)}$, and $\varphi^{(2)}$ are Y-periodic with the local coordinate y. The strain $e_{ij}\ (u_i^{\varepsilon})$ and electric field vector $E_i\ (\varphi^{\varepsilon})$ can be expanded similarly:

$$e_{ij}(u_i^{\varepsilon}) = \varepsilon^{-1}e_{ij}^{(-1)} + e_{ij}^{(0)} + \varepsilon e_{ij}^{(1)} + \ldots \tag{9.138a}$$

$$E_i(\varphi^{\varepsilon}) = \varepsilon^{-1}E_i^{(-1)} + E_i^{(0)} + \varepsilon E_i^{(1)} + \ldots \tag{9.138b}$$

to obtain the following definitions from Equation 9.128 and Equation 9.130:

$$e_{ij}^{(-1)} = \frac{1}{2}\left(\frac{\partial u_i^{(0)}}{\partial y_j} + \frac{\partial u_j^{(0)}}{\partial y_i}\right) \tag{9.139a}$$

$$e_{ij}^{(n)} = \frac{1}{2}\left(\frac{\partial u_i^{(n)}}{\partial x_j} + \frac{\partial u_j^{(n)}}{\partial x_i} + \frac{\partial u_i^{(n+1)}}{\partial y_j} + \frac{\partial u_j^{(n+1)}}{\partial y_i}\right), n = 0,1,\ldots \tag{9.139b}$$

$$E_i^{(-1)} = -\frac{\partial \varphi^{(0)}}{\partial y_i}, \tag{9.139c}$$

$$E_i^{(n)} = -\left(\frac{d\varphi^{(n)}}{dx_i} + \frac{d\varphi^{(n+1)}}{dy_i}\right), n = 0.1,\ldots, \tag{9.139d}$$

The stress $\sigma_{ij}^{\varepsilon}$ and electric displacement D_i^{ε} are treated similarly:

$$\sigma_{ij}^{\varepsilon} = \varepsilon^{-1}\sigma_{ij}^{(-1)} + \sigma_{ij}^{(0)} + \varepsilon\sigma_{ij}^{(1)} + \ldots \tag{9.140a}$$

$$D_i^{\varepsilon} = \varepsilon^{-1}D_i^{(-1)} + D_i^{(0)} + \varepsilon D_i^{(1)} + \ldots \tag{9.140b}$$

The following definitions are derived from Equation 9.135:

$$\sigma_{ij}^{(n)} = C_{ijkl}e_{kl}^{(n)}(x,y) - e_{ijk}^{T}E_k^{(n)}(x,y), n = -1,0,1,\ldots \tag{9.141a}$$

$$D_i^{(n)} = e_{ijk}e_{ik}^{(n)}(x,y) + \varepsilon_{ij}E_j^{(n)}(x,y), n = -1,0,1,\ldots \tag{9.141b}$$

9.5.4 Microscale Independent Terms

The following results are deduced from Equation 9.127 and Equation 9.129, and expansions (Equation 9.140a and Equation 9.140b):

$$\frac{\partial \sigma_{ij}^{(-1)}}{\partial y_j} = 0, \qquad \text{and} \qquad \frac{\partial \mathrm{D}_{\mathrm{j}}^{(\ 1)}}{\partial y_j} = 0 \tag{9.142a}$$

$$\frac{\partial \sigma_{ij}^{(0)}}{\partial y_j} + \frac{\partial \sigma_{ij}^{(-1)}}{\partial x_j} = 0, \quad \text{and} \quad \frac{\partial D_j^{(0)}}{\partial y_j} + \frac{\partial D_j^{(-1)}}{\partial x_j} = 0 \tag{9.142b}$$

$$\frac{\partial \sigma_{ij}^{(1)}}{\partial y_j} + \frac{\partial \sigma_{ij}^{(0)}}{\partial x_j} + f_i = 0, \ldots \tag{9.142c}$$

$$\frac{\partial D_j^{(1)}}{\partial y_j} + \frac{\partial D_j^{(0)}}{\partial x_j} + \rho_b = 0, \ldots \tag{9.142d}$$

From Equation 9.142a, we get

$$\frac{\partial}{\partial y_j}\left[C_{ijkl} e_{kl}^{(-1)} - \breve{e}_{kl}^{(-1)} E_k^{(-1)}\right] = 0 \tag{9.143a}$$

$$\frac{\partial}{\partial y_j}\left[\breve{e}_{ijk} e_{jk}^{(-1)} + \varepsilon_{ij} E_j^{(-1)}\right] = 0 \tag{9.143b}$$

Inserting Equation 9.139a and Equation 9.139c into the above equations, multiplying each equation by $u_i^{(0)}$ and $\varphi^{(0)}$, respectively, integrating on the Y domain, and adding the resulting equations, we obtain the following orthogonal condition:

$$\int_Y \frac{\partial \varphi^{(0)}}{\partial y_i} \varepsilon_{ij} \frac{\partial \varphi^{(0)}}{\partial y_j} dY + \int_Y \frac{\partial u_i^{(0)}}{\partial y_j} C_{ijkl} \frac{\partial u_k^{(0)}}{\partial y_l} dY = 0 \tag{9.144}$$

As the constitutive matrices C_{ijkl} and ε_{ij} are always positive, the result above proves that $u_i^{(0)}$ and $\varphi^{(0)}$ are independent of the local coordinate y:

$$\frac{\partial u_i^{(0)}}{\partial y_j} = 0 \quad \text{and} \quad u_i^{(0)}(x, y) = u_i^{(0)}(x) \tag{9.145a}$$

$$\frac{\partial \varphi^{(0)}}{\partial y_j} = 0 \quad \text{and} \quad \varphi^{(0)}(x, y) = \varphi^{(0)}(x) \tag{9.145b}$$

Considering this condition, the strain and stress components of order $n = -1$ defined by Equation 9.139a, Equation 9.139c, and Equation 9.141b are:

$$e_{ij}^{(-1)} = 0 \quad \text{and} \quad E_{ij}^{(-1)} = 0 \tag{9.146a}$$

$$\sigma_{ij}^{(-1)} = 0 \quad \text{and} \quad D_{ij}^{(-1)} = 0 \tag{9.146b}$$

9.5.5 Homogenized Problem

The objective of the homogenization method is to replace the problem defined by Equation 9.127 to Equation 9.135 with the following equivalent homogeneous problem:

$$\frac{\partial \sigma_{xij}}{\partial x_j} + f_{xi} = 0 \quad \text{in} \quad \Omega \tag{9.147}$$

$$e_{xij} = \frac{1}{2}\left(\frac{\partial u_i^{(0)}}{\partial x_j} + \frac{\partial u_j^{(0)}}{\partial x_i}\right) \quad \text{in} \quad \Omega \tag{9.148}$$

$$\frac{\partial D_{xi}}{\partial x_i} + \rho_{bx} = 0 \quad \text{in} \quad \Omega \tag{9.149}$$

$$\phi_{xi} = -\frac{\partial \varphi^{(0)}}{\partial x_i} \quad \text{in} \quad \Omega \tag{9.150}$$

$$\begin{Bmatrix} \sigma_x \\ D_x \end{Bmatrix} = D^H \begin{Bmatrix} e_x \\ \phi_x \end{Bmatrix} \quad \text{in} \quad \Omega \tag{9.151}$$

$$\sigma_{xij} n_j = \overline{T}_{xi} \ \text{on} \ \Gamma_T \tag{9.152}$$

$$D_{xj} n_j = \omega_x \ \text{on} \ S_\omega \tag{9.153}$$

$$u_i^{(0)} = \overline{u}_i \ \text{on} \ \Gamma_u \tag{9.154}$$

$$\varphi^{(0)} = \overline{\varphi} \ \text{on} \ S_\varphi \tag{9.155}$$

In the equations above, D^H is the homogenized constitutive matrix, and σ_x and D_x represent the average values of $\sigma^{(0)}$ and $D^{(0)}$ in the Y domain:

$$\sigma_x = \frac{1}{|y|}\int \sigma^{(0)} dy \quad \text{and} \quad D_x = \frac{1}{|y|}\int S^{(0)} dy \tag{9.156}$$

Equation 9.147 and Equation 9.149 represent the average enforcement of the local conditions (Equation 9.142c and Equation 9.142d). The same Equation 9.152 and Equation 9.153, regard now to the local conditions (Equation 9.131 and Equation 9.132, respectively). Hence, f_x, ρ_{bx}, $\overline{T}_x$, and $\overline{\omega}_x$ represent the constituent average values for body force, extrinsic bulk charge, and applied tractions and potential, respectively.

Hence, according to the results (Equation 9.146), Equation 9.142b presented in Section 9.5.4 that are changed to the following:

$$\frac{\partial \sigma_{ij}^{(0)}}{\partial y_j} = 0 \quad \text{and} \quad \frac{\partial \mathrm{D}_{\mathrm{j}}^{(0)}}{\partial y_j} = 0 \tag{9.157}$$

$$e_{ij}^{(0)} = e_{xij} + e_{yij} \text{ and } E_i^{(0)} = \phi_{xi} + \phi_{yi} \tag{9.158}$$

$$\sigma_{ij}^{(0)} = C_{ijkl}e_{kl}^{(0)} - e_{ijk}^T E_k^{(0)} \quad \text{and} \quad D_i^{(0)} = e_{ijk}e_{jk}^{(0)} + \varepsilon_{ij}E_j^{(0)} \tag{9.159}$$

where

$$e_{yij} = \frac{1}{2}\left(\frac{\partial u_i^{(1)}}{\partial y_j} + \frac{\partial u_j^{(1)}}{\partial y_i}\right) \tag{9.160a}$$

$$\phi_{yi} = -\frac{\partial \varphi^{(1)}}{\partial y_i} \tag{9.160b}$$

We assume the following approximations for the first-order displacement and potential:

$$\begin{Bmatrix} u_i^{(1)}(x,y) \\ \varphi^{(1)}(x,y) \end{Bmatrix} = \begin{Bmatrix} \chi_i^{kl}(y) & \Pi_i^j(y) \\ \Gamma^{kl}(y) & \Lambda^j(y) \end{Bmatrix} \begin{Bmatrix} e_{xkl}(u_i^{(0)}) \\ \phi_{xj}(\varphi^{(0)}) \end{Bmatrix} \tag{9.161}$$

where $\chi_i^{kl}(y)$, $\Pi_i^j(y)$, $\Pi^{kl}(y)$, and $\Lambda^k(y)$ are Y-periodic functions defined in the unit cell Y. These functions are constrained to satisfy Equation 9.157, which can be written as follows, using Equation 9.158 to Equation 9.160:

$$\frac{\partial}{\partial y_j}\left[C_{ijkl}e_{ykl}(u_i^{(1)} - \breve{e}_{ijk}^T\phi_{xk}(\varphi^{(1)})\right] = -\frac{\partial}{\partial y_j}\left[C_{ijkl}e_{xkl}(u_i^{(0)}) - \breve{e}_{ijk}^T\phi_{xk}(\varphi^{(0)})\right] \tag{9.162a}$$

$$\frac{\partial}{\partial y_j}\left[\breve{e}_{jkl}e_{ykl}(u_i^{(1)}) + \varepsilon_{jk}\phi_{yk}(\varphi^{(1)})\right] = -\frac{\partial}{\partial y_j}\left[\breve{e}_{jkl}e_{xkl}(u_i^{(0)}) + \varepsilon_{jk}\phi_{xk}(\varphi^{(0)})\right] \tag{9.162b}$$

The definitions below for the zero-th order stress and electric displacement terms are obtained by inserting the approximation (Equation 9.161) into the definitions (Equation 9.158 and Equation 9.159):

$$\sigma_{ij}^{(0)}(x,y) = C_{ijmn}\left\{\left[T_{mn}^{kl} + e_{ymn}(\chi_i^{kl})\right]e_{xkl}(u_i^{(0)}) + e_{ymn}(\Pi_i^j)\phi_{xj}(\varphi^{(0)})\right\} - \breve{e}_{ijp}^T\left\{\phi_{yp}(\Gamma^{kl})e_{xkl}(u_i^{(0)}) + [M_p^k + \phi_{yp}(\Lambda^k)]\phi_{xk}(\varphi^{(0)})\right\} \tag{9.163a}$$

$$D_i^{(0)}(x,y) = \breve{e}_{ipq}\left\{[N_{pq}^{kl} + e_{ypq}(\chi_i^{kl})]e_{xkl}(u_i^{(0)}) + e_{ypq}(\Pi_i^j)\phi_{xj}(\varphi^{(0)})\right\} + \varepsilon_{ip}\left\{\phi_{yp}(\Gamma^{kl})e_{xkl}(u_i^{(0)}) + [H_p^k + \phi_{yp}(\Lambda^k)]\phi_{xk}(\varphi^{(0)})\right\} \tag{9.163b}$$

They can be written in matrix form:

$$\begin{Bmatrix}\sigma^{(0)}\\ D^{(0)}\end{Bmatrix} = \begin{bmatrix} C\left[T + e_y(\chi)\right] - \breve{e}^T\phi_y(\Gamma) & -\breve{e}^T\left[M + \phi_y(\Lambda)\right] + Ce_y(\Pi) \\ \breve{e}\left[N + e_y(\chi)\right] + \varepsilon\,\phi_y(\Gamma) & \varepsilon\left[H + \phi_y(\Lambda)\right] + \breve{e}e_y(\Pi)\end{bmatrix}$$

$$\begin{Bmatrix} e_x \\ \phi_x \end{Bmatrix} = D\begin{Bmatrix} e_x \\ \phi_x \end{Bmatrix} \tag{9.164}$$

$$e_y(\chi) = e_{yij}(\chi_k^{pq}), e_y(\Pi) = e_{yij}(\Pi_k'), e_x = e_{xpq}(u_k^{(0)})$$

$$\phi_y(\Gamma) = \phi_{yp}(\Gamma^{kl}), \phi_y(\Lambda) = \phi_{yp}(\Lambda^k), \phi_x = \phi_{xp}(\varphi^{(0)})$$

In the equations above, T_{ij}^{kl} is a fourth-order unit tensor $T_{ij}^{kl} = \frac{1}{2}(\delta_{ik}\delta_{jl} + \delta_{il}\delta_{jk})$, where ij is the Kronecker symbol. N_{pq}^{kl} is also a fourth-order unit tensor, and M_p^k and H_p^k are second-order unit tensors.

$\sigma^{(0)}(x,y)$ and $D_i^{(0)}(x,y)$ are the second asymptotic expansion patterns of $\sigma^{(\varepsilon)}(x)$ and $D^{(\varepsilon)}(x)$, respectively. Therefore, Equation 9.163a and Equation 9.163b together result in Equation 9.146b, which can be used to determine the microscopic stress and electric displacement field:

$$\lim_{\varepsilon\to 0} \sigma_{ij}^{\varepsilon} = \sigma_{ij}^{(0)}$$

$$\lim_{\varepsilon\to 0} D_i^{\varepsilon}(x) = D_i^{(0)}$$

The homogenized equivalent elastic-electric constant present in Equation 9.151 is independent of the y coordinator system. It is obtained by inserting the definition (Equation 9.164) for the microscopic stress and electric displacement fields into the average definition (Equation 9.156):

$$\boldsymbol{D}^H = \frac{1}{|y|}\int \boldsymbol{D}dy \tag{9.165}$$

9.5.6 Finite Element Variational Statements

In order to obtain the homogenized constant, it is necessary to determine $\chi_i^{kl}, \Pi_i^j, \Gamma^{kl}$, and Λ^k, using Equation 9.162a and Equation 9.162b. According to the approximation (Equation 9.161), we obtain

$$\begin{bmatrix} C_{ijkl} & -\breve{e}_{ijk}^T \\ \breve{e}_{ijk} & \varepsilon_{ij} \end{bmatrix} \frac{\partial}{\partial y_j} \left\{ \begin{bmatrix} e_{ykl} & 0 \\ 0 & \phi_{yl} \end{bmatrix} \begin{bmatrix} \chi_i^{kl} & \Pi_i^j \\ \Gamma^{kl} & \Lambda^j \end{bmatrix} \right\} + \frac{\partial}{\partial y_j} \begin{bmatrix} C_{ijkl} & -\breve{e}_{ijk}^T \\ \breve{e}_{ijk} & \varepsilon_{ij} \end{bmatrix} = 0 \quad (9.166)$$

The virtual work principle states that

$$\int_Y \begin{bmatrix} e_{ykl} & 0 \\ 0 & \phi_{yl} \end{bmatrix} \delta \begin{bmatrix} \chi_i^{kl} & \Pi_i^j \\ \Gamma^{kl} & \Lambda^j \end{bmatrix} \begin{bmatrix} C_{ijkl} & -\breve{e}_{ijk}^T \\ \breve{e}_{ijk} & \varepsilon_{ij} \end{bmatrix} \begin{bmatrix} e_{ykl} & 0 \\ 0 & \phi_{yl} \end{bmatrix} \begin{bmatrix} \chi_i^{kl} & \Pi_i^j \\ \Gamma^{kl} & \Lambda^j \end{bmatrix} dY + \int_Y \begin{bmatrix} e_{ykl} & 0 \\ 0 & \phi_{yl} \end{bmatrix} \delta \begin{bmatrix} \chi_i^{kl} & \Pi_i^j \\ \Gamma^{kl} & \Lambda^j \end{bmatrix} \begin{bmatrix} C_{ijkl} & -\breve{e}_{ijk}^T \\ \Gamma^{kl} & \Lambda^j \end{bmatrix} \begin{bmatrix} C_{ijkl} & -\breve{e}_{ijk}^T \\ \breve{e}_{ijk} & \varepsilon_{ij} \end{bmatrix} dY = 0 \quad (9.167)$$

It is easy to prove that this equation is the first-order variation of the following potential functional:

$$\Pi_P\left(\begin{bmatrix} \chi_i^{kl} & \Pi_i^j \\ \Gamma^{kl} & \Lambda^j \end{bmatrix}\right) = \sum_e \int_Y \frac{1}{2} \begin{bmatrix} e_{ykl} & 0 \\ 0 & \Phi_{yl} \end{bmatrix} \begin{bmatrix} \chi_i^{kl} & \Pi_i^j \\ \Gamma^{kl} & \Lambda^j \end{bmatrix} \begin{bmatrix} C_{ijkl} & -\breve{e}_{ijk}^T \\ \breve{e}_{ijk} & \varepsilon_{ij} \end{bmatrix} \begin{bmatrix} e_{ykl} & 0 \\ 0 & \Phi_{yl} \end{bmatrix} \begin{bmatrix} \chi_i^{kl} & \Pi_i^j \\ \Gamma^{kl} & \Lambda^j \end{bmatrix} dY + \int_Y \begin{bmatrix} e_{ykl} & 0 \\ 0 & \Phi_{yl} \end{bmatrix} \begin{bmatrix} \chi_i^{kl} & \Pi_i^j \\ \Gamma^{kl} & \Lambda^j \end{bmatrix} \begin{bmatrix} C_{ijkl} & -\breve{e}_{ijk}^T \\ \breve{e}_{ijk} & \varepsilon_{ij} \end{bmatrix} dY \quad (9.168)$$

If we define the extensive strain and the extensive electric field vector

$$\begin{Bmatrix} \bar{\varepsilon}^{kl} \\ \bar{E}^{kl} \end{Bmatrix} = \begin{bmatrix} e_{ykl} & 0 \\ 0 & \Phi_{yl} \end{bmatrix} \begin{bmatrix} \chi_i^{kl} & \Pi^j \\ \Gamma^{kl} & \Lambda^j \end{bmatrix} \quad (9.169)$$

$$\begin{Bmatrix} \bar{\varepsilon}^{kl} \\ \bar{E}^{kl} \end{Bmatrix} = \begin{bmatrix} C_{ijkl} & -\breve{e}_{ijk}^T \\ \breve{e}_{ijk} & \varepsilon_{ij} \end{bmatrix}^{-1} \begin{Bmatrix} \tilde{\sigma}^{kl} \\ \tilde{D}^{kl} \end{Bmatrix} = \mathbf{C}^{-1} \begin{Bmatrix} \tilde{\sigma}^{kl} \\ \tilde{D}^{kl} \end{Bmatrix} = \mathbf{S} \begin{Bmatrix} \tilde{\sigma}^{kl} \\ \tilde{D}^{kl} \end{Bmatrix} \quad (9.170)$$

which are Y-periodic functions in the unit cell, we obtain a two-field Hellinger–Reissner functional:

$$\Pi_{HR}\left(\begin{bmatrix}\chi_i^{kl} & \Pi_i^j \\ \Gamma^{kl} & \Lambda^j\end{bmatrix}, \begin{Bmatrix}\tilde{\sigma}^{kl} \\ \tilde{D}^{kl}\end{Bmatrix}\right)$$

$$= \int_Y \left[-\frac{1}{2}\begin{Bmatrix}\tilde{\sigma}^{kl} \\ \tilde{D}^{kl}\end{Bmatrix}^T \mathbf{S} \begin{Bmatrix}\tilde{\sigma}^{kl} \\ \tilde{D}^{kl}\end{Bmatrix} + \begin{Bmatrix}\tilde{\sigma}^{kl} \\ \tilde{D}^{kl}\end{Bmatrix}^T \begin{bmatrix} e_{ykl} & 0 \\ 0 & \Phi_{yl}\end{bmatrix}\begin{bmatrix}\chi_i^{kl} & \Pi_i^j \\ \Gamma^{kl} & \Lambda^j\end{bmatrix} \right. \tag{9.171}$$

$$\left. + \begin{bmatrix} e_{ykl} & 0 \\ 0 & \Phi_{yl}\end{bmatrix}\begin{bmatrix}\chi_i^{kl} & \Pi_i^j \\ \Gamma^{kl} & \Lambda^j\end{bmatrix}\mathbf{C}\right] dY$$

Using the Lagrange multiplier method and relaxing the compatibility condition in the potential (Equation 9.168), or by employing Legendre transformation on the Hellinger–Reissner potential (Equation 9.171), we obtain the three-field Hu–Washizu functional:

$$\Pi_{HW}\left(\begin{bmatrix}\chi_i^{kl} & \Pi_i^j \\ \Gamma^{kl} & \Lambda^j\end{bmatrix}, \begin{Bmatrix}\bar{\varepsilon}^{kl} \\ \bar{E}^{kl}\end{Bmatrix}, \begin{Bmatrix}\tilde{\sigma}^{kl} \\ \tilde{D}^{kl}\end{Bmatrix}\right)$$

$$= \int_Y \left[-\frac{1}{2}\begin{Bmatrix}\tilde{\sigma}^{kl} \\ \tilde{D}^{kl}\end{Bmatrix}^T \mathbf{S} \begin{Bmatrix}\tilde{\sigma}^{kl} \\ \tilde{D}^{kl}\end{Bmatrix} - \begin{Bmatrix}\tilde{\sigma}^{kl} \\ \tilde{D}^{kl}\end{Bmatrix}^T \left(\begin{Bmatrix}\tilde{\varepsilon}^{kl} \\ \tilde{E}^{kl}\end{Bmatrix} - \begin{bmatrix} e_{ykl} & 0 \\ 0 & \Phi_{yl}\end{bmatrix}\begin{bmatrix}\chi_i^{kl} & \Pi_i^j \\ \Gamma^{kl} & \Lambda^j\end{bmatrix} \right) \right. \tag{9.172}$$

$$\left. + \begin{bmatrix} e_{ykl} & 0 \\ 0 & \Phi_{yl}\end{bmatrix}\begin{bmatrix}\chi_i^{kl} & \Pi_i^j \\ \Gamma^{kl} & \Lambda^j\end{bmatrix}\mathbf{C}\right] dY$$

Although hybrid elements based on the Hellinger–Reissner principle or the Hu–Washizu principle can improve the accuracy of the approximate displacement and stress solutions, they will not be used here. In this section we will instead introduce displacement-incompatible elements based on the potential (Equation 9.168).

9.5.7 Finite Element Implementation

Subdivide the unit cell domain Y into the finite element subdomains Y_e, such that $\cup Y_e = Y, Y_a \cap Y_b = \varnothing$ and $\partial Y_a \cap \partial Y_b = S_{ab}$ (a and b are arbitrary elements).

In each element, the term $\begin{bmatrix}\chi_i^{kl} & \Pi_i^j \\ \Gamma^{kl} & \Lambda^j\end{bmatrix}$ is divided into a compatible part $\begin{bmatrix}\chi_{iq}^{kl} & \Pi_{iq}^j \\ \Gamma_q^{kl} & \Lambda_q^j\end{bmatrix}$, so that the functional (Equation 9.168) can be rewritten as

$$\Pi_p\left(\begin{bmatrix}\chi_i^{kl} & \Pi_i^j\\ \Gamma^{kl} & \Lambda^j\end{bmatrix}=\begin{bmatrix}\chi_{iq}^{kl} & \Pi_{i\lambda}^j\\ \Gamma^{kl} & \Lambda_q^j\end{bmatrix}+\begin{bmatrix}\chi_{i\lambda}^{kl} & \Pi_{i\lambda}^j\\ \Gamma_\lambda^{kl} & \Lambda_\lambda^j\end{bmatrix}\right)$$

$$=\sum_e\int_{Y_e e}\frac{1}{2}\begin{bmatrix}e_{ykl} & 0\\ 0 & \Phi_{yl}\end{bmatrix}\begin{bmatrix}\chi_i^{kl} & \Pi_i^j\\ \Gamma^{kl} & \Lambda^j\end{bmatrix}\mathbf{C}\begin{bmatrix}e_{ykl} & 0\\ 0 & \Phi_{yl}\end{bmatrix}\begin{bmatrix}\chi_i^{kl} & \Pi_i^j\\ \Gamma^{kl} & \Lambda^j\end{bmatrix}dY \tag{9.173}$$

$$+\int_{Y_e}\begin{bmatrix}e_{ykl} & 0\\ 0 & \Phi_{yl}\end{bmatrix}\begin{bmatrix}\chi_{iq}^{kl} & \Pi_{iq}^j\\ \Gamma_q^{kl} & \Lambda_q^j\end{bmatrix}\mathbf{C}dY$$

The stationary condition of the above functional recovers the equilibrium (Equation 9.162a and Equation 9.162b) and the equilibrium of interelement tractions if the following PTC is met *a priori*:

$$\int_{Y_e}\begin{bmatrix}e_{ykl} & 0\\ 0 & \Phi_{yl}\end{bmatrix}\begin{bmatrix}\chi_{i\lambda}^{kl} & \Pi_{i\lambda}^j\\ \Gamma_\lambda^{kl} & \Lambda_\lambda^j\end{bmatrix}dY=0$$

or equivalently:

$$\oint_{\partial Y_e}\begin{bmatrix}\chi_{i\lambda}^{kl} & \Pi_{i\lambda}^j\\ \Gamma_\lambda^{kl} & \Lambda_\lambda^j\end{bmatrix}n_j\,ds=0 \tag{9.174}$$

The incompatible functions meeting the PTC can now be easily formulated. If we refer to the eight-node 3-D isoparametric element, the compatible field value is related to the nodal values ${}^e q^k$ (relating to the mechanical displacement) and ${}^e\Lambda^k$ (relating to electric potential) via the bilinear interpolation functions:

$$\begin{bmatrix}\chi_{i\lambda}^{kl} & \Pi_{i\lambda}^j\\ \Gamma & \Lambda_\lambda^j\end{bmatrix}=\mathbf{N}\begin{bmatrix}{}^e q^{kl} & {}^e\theta^j\\ {}^e\vartheta^{kl} & {}^e o^k\end{bmatrix} \tag{9.175}$$

where

$$\mathbf{N}=[N_1 \quad N_2 \quad N_3 \quad N_4 \quad N_5 \quad N_6 \quad N_7 \quad N_8]$$

$$N_i=\frac{1}{8}(1+\xi_i\xi)(1+\eta_i\eta)(1+\varsigma_i\varsigma)$$

(ξ,η,ζ) represents the isoparametric coordinates, (ξ_i,η_i,ζ_i) are the isoparametric coordinates of point i with the global coordinates (x_i,y_i,z_i), and $i=1\sim 8$.

The incompatible term $\begin{bmatrix} \chi_{i\lambda}^{kl} & \Pi_{i\lambda}^{j} \\ \Gamma_{\lambda}^{kl} & \Lambda_{\lambda}^{j} \end{bmatrix}$ is related to the element inner parameters $\begin{bmatrix} {}^{e}q_{\lambda}^{kl} & {}^{e}\theta_{\lambda}^{j} \\ {}^{e}\upsilon_{\lambda}^{kl} & {}^{e}o_{\lambda}^{k} \end{bmatrix}$ via the shape functions N_{λ}^{*}:

$$\begin{bmatrix} \chi_{i\lambda}^{kl} & \Pi_{i\lambda}^{j} \\ \Gamma_{\lambda}^{kl} & \Lambda_{\lambda}^{j} \end{bmatrix} = N_{\lambda}^{*} \begin{bmatrix} {}^{e}q_{\lambda}^{kl} & {}^{e}\theta_{\lambda}^{j} \\ {}^{e}\vartheta_{\lambda}^{kl} & {}^{e}o_{\lambda}^{k} \end{bmatrix} \tag{9.176}$$

Here, incompatible terms are employed in each element as derived in References 78 and 79:

$$\begin{aligned} \chi_{i\lambda}^{kl} &= ([\xi^2 \quad \eta^2 \quad \zeta^2] - [\xi \quad \eta \quad \zeta] P^{*} P_{\lambda}) \begin{bmatrix} {}^{e}q_{\lambda}^{kl} & {}^{e}\theta_{\lambda}^{j} \\ {}^{e}\vartheta_{\lambda}^{kl} & {}^{e}o_{\lambda}^{k} \end{bmatrix} \\ &= (N_{\lambda} - N^{*} P^{*} P_{\lambda}) \begin{bmatrix} {}^{e}q_{\lambda}^{kl} & {}^{e}\theta_{\lambda}^{kl} \\ {}^{e}\vartheta_{\lambda}^{kl} & {}^{e}o_{\lambda}^{kl} \end{bmatrix} \end{aligned} \tag{9.177}$$

In the equation above,

$$P^{*} = \oint_{\partial Y_e} \begin{Bmatrix} l \\ m \\ n \end{Bmatrix} N^{*} ds = \int_{Y_e} \begin{Bmatrix} \dfrac{\partial}{\partial y_1} \\ \dfrac{\partial}{\partial y_2} \\ \dfrac{\partial}{\partial y_3} \end{Bmatrix} N^{*} dY = \int_{Y_e} J^{-1} dY$$

$$P_{\lambda} = \oint_{\partial Y_e} \begin{Bmatrix} l \\ m \\ n \end{Bmatrix} N_{\lambda} ds = \int_{Y_e} \begin{Bmatrix} \dfrac{\partial}{\partial y_1} \\ \dfrac{\partial}{\partial y_2} \\ \dfrac{\partial}{\partial y_3} \end{Bmatrix} N_{\lambda} dY = 2\int_{Y_e} J^{-1} \begin{bmatrix} \xi & & \\ & \eta & \\ & & \zeta \end{bmatrix} dY$$

J is related to the element Jacobian:

$$J = \begin{bmatrix} a_1 + a_4\eta + a_5\zeta + a_7\eta\zeta & b_1 + b_4\eta + b_5\zeta + b_7\eta\zeta \\ a_2 + a_4\xi + a_6\zeta + a_7\xi\zeta & b_2 + b_4\xi + b_6\zeta + b_7\xi\zeta \\ a_3 + a_5\xi + a_6\eta + a_7\xi\eta & b_3 + b_5\xi + b_6\eta + b_7\xi\eta \\ c_1 + c_4\eta + c_5\zeta + c_7\eta\zeta & \\ c_2 + c_4\xi + c_6\zeta + c_7\xi\zeta & \\ c_3 + c_5\xi + c_6\eta + c_7\xi\eta & \end{bmatrix} \tag{9.178}$$

and the coefficients a_i and b_i , c_i (I = 1–17) are dependent on the element nodal coordinates y_j^i:

$$\begin{bmatrix} a_1 & b_1 & c_1 \\ \vdots & \vdots & \vdots \\ a_7 & b_7 & c_7 \end{bmatrix} =$$

$$\begin{bmatrix} -1 & 1 & 1 & -1 & -1 & 1 & 1 & -1 \\ -1 & -1 & 1 & 1 & -1 & -1 & 1 & 1 \\ 1 & 1 & 1 & 1 & -1 & -1 & -1 & -1 \\ 1 & -1 & 1 & -1 & 1 & -1 & 1 & -1 \\ -1 & 1 & 1 & -1 & 1 & -1 & -1 & 1 \\ -1 & -1 & 1 & 1 & 1 & 1 & -1 & -1 \\ 1 & -1 & 1 & -1 & -1 & 1 & -1 & -1 \end{bmatrix} \tag{9.179}$$

$$\begin{bmatrix} y_1^1 & y_1^2 & y_1^3 \\ \vdots & \vdots & \vdots \\ y_8^1 & y_8^2 & y_8^3 \end{bmatrix}$$

With the above assumed field, we obtain

$$\begin{aligned} \begin{bmatrix} e_{ykl} & 0 \\ 0 & \Phi_{yl} \end{bmatrix} \begin{bmatrix} \chi_i^{kl} & \Pi_i^j \\ \Gamma^{kl} & \Lambda^j \end{bmatrix} &= \begin{bmatrix} e_{ykl} & 0 \\ 0 & \Phi_{yl} \end{bmatrix} [N] \begin{bmatrix} {}^e q^{kl} & {}^e\theta_i^j \\ {}^e\vartheta^{kl} & {}^e o^k \end{bmatrix} \\ &= [B] \begin{bmatrix} {}^e q^{kl} & {}^e\theta_i^j \\ {}^e\vartheta^{kl} & {}^e o^k \end{bmatrix} \end{aligned} \tag{9.180}$$

where

$$B = \begin{bmatrix} e_{ykl} & 0 \\ 0 & \Phi_{yl} \end{bmatrix} [N]$$

The equation for the finite element can be written as

$$K \begin{bmatrix} q^{kl} & \theta_i^j \\ \vartheta^{kl} & o^k \end{bmatrix} = P \tag{9.181}$$

where

$$K = \int_{-1}^{1} \int_{-1}^{1} \int_{-1}^{1} B^T C B J d\xi d\eta d\zeta \tag{9.182a}$$

$$P = -\int_{-1}^{1}\int_{-1}^{1}\int_{-1}^{1} B^T C J d\xi d\eta d\zeta \tag{9.182b}$$

$\mathbf{C}$ is definied by Equation 9.136 and J is the determinant of the Jacobian matrix. The term $\begin{bmatrix} q_\lambda^{kl} & \theta_\lambda^j \\ \upsilon_\lambda^{kl} & o_\lambda^k \end{bmatrix}$ is solved from Equation 9.181, and the homogenized equivalent elastic, piezoelectric, and dielectric constants are obtained from Equation 9.164 and Equation 9.165.

9.5.8 Numerical Results

A cubic finite-element mesh is chosen in numeral analysis, and the constitutive properties $\mathbf{C}$ are computed at the Gauss integral points. If the Gauss point belongs to the yarn's region, the constitutive properties $\mathbf{C}$ are those of the yarn. Otherwise the matrix properties are selected. Using this information on the RVE–unit cell, the stiffness matrix **K** and the force matrix **P** are calculated using Equation 9.182a and Equation 9.182b.

First, we consider a 3-D unidirectional composite (Figure 9.43). The material is a PZT-7A continuous fiber-reinforced epoxy matrix. The fiber and matrix constants are shown in Table 9.8. The results are shown in Figure 9.44, which shows that the solution obtained with the homogenization method is closer to the experimental results reported in Reference 80 than those obtained using the Mori–Tanaka method.

The 3-D piezoelectric braided composites have 12 straight fibers in the RVE or a unit-cell model [82]. We assume that the braided direction along coordination axis y_3 (Figure 9.45). The average braided angle $\bar{\beta}$ can be calculated when the fiber equation in the RVE–unit cell is known. The volume fraction of fibers (V_f) in the RVE is maintained by a suitable choice of the yarn diameter. This model is more suitable for the topogical structure in the 3-D braided performs.

The 3-D four-step braided composite is built with piezoceramic fibers in a polymeric matrix. The properties of the PZT-5 piezoceramic fiber are given in Table 9.9 The epoxy material has the following isotropic properties: $C_{11} = 3.86$ (Gpa), $C_{12} = 2.57$ (Gpa), $C_{44} = 0.64$ (Gpa), and $\epsilon_{11} = 0.079$ (10^{-9} C^2/Nm^2). The fiber average braided angle $\bar{\beta}$ is 30°. The volume fraction of fibers (V_f) in the RVE–unit cell changes from 15 to 60%.

The variations of the effective elastic, piezoelectric constants and dielectric parameters of the 3-D four-step braided piezoceramic in epoxy material with the volume fraction of fibers (V_f) in the RVE–unit cell are illustrated in Figure 9.46 to Figure 9.48.

Figure 9.46 shows that the variation of the effective elastic constants C_{11}, C_{13}, C_{12}, and C_{33} increase with an increase of V_f, as in regular braided composites. The variation of the effective piezoelectric constants e_{13}, e_{33}, and e_{15} with V_f is shown in Figure 9.47. The values of e_{15}

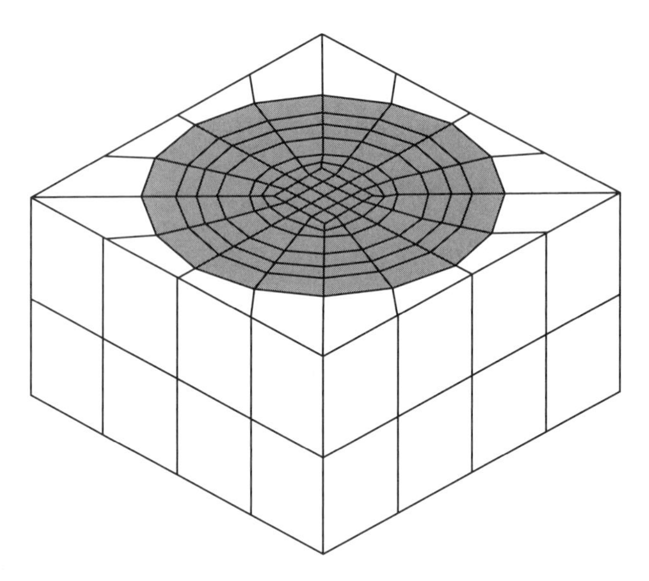

FIGURE 9.43
Unidirectional composites.

TABLE 9.8
Material Properties of PZT-7A and Epoxy

Material	C_{11}(GPa)	C_{12}(GPa)	C_{13}(GPa)	C_{33}(GPa)	C_{44}(GPa)
PZT-7A	148	76.2	74.2	131	25.4
	e_{31}(C/m)	e_{33}(C/m)	e_{15}(C/m)	$\in_{11}/\in_0^a$	$\in_{33}/\in_0$
	−2.1	9.5	9.2	460	235
Epoxy	8.0	4.4	4.4	8.0	1.8
	e_{31}(C/m)	e_{33}(C/m)	e_{15}(C/m)	$\in_{11}/\in_0$	$\in_{33}/\in_0$
	0	0	0	4.2	4.2

Source: Data from Chen, L., Tao, X.M., and Choy, C.L., *Compos. Sci. Technol.*, 59, 391, 1999.

and e_{33} rise with the increase of V_f. e_{13} is quasi-independent of V_f. The parameters $\in_{11}$ and $\in_{33}$ decrease with an increase of V_f, as shown in Figure 9.48.

9.5.9 Conclusion

Based on homogenization theory, the potential functional, the two-field Hellinger–Reissner functional, and the three-field Hu–Washizu functional for piezoelectric systems, the micro analysis for the properties of a piezoceramic material is established in this section.

The general PTC is satisfied, and an incompatible displacement finite-element method for the analysis of elastic and electric coupled systems with periodic microscopic structures is developed.

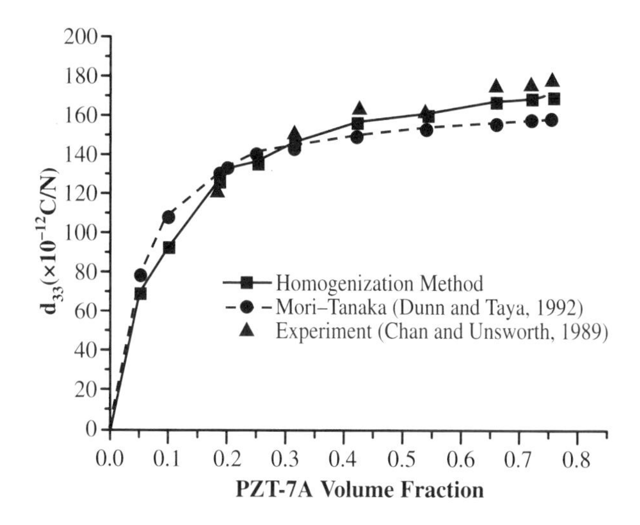

FIGURE 9.44
Results of unidirectional composites compared with experiment and the Mori–Tanaka method.

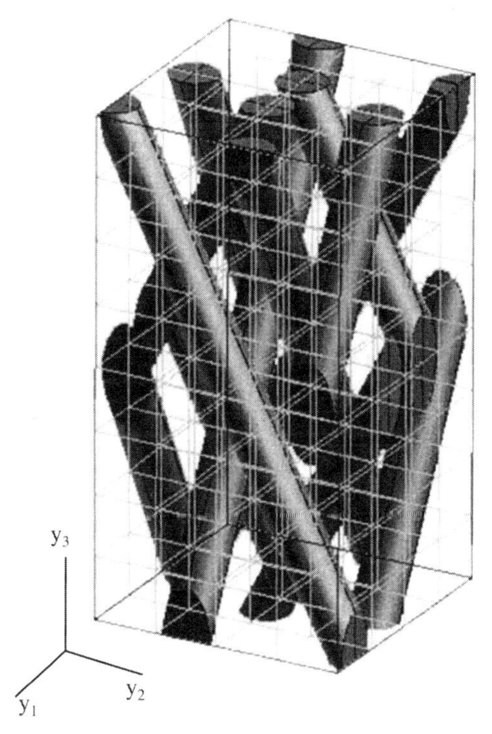

FIGURE 9.45
The RVE model and finite element method mesh of piezoelectric braided composites.

TABLE 9.9
Material Properties of PZT-5

Elastic compliance	C_{11}	C_{33}	C_{12}	C_{13}	C_{44}
Values (GPa)	121	111	75.4	75.2	21.1
Piezoelectric constants	e_{15}	e_{24}	e_{13}	e_{23}	e_{33}
Values (C/m^2)	12.3	12.3	–5.4	–5.4	15.8
Dielectric permittivity	$\in_{11}$	$\in_{22}$	$\in_{33}$		
Values ($\times 10^{-9}$ C^2/Nm^2)	8.11	8.11	7.35		

Source: Data from Chen, L., Tao, X.M., and Choy, C.L., *Compos. Sci. Technol.*, 59, 391, 1999.

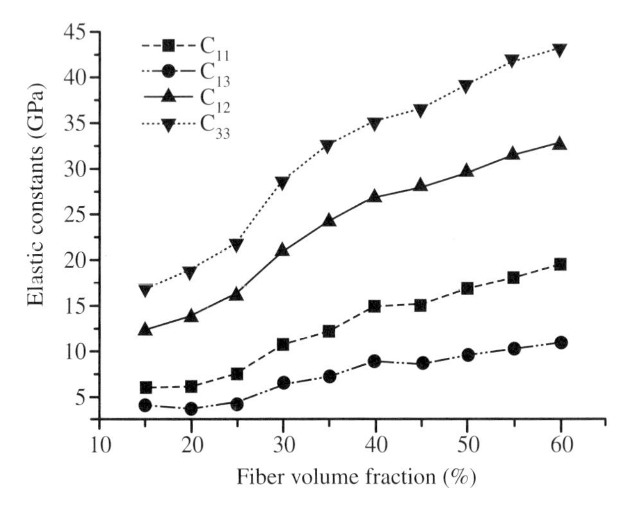

FIGURE 9.46
Variation of effective elastic constants with the fiber volume fraction.

The variation of effective elastic parameter, piezoelectric, and dielectric properties of 3-D four-step braided composites with the volume fraction of the fibers V_f is studied.

The numerical example shows that the variation of the elastic constants of 3-D piezoelectric braided composites with the fiber volume fraction is the same as those found for regular braided composites. A different behavior is found for the piezoelectric and dielectric constants.

The hybrid finite element method for elastic-electric coupled system analysis of periodic microscopic structures can be obtained from the two-field Hellinger–Reissner functional, and the stress boundary condition on a unit cell can be satisfied easily.

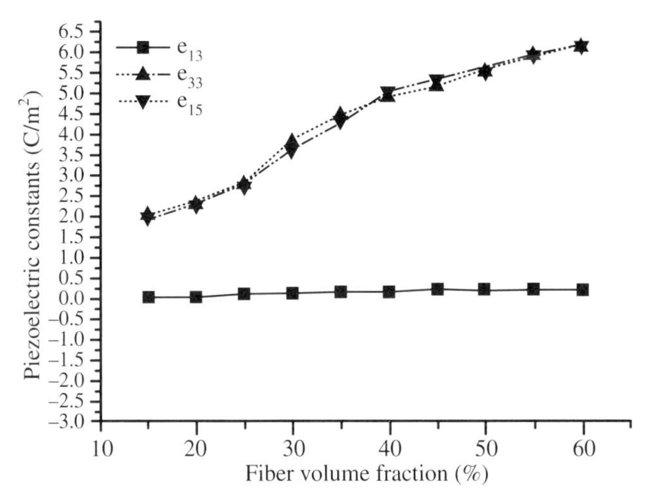

FIGURE 9.47
Variation of effective piezoelectric constants with the fiber volume fraction.

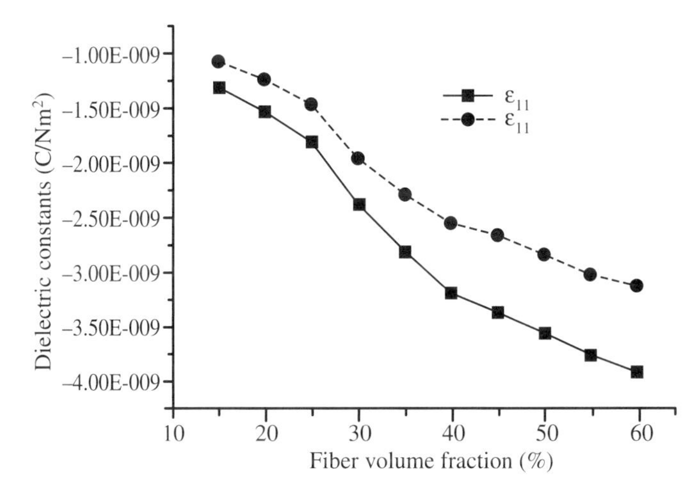

FIGURE 9.48
Variation of effective dielectric constants with the fiber volume fraction.

References

1. Reddy, J.N., A simple higher-order theory for laminated composite plates, *J. Appl. Mech.*, 51, 745, 1984.
2. Reddy, J.N. and Robbins, D.H., Theories and computational models for composite laminates, *Appl. Mech. Rev.*, 47, 147, 1994.
3. Di, S.M., A third-order triangular multilayered plate finite element with continuous interlaminar stresses, *Int. J. Numer. Methods Eng.*, 38, 1, 1995.

4. Mau, S.T., Tong, P., and Pian, T.H.H., Finite element solutions for laminated thick plates, *J. Compos. Mater.*, 6, 304, 1972.
5. Spilker, R.L., Chou, S.C., and Orringer, O., Alternate hybrid-stress elements for analysis of multilayer composite plates, *J. Compos. Mater.*, 11, 51, 1977.
6. Spilker, R.L., A hybrid-stress finite element formulation for thick multilayer laminates, *Comput. Struct.*, 11, 507, 1980.
7. Spilker, R.L., Hybrid-stress eight-node elements for thin and thick multilayer laminated plates, *Int. J. Numer. Methods Eng.*, 18, 801, 1982.
8. Spilker, R.L., Hybrid stress formulations for multilayer isoparametric plate element, in *Finite Element Methods for Plate and Shell Structures*, Vol. 1, *Element Technology*, Pineridge Press, Swansea, U.K., 1986.
9. Pian, T.H.H. and Li, M.S., Stress analysis of laminated composites by hybrid finite elements, IUTAM/IACM Symposium, Vienna, 1989.
10. Huang, Y.Q., Hybrid Approach to Laminated Composite and Piezoelectric Plates, PhD thesis, University of Science and Technology of China, Hefei, 1997.
11. Jing, H.S. and Liao, M.L., Partial hybrid stress element for the analysis of thick laminated composite plates, *Int. J. Numer. Methods Eng.*, 28, 2813, 1989.
12. Di, S.L. and Ramm, E., Hybrid stress formulation for higher-order theory of laminated shell analysis, *Comput. Methods Appl. Mech. Eng.*, 109, 359, 1993.
13. Cheung, Y.K. and Di, S.L., Analysis of laminated composite plates by hybrid stress isoparametric element, *Int. J. Solids Struct.*, 30, 2843, 1993.
14. Pian, T.H.H. and Wu, C.C., A rational approach for choosing stress term of hybrid finite element formulations, *Int. J. Numer. Methods Eng.*, 26, 2331, 1988.
15. Wu, C.C. and Pian, T.H.H., *Incompatible Numerical Analysis and Hybrid Element Method*, Science Press, Beijing, 1997.
16. Reissner, E., On a certain mixed variational theorem and a proposed application, *Int. J. Numer. Methods Eng.*, 20, 1366, 1984.
17. Reissner, E., On a mixed variational theorem and on deformable plate theory, *Int. J. Numer. Methods Eng.*, 23, 193, 1986.
18. Moriya, K., Laminated plate and shell elements for finite element analysis of advanced fiber reinforced composite structures, *Nippon Kikai Gakkai Ronbunshu*, 52, 1600, 1986.
19. Huang, Q., The basis of variational theorem of hybrid element and three-dimensional laminate theory: a new method for analyzing interlaminar stress of composite materials, *Appl. Math. Mech.*, 9, 599, 1988.
20. Zhong, W.X., *A New System of Elasticity Solutions*, Press of Dalian University of Technology, 1995.
21. Huang, Y.Q., Liu, M., and Wu, C.C., Bimaterial interface hybrid element for piezoelectric laminated analysis, *J. Strain Anal. Eng. Des.*, 34, 97, 1999.
22. Pagano, N.J., Exact solutions for composite laminates in cylindrical bending, *J. Compos. Mater.*, 3, 398, 1969.
23. Pagano, N.J., Exact solutions for rectangular bidirectional composites and sandwich plates, *J. Compos. Mater.*, 4, 20, 1970.
24. Huang, Y.Q., Liu, M., and Wu, C.C., A 3D penalty-hybrid finite element method for the analysis of laminated system, *Chin. J. Comput. Mech.*, 14, 83, 1997.
25. Heyliger, P., Static behavior of laminated elastic/piezoelectric plates, *Am. Inst. Aeronaut. Astronaut. J.*, 32, 2481, 1994.
26. Heyliger, P. and Brooks, S., Free vibration of piezoelectric laminates in cylindrical bending, *Int. J. Solids Struct.*, 32, 2945, 1995.

27. Lee, J.S. and Jiang, L.Z., Exact electroelastic analysis of piezoelectric laminae via state space approach, *Int. J. Solids Struct.*, 33, 977, 1996.
28. Ray, M.C., Bhattacharya, R., and Samanta, B., Exact solutions for static analysis of intelligent structures, *Am. Inst. Aeronaut. Astronaut. J.*, 31, 1684, 1993.
29. Ray, M.C., Rao, K.M., and Samanta, B., Exact solution for static analysis of an intelligent structure under cylindrical bending, *Comput. Struct.*, 47, 1031, 1993.
30. Sun, K. and Zhang, F.X., *Piezoelectricity*, National Defense Industry Press, Beijing, 1984.
31. Jiarang, E. and Jiangiao, Y., An exact solution for the statics and dynamics of laminated thick plates, *Int. J. Solids & Structures*, 26, 655–622, 1990.
32. Deeg, W.F., The Analysis of Dislocation, Crack and Inclusion Problems in Piezoelectric Solids, Ph.D. dissertation, Stanford University, Stanford, CA.
33. Dunn, M.L., The effects of crack face boundary conditions on the fracture mechanics of piezoelectric solids, *Eng. Fracture Mech.*, 48, 25, 1994.
34. EerNisse, E.P., Variational method for electroelastic vibration analysis, *IEEE Trans. J. Sonics Ultrason.*, 14, 59, 1983.
35. Jackson, J.D., *Classical Electrodynamics*, Wiley, New York, 1976.
36. Kumar, S. and Singh, R.N., Crack propagation in piezoelectric materials under combined mechanical and electrical loadings, *Acta Mater.*, 44, 173, 1996.
37. Kumar, S. and Singh, R.N., Energy release rate and crack propagation in piezoelectric materials. Part I. Mechanical/electrical load. Part II. Combined mechanical and electrical loads, *Acta Mater.*, 45, 849, 1997.
38. Liu, M., Finite Element Analysis for Cracked Piezoelectric Medium, M.Sc. thesis, Department of Modern Mechanics, University of Science and Technology of China, Hefei, 1998.
39. Pak, Y.E., Crack extension force in a piezoelectric material, *J. Appl. Mech.*, 57, 47, 1990.
40. Pak, Y.E., Linear electro-elastic fracture mechanics of piezoelectric materials, *Int. J. Fracture*, 54, 79, 1992.
41. Park, S.B. and Sun, C.T., Effect of electric field on fracture of piezoelectric ceramics, *Int. J. Fracture*, 70, 203, 1995.
42. Parton, V.Z., Fracture mechanics of piezoelectric materials, *Acta Astronaut.*, 3, 671, 1976.
43. Pian, T.H.H. and Sumihara, K., Rational approach for assumed stress finite elements, *Int. J. Numer. Methods Eng.*, 20, 1685, 1984.
44. Sosa, H., On the fracture mechanics of piezoelectric solids, *Int. J. Solids Struct.*, 29, 2613, 1992.
45. Sosa, H. and Khutoryansky, N., New developments concerning piezoelectric materials with defects, *Int. J. Solids Struct.*, 33, 3399, 1996.
46. Sze, K.Y., Efficient formulation of robust hybrid elements using orthogonal stress/strain interpolants and admissible matrix formulation, *Int. J. Numer. Methods Eng.*, 35, 1, 1992.
47. Wu, C.C. and Cheung, Y.K., On optimization approach of hybrid stress elements, *Finite Elements Anal. Des.*, 21, 111, 1995.
48. Hashin Z., Analysis of composite materials: a survey, *J. Appl. Mech.*, 50, 481, 1983.
49. Christensen, R.M., A critical evaluation for a class of micromechanics models. *J. Mech. Phys. Solids*, 38, 379, 1990.
50. Tolonen, H. and Sjolind, S.G., Effect of mineral fibers on properties of composite matrix materials, *Mech. Compos. Mater.*, 31, 435, 1995.

51. Du, S.Y. and Wang, B., *Micromechanics of Composite Materials*, Science Press, Beijing, 1998.
52. Xiao, Q.Z., Karihaloo, B.L., Li, Z.R., and Williams, F.W., An improved hybrid-stress element approach to torsion of shafts, *Comput. Struct.*, 71, 535, 1999.
53. Bensoussan, A., Lions, J.L., and Panaicolaou, G., Asymptotic analysis for periodic structures, North-Holland, New York, 1978.
54. Bakhvalov, A. and Panassenko, G.P., Homogenization: averaging process in periodic media, Kluwer Academic, Dordrecht, The Netherlands, 1989.
55. Lene, F. and Leguillon, D., Homogenized constitutive law for a partially cohesive composite material, *Int. J. Solids Struct.*, 18, 443, 1982.
56. Guedes, J.M. and Kikuchi, N., Preprocessing and postprocessing for materials based on the homogenization method with adaptive finite element methods, *Comp. Methods Appl. Mech. Eng.*, 83, 143, 1990.
57. Jansson, S., Homogenized nonlinear constitutive properties and local stress concentrations for composites with periodic internal structure, *Int. J. Solids Struct.*, 29, 2181, 1992.
58. Kalamkarov, A.L., Composite and reinforced elements of construction, Wiley, New York, 1992.
59. Fish, J., Nayak, M., and Holmes, M.H., Microscale reduction error indicators and estimators for a periodic heterogeneous medium, *Comput. Mech.*, 14, 323, 1994.
60. Lukkassen, D., Persson, L.E., and Wall, P., Some engineering and mathematical aspects on the homogenization method, *Compos. Eng.*, 5, 519, 1995.
61. Sun, H.Y., Di, S.L., Zhang, N., and Wu, C.C., Micromechanics of composite materials using multivariable finite element method and homogenization theory, *Int. J. Solids Struct.*, 38, 3007, 2001.
62. Mascarenhas, M.L. and Trabucho, L., Homogenised behavior of a beam with multicellular cross section, *Appl. Anal.*, 38, 97, 1990.
63. Mascarenhas, M.L. and Polisevski, D., The warping, the torsion and the Neumann problems in a quasi-periodically perforated domain, *Math. Model. Numer. Anal.*, 28, 37, 1994.
64. Lin, C.C. and Segel, L.A., *Mathematics Applied to Deterministic Problems in the Natural Science,* Macmillan, New York, 1974.
65. Timoshenko, S.P. and Goodier, J.N., *Theory of Elasticity,* 3rd ed., McGraw-Hill, New York, 1970.
66. Wu, C.C., Huang, M.G., and Pian, T.H.H., Consistency condition and convergence criteria of incompatible elements, general formulation of incompatible functions and its application, *Comput. Struct.*, 27, 639, 1987.
67. Simo, J.C. and Hughes, T.R.J., On the variational foundations of assumed strain methods, *J. Appl. Mech.*, 53, 51, 1986.
68. Simo, J.C. and Rifai, M.S., A class of mixed assumed strain methods and the method of incompatible modes, *Int. J. Numer. Methods Eng.*, 29, 1595, 1990.
69. Wu, C.C. and Bufler, H., Multivariable finite elements: consistency and optimization, *Sci. China A,* 34, 284, 1991.
70. Bathe, K.J., *Finite Element Procedures in Engineering Analysis,* Prentice-Hall, Englewood Cliffs, NJ, 1982.
71. Zhao, Y.H. and Weng, G.J., Effective elastic moduli of ribbon reinforced composites, *J. Appl. Mech.*, 57, 158, 1990.
72. Safari, A., Jadidian, B., and Chou, T.W., Three-dimensionally braided PZT fiber/polymer composites, in Patent Disclosure, 1998.

73. Ruan, X., Safari, A., and Chou, T.W., Effective elastic, piezoelectric and dielectric properties of braided fabric composites, *Composites A*, 30, 1435, 1999.
74. Lions, J.L., *Some Methods in the Mathematical Analysis of Systems and Their Control,* Science Press, Beijing, 1981.
75. Bakhvalov, A. and Panassenko, G.P., *Homogenization: Averaging Process in Periodic Media,* Kluwer Academic, Dordrecht, The Netherlands, 1989.
76. Guedes, J.M. and Kikuchi, N., Preprocessing and postprocessing for materials based on the homogenization method with adaptive finite element methods, *Comput. Methods Appl. Mech. Eng.*, 83, 143, 1990.
77. Fish, J., Shek, K., Pandheeradi, M., and Shephard, M.S., Computational plasticity for composite structures based on mathematical homogenization: theory and practice, *Comp. Methods Appl. Mech. Eng.*, 148, 53, 1997.
78. Pian, T.H.H. and Wu, C.C., General formulation of incompatible shape function and incompatible isoparametrics element, in *Proc. Invitational China American Workshop on FEM,* Chende, China, 1986, pp. 159–165.
79. Wu, C.C., Sze, K.Y., and Huang, Y.Q., Numerical solutions on fracture of piezoelectric materials by hybrid element, *Int. J. Solids & Structures*, 38, 4315–4329, 2001.
80. Chan, H.L.W. and Unsworth, J., Simple model for piezoelectric ceramic/polymer 1–2 composites used in ultrasonic transducer applications, *IEEE Trans. Uktrason. Ferroelectr. Frequency Control*, 36, 434, 1989.
81. Dun, M.L. and Taya, M., Micromechanics predictions of the effective electroelastic moduli of piezoelectric composites, *Comput. Struct.*, 30, 161, 1993.
82. Chen, L., Tao, X.M., and Choy, C.L., On the microstructure of three-dimensional braided perform, *Compos. Sci. Technol.*, 59, 391, 1999.

10

Finite Element Implementation

10.1 Overview

This chapter develops a finite element program written in Fortran 90 for the linear elastic static analysis of plane stress, plane strain, axisymmetric, and three-dimensional (3-D) solid problems. Plane problems with piezoelectricity are also included. Various shapes of elements including three- and six-node triangular elements, four-, eight-, and nine-node quadrilateral elements, and eight-node brick elements are considered. About 30 kinds of isoparametric displacement elements, incompatible elements, and hybrid stress elements are integrated into this program. The program can output nodal displacements and reactions, and stresses at Gaussian and nodal points by shape function interpolation. In addition, stresses at nodal points of an incompatible element can also be computed by bi- and tri-extrapolation through the values at Gaussian points.

Dynamic arrays are allocated in the program and therefore the maximum nodes and elements permitted are only dependent on the computer memory. Furthermore, since the program is designed with various function modules, it is convenient to add new elements, such as plate and shell elements. Smart design also guarantees that elements with different nodes or different model numbers can be adopted together.

10.2 Description of Variables and Subroutines

10.2.1 Definitions of Main Variables and Arrays

nelem	Total number of elements.
npoin	Total number of nodes used.
mpoin	Maximum node number.

nnode	Number of element nodes (three, four, eight, or nine).
ngroup	Number of element groups.
nmats	Number of material kinds.
nload	Number of load cases.
nprop	Number of material constants used (3 for plane stress and strain, 2 for axisymmetric and 3-D solids, 26 for 2-D piezoelectric problems).
nevab	Number of nodal degrees of freedom (DOFs) in an element.
noutp	Option of output (one for displacements; two for displacements and reactions; three for displacements, reactions, and stresses).
nvfix	Number of prescribed nodes.
ntype	Problem type identifier (one, two, three, four, or five for plane stress, plane strain, axisymmetric, 3-D, plane strain with piezoelectricity).
ngaus	Number of Gaussian integration order or hammer integration points (one, two, three, or four for quadrilateral and brick elements; one, three, four, six, or seven for triangular elements).
ngaup	Number of integration points used.
nbeta	Number of stress parameters in an element.
ndime	Number of dimensions (two or three).
ndofn	Number of DOFs per node (two or three).
nstre	Number of stress components.
ntotv	Total number of DOFs.
model	Element model identifier.
nlambda	Number of incompatible internal DOFs in an element.
mstif	Length of the global stiffness matrix.
coord(ndime,npoin)	coord(I,J) = Ith coordinate of node J.
elements(nelem)%lnods (nnode)	Elements(I)%lnods(J) = Jth node number of element I.
estif(nevab,nevab)	Element stiffness matrix.

grpno(ngroup)	grpno(I) = Ith group number.
nofix(nvfix)	nofix(I) = Ith prescribed node number.
fixed(ntotv)	Store prescribed displacement values of fixed nodes initially and nodal reactions finally.
asdis(ntotv)	Store nodal displacements.
iffix(ntotv)	Store nodal constraint code.
props(nprop,nmats)	props(I,J) = Ith material constant of material J.
presc(nvfix,ndofn)	presc(I,J) = Jth prescribed displacement component of the Ith prescribed node.
posgp(ndime,ngaup)	posgp(I,J) = Ith local coordinate of the Jth integration point.
weigp(ngaup)	weigp(J) = weighting coefficient of the Jth integration point.
aload(ntotv,nload)	aload(:,I) = global load vector of load case I.
astif(mstif)	Global stiffness matrix with 1-D variable bandwidth storage.
jffix(nvfix*ndofn)	Store nodal constraint code.

10.2.2 Functions of Subroutines Used

openfile	Open files for input and output.
timeanddate	Get system date and time, output elapsed time when finished.
input	Read element connections, nodal coordinates, constraints, material information, etc.; write to output file simultaneously; interpolate coordinates of midside and central nodes if necessary; allocate and initialize global variables and arrays.
load	Read load information and compute the global load vector.
bandwidth	Compute half bandwidth and column heights.
gaussq	Evaluate local coordinates and weights of each integration point.
kmatx	Form global stiffness matrix.
solve	Solve global algebraic equations by the LDL^T method.

output	Output nodal displacements and reactions and elemental stresses.
nodexy	Interpolate midside and central node coordinates for six-, eight-, or nine-node 2-D elements if necessary.
bmatrix	Compute strain-displacement matrix and stress interpolation matrix if necessary.
stiff	Evaluate elemental stiffness matrix.
fimatrix	Calculate stress interpolation matrix for hybrid elements.
jacob	Evaluate Jacobian matrix and Cartesian derivatives of shape functions.
shapefun	Evaluate shape functions and their local derivatives.
modps	Evaluate material stiffness matrix for displacement elements or flexibility matrix for hybrid stress elements.
materialpiezo	Evaluate material matrix for 2-D piezoelectric problems.
assemble	Assemble the elemental stiffness into the global matrix.
columnheight	Evaluate the column height.
aibi	Calculate element geometry coefficients for four-node 2-D element.
inverse	Inverse a symmetric matrix.
condensation	Static condensation.
restore	Restore the internal parameters.
NH11OPH18	Calculate coefficients for 3-D incompatible element NH11, Jacobian at element center for hybrid element OPH-18*b*, and equilibrium matrix for element OPH-18*b*(*a*) by numerical integration.
computestress	Compute stresses at Gaussian and nodal points within each element.
reviseh	Modify H matrix for hybrid stress elements with penalty equilibrium constraint.
readio	Evaluate the number of data in the input string and append zeros when necessary.

stress	Compute stresses at the sampling point.
extrapolation	Evaluate the bi- or trilinear extrapolation matrix.
stretransmatrix	Evaluate stress transformation matrix.
stratransmatrix	Evaluate strain transformation matrix.

(a) 3-node 2D element

(b) 6-node 2D element

(c) 4-node 2D element

(d) 8-node 2D element

(e) 9-node 2D element

(f) 8-node brick element

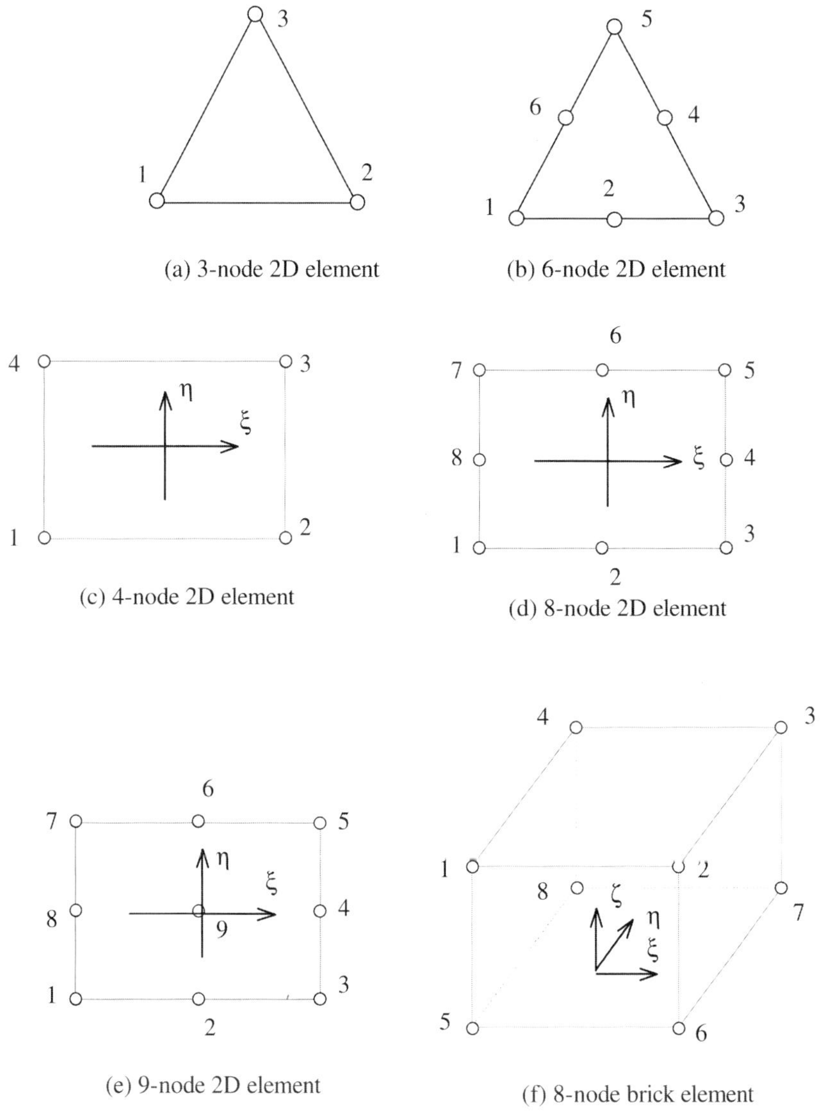

FIGURE 10.1
Coordinate configurations and node arrangements of 2-D and 3-D elements.

10.3 Instructions for Input Data

The input data can be prepared in the free format; that is, integer and real numbers are separated by spaces, tabs, and commas. Instructions for the formatted input file are given below. Note that default data are not required.

Section 1: Problem title. Limited to 80 characters within the first line

Section 2: Governing variables. The following variables are required in one line:

nelem:	Total number of elements
nvfix:	Number of prescribed nodes
ngroup:	Number of element groups
nmats:	Number of material kinds
ntype:	Problem type identifier (1, 2, 3, 4, 5 for plane stress, plane strain, axisymmetric, 3-D, plane strain with piezoelectricity)
noutp:	Output option (1 for displacements, 2 for displacements and reactions, 3 for displacements, reactions and stresses)
nload:	Number of load cases (default is zero)

Section 3: Total nelem lines in this section. Each line contains:

numel:	Element number
grpno(numel):	Group number
elements(numel)%lnods(1:nnode):	Node arrangement (see Figure 10.1 for details of various elements)

Section 4: Nodal coordinates. All nodes at the element corners are required. This section ends at the node with the maximum number. Each line contains:

Inode:	Node number
coord(1:ndime,inode):	Nodal coordinates

Section 5: Constraint information. Total nvfix lines. Each line contains:

nofix(I):	Ith prescribed node number
ifpre:	Constraint code

presc(I,1:ndofn): Prescribed values of the Ith fixed node

Note that the constraint codes are varied with different problem types. Details are:

ntype = 1, 2, 3 for plane and axisymmetric problems.
ifpre = 10: DOF in the x or r directions is prescribed.
ifpre = 01: DOF in the y or z directions is prescribed.
ifpre = 11: Both DOFs are prescribed.
ntype = 4, 5 for 3-D solids and 2-D plane problems with piezoelectricity:
ifpre = 100: DOF in the x direction is prescribed.
ifpre = 010: DOF in the y direction is prescribed.
ifpre = 001: DOF in the z direction or potential is prescribed.
ifpre = 101: DOFs in the x direction and z direction or potential are prescribed.
ifpre = 110: DOFs in the x and y directions are prescribed.
ifpre = 111: All nodal DOFs are prescribed.

Section 6: Element group information. Total ngroup lines. Each line contains:

lgroup: Group number
lprop: Material number
model: Element model (see Table 10.1 for details)
ngaus: Number of Gaussian integration order or hammer integration points (1, 2, 3, 4 for quadrilateral and brick elements; 1, 3, 4, 6, 7 for triangular elements)

Section 7: Material constants. The following variables are required for each kind of material:

numat: Material number
props(1:nprop,numat): constants of the material numat, where for ntype = 1,2 with nprop = 3, and ntype = 3,4 with nprop = 2
props(1,numat) = Young's modulus for ntype = 1, 2, 3, 4
props(2,numat) = Poisson's ratio for ntype = 1, 2, 3, 4
props(3,numat) = thickness for ntype = 1, 2 for 2-D piezoelectric problems with ntype = 5 and nprop = 26

TABLE 10.1

List of 2- and 3-D Elements Used

Model Number	Identifier	Description and Notes
Plane Stress and Plane Strain Elements		
0	T3, T6 / Q4, Q8, Q9	Triangular/quadrilateral isoparametric elements
1	Q6	Wilson element (1973)
2	QM6	Taylor element (1976)
3	NQ6	Wu et al.'s incompatible element (W–H–P 1987)
5	R-u:Q6 $\in P_2$	Revised displacement (R-u)–based quadratic incompatible element (W–H–R 2001)
6	R-S:Q6 $\in P_2$	Revised stiffness (R-s)–based quadratic incompatible element (W–H–R 2001)
100	PS	Pian–Sumihara hybrid stress element (P–S 1984)
101	PS(α)	Penalty-equilibrium hybrid element (W–C 1995)
Axisymmetric Elements		
0	Axi-T3, Axi-T6	Axisymmetric three- and six-node triangular isoparametric elements
0	Axi-Q4, Axi-Q8, Axi-Q9	Axisymmetric four-, eight-, and nine-node quadrilateral isoparametric elements
3	AQ6	Axisymmetric four-node incompatible element (W–C 1992)
4	R-S:Axi-Q6	Axisymmetric R-S–based incompatible element (W–H–R 2001)
101	Axi-8β(α)	Axisymmetric penalty-equilibrium hybrid element (W–C 1995)
Plane Strain Elements with Piezoelectricity		
0	PZT-T3, PZT-T6	2-D piezoelectric three- and six-node triangular isoparametric elements
0	PZT-Q4, PZT-Q8, PZT-Q9	2-D piezoelectric four-, eight-, and nine-node quadrilateral isoparametric elements
100	PZT-9β	Piezoelectric hybrid element (Liu–Wu 1999)
3-D elements		
0	H8	Eight-node isoparametric brick element
1	H11	Wilson's 3-D element (1973)
3	NH11	3-D incompatible element (W–H–P 1987)
100	OPH-18β	3-D optimal 18β hybrid element (Wu–Bufler 1991)
101	OPH-18β(α)	Penalty-equilibrium 18β hybrid element (W–C 1995)

props(1,numat) = thickness

props(2:11,numat) = $E_1, E_2, E_3, \mu_{12}, \mu_{13}, \mu_{23}, G_{12}, G_{13}, G_{23}, 0$

or

props(2:11,numat) = $C_{11}, C_{22}, C_{33}, C_{12}, C_{13}, C_{23}, C_{44}, C_{55}, C_{66}, 0$

props(12:17,numat) = $e_{31}, e_{33}, e_{15}, \in_1, \in_2, \in_3$

props(18:26,numat) = angles (in degrees) between the physical space (x,y,z) and the material principal axes (1,2,3). It can be inputted in the 3 × 3 matrix form below:

	x	y	z
1	θ_{11}	θ_{12}	θ_{13}
2	θ_{21}	θ_{22}	θ_{23}
3	θ_{31}	θ_{32}	θ_{33}

In other words, props(18:26,numat) = $\theta_{3\times3}$.

Section 8: Load title. Limited to 80 characters within one line

Section 9: Load governing variables

nplod: Number of nodes subjected to concentrated forces

ibody: 0 for no body force; 1 for body force

nedge: Number of element edges subjected to distributed forces (inapplicable to 3-D problem)

Section 10: Nodal point loads. Total nplod lines. Each line contains:

lodpt: Node number

point(1:ndofn): Load components

Section 11: Body force. This section is unnecessary if ibody = 0.

bodyf(1:ndofn): Components of the body force

Section 12: Distributed edge loads. Total edge × two lines.

For each edge, the first line is:

neass: Element number

noprs(1:nodeg): Node numbers along the edge (counterclockwise)

The second line is:

press(1:ndofn, 1:nodeg): Distributed loads acting at the edge nodes

Note that nodeg = 2 for three- and four-node 2-D elements; nodeg = 3 for six-, eight-, and nine-node 2-D elements; press(1, I) represents the normal load component acting at the Ith edge node (in normal to the edge is positive); press(2, I) represents the tangential load component acting at the Ith edge node (counterclockwise is positive); press(3, I) represents the electric charge acting at the Ith edge node for 2-D piezoelectric problems.

Sections 8 to 12: repeated if more load cases are considered.

10.4 Examples

Figure 10.2 shows a cantilever beam subjected to pure bending loads. Five irregular elements are used. The sampling input and output files of this problem are given provided that the NQ6 element is adopted.

The input file:

```
5 element cantilever beam under bending
5 2 1 1 1 3 1
1 1 1 3 4 2
2 1 3 5 6 4
3 1 5 7 8 6
4 1 7 9 10 8
5 1 9 11 12 10
1 0. 0.
2 0. 2.
3 1. 0.
4 2. 2.
5 2. 0.
6 4. 2.
7 4. 0.
8 5. 2.
9 7. 0.
10 6. 2.
11 10. 0.
12 10. 2.
1 11 0. 0.
2 10 0. 0.
1 1 3 2
1 1500. 0.25 1.
loads
2 0 0
11 1000. 0.
12 -1000. 0.
```

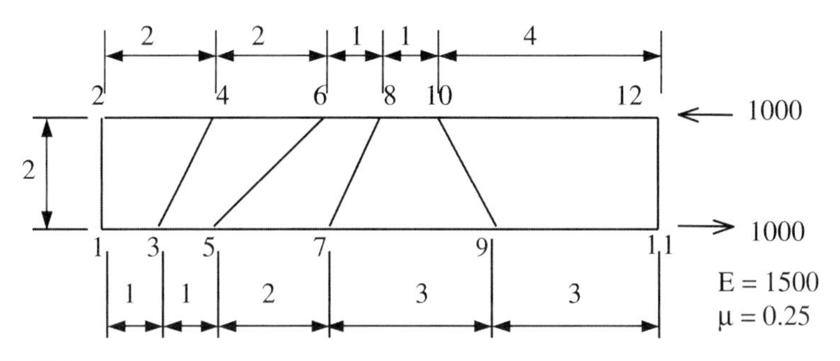

FIGURE 10.2
Four-node finite element meshes of a cantilever beam under pure bending.

The output file is

```
    Start at 15:44:20.00 03/20/2003

               PLANE STRESS

five-element cantilever beam under bending

 nelem = 5   nvfix  = 2   ntype = 1   nmats = 1   ngroup = 1
 ndime = 2   nstre  = 3   ndofn = 2   noutp = 3   nprop  = 3   nload = 1

           Element      Group No.        Node Numbers
              1             1           1    3    4    2
              2             1           3    5    6    4
              3             1           5    7    8    6
              4             1           7    9   10    8
              5             1           9   11   12   10

Node       X                 Y
 1    0.000000E+00    0.000000E+00
 2    0.000000E+00    2.000000E+00
 3    1.000000E+00    0.000000E+00
 4    2.000000E+00    2.000000E+00
 5    2.000000E+00    0.000000E+00
 6    4.000000E+00    2.000000E+00
 7    4.000000E+00    0.000000E+00
 8    5.000000E+00    2.000000E+00
 9    7.000000E+00    0.000000E+00
10    6.000000E+00    2.000000E+00
11    1.000000E+01    0.000000E+00
12    1.000000E+01    2.000000E+00
```

```
Node      Code        Fixed Values
1          11      0.000000 0.000000
2          10      0.000000 0.000000


===================== Element Properties =====================
Element Group No. = 1
Material No. = 1   Element Model: NQ6      ngaus = 2


===================== Material Properties ====================
 Material No. = 1
Young Modulus = 1.500000E+03  Poisson Ratio = 2.500000E-01
     Thickness = 1.000000E+00
Load case 1

loads

         2            0 0
11    1000.000      0.000
12   -1000.000      0.000


Load case 1

Displacements

Node     X-coord        Y-coord         X-disp.          Y-disp.
 1   0.000000E+00   0.000000E+00   0.000000E+00     0.000000E+00
 2   0.000000E+00   0.200000E+01   0.000000E+00    -0.554241E+00
 3   0.100000E+01   0.000000E+00   0.267706E+01    0.792288E+00
 4   0.200000E+01   0.200000E+01   -0.334373E+01   0.295441E+01
 5   0.200000E+01   0.000000E+00   0.533104E+01    0.276435E+01
 6   0.400000E+01   0.200000E+01   -0.658193E+01   0.143173E+02
 7   0.400000E+01   0.000000E+00   0.863064E+01    0.143657E+02
 8   0.500000E+01   0.200000E+01   -0.929731E+01   0.254954E+02
 9   0.700000E+01   0.000000E+00   0.125368E+02    0.478040E+02
10   0.600000E+01   0.200000E+01   -0.118701E+02   0.353039E+02
11   0.100000E+02   0.000000E+00   0.186052E+02    0.960673E+02
12   0.100000E+02   0.200000E+01   -0.186052E+02   0.938902E+02

Reactions

Node     X-coord        Y-coord         X-reac.          Y-reac.
1    0.000000E+00   0.000000E+00  -0.100000E+04    -0.147367E-10
2    0.000000E+00   0.200000E+01   0.100000E+04     0.000000E+00

Stresses at Gaussian and/or nodal points
No. X-coord  Y-coord    XX-stress        YY-stress      XY-stress

Element No. = 1
Stresses at Gaussian points by shape function interpolation
1  0.2560     0.4226     0.173716E+04    0.204487E+02   0.313742E+00
2  0.3780     1.5774    -0.174021E+04    0.765288E+01  -0.288521E+01
```

```
3  0.9553     0.4226     0.173221E+04  -0.588362E+02  -0.195075E+02
4  1.4107     1.5774    -0.174356E+04  -0.460403E+02  -0.163085E+02

Stresses at nodal points by shape function interpolation
1  0.0000     0.0000     0.394938E+04  -0.202474E+03  -0.179047E+03
2  0.0000     1.0000    -0.455289E+03  -0.446236E+03   0.666400E+02
3  0.0000     2.0000    -0.254330E+04  -0.110834E+03  -0.943219E+02
4  0.5000     0.0000     0.415842E+04   0.571316E+03  -0.198281E+03
5  0.7500     1.0000    -0.315927E+03   0.696241E+02   0.538175E+02
6  1.0000     2.0000    -0.243878E+04   0.276061E+03  -0.103939E+03
7  1.0000     0.0000     0.393898E+04  -0.368820E+03  -0.220633E+03
8  1.5000     1.0000    -0.462220E+03  -0.557133E+03   0.389157E+02
9  2.0000     2.0000    -0.254850E+04  -0.194007E+03  -0.115115E+03

Stresses at nodal points by bi-/tri- linear extrapolation
1  0.0000     0.0000     0.301200E+04   0.575812E+02   0.959687E+01
2  0.0000     1.0000     0.994595E-08   0.383875E+02   0.479843E+01
3  0.0000     2.0000    -0.301200E+04   0.191937E+02   0.106007E-06
4  0.5000     0.0000     0.300720E+04  -0.191937E+02  -0.959687E+01
5  0.7500     1.0000    -0.359883E+01  -0.191937E+02  -0.959687E+01
6  1.0000     2.0000    -0.301440E+04  -0.191937E+02  -0.959687E+01
7  1.0000     0.0000     0.300240E+04  -0.959687E+02  -0.287906E+02
8  1.5000     1.0000    -0.719765E+01  -0.767749E+02  -0.239922E+02
9  2.0000     2.0000    -0.301679E+04  -0.575812E+02  -0.191937E+02

Element No. = 2
Stresses at Gaussian points by shape function interpolation
1  1.4673     0.4226     0.172154E+04   0.277172E+00   0.915073E+01
2  2.1667     1.5774    -0.168098E+04   0.241248E+02   0.270364E+02
3  2.1667     0.4226     0.180465E+04   0.148041E+03   0.119973E+03
4  3.1994     1.5774    -0.162469E+04   0.124193E+03   0.102088E+03

Stresses at nodal points by shape function interpolation
1  1.0000     0.0000     0.398625E+04   0.214899E+03  -0.693283E+03
2  1.5000     1.0000    -0.513312E+03  -0.755636E+03   0.197691E+03
3  2.0000     2.0000    -0.241674E+04   0.144529E+03  -0.314361E+03
4  1.5000     0.0000     0.430890E+04   0.131172E+04  -0.572182E+03
5  2.2500     1.0000    -0.298216E+03  -0.244242E+02   0.278425E+03
6  3.0000     2.0000    -0.225541E+04   0.692938E+03  -0.253810E+03
7  2.0000     0.0000     0.416064E+04   0.524918E+03  -0.460769E+03
8  3.0000     1.0000    -0.397055E+03  -0.548957E+03   0.352700E+03
9  4.0000     2.0000    -0.232954E+04   0.299538E+03  -0.198103E+03

Stresses at nodal points by bi-/tri- linear extrapolation
1  1.0000     0.0000     0.293293E+04  -0.689268E+02  -0.427522E+02
2  1.5000     1.0000    -0.523561E+01  -0.331554E+02  -0.159237E+02
3  2.0000     2.0000    -0.294340E+04   0.261606E+01   0.109049E+02
4  1.5000     0.0000     0.301341E+04   0.741589E+02   0.645620E+02
5  2.2500     1.0000     0.551287E+02   0.741589E+02   0.645620E+02
6  3.0000     2.0000    -0.290316E+04   0.741589E+02   0.645620E+02
7  2.0000     0.0000     0.309390E+04   0.217245E+03   0.171876E+03
8  3.0000     1.0000     0.115493E+03   0.181473E+03   0.145048E+03
9  4.0000     2.0000    -0.286291E+04   0.145702E+03   0.118219E+03
```

```
Element No. = 3
Stresses at Gaussian points by shape function interpolation
1  2.8006   0.4226   0.159358E+04  -0.169388E+03  -0.164321E+03
2  3.8333   1.5774  -0.195190E+04  -0.268802E+03  -0.238882E+03
3  3.8333   0.4226   0.182823E+04   0.247770E+03   0.148547E+03
4  4.5327   1.5774  -0.160541E+04   0.347185E+03   0.223108E+03

Stresses at nodal points by shape function interpolation
1  2.0000   0.0000   0.225209E+04  -0.486566E+03   0.156069E+03
2  3.0000   1.0000   0.299448E+03   0.567849E+03  -0.588374E+03
3  4.0000   2.0000  -0.452808E+04  -0.101232E+04   0.320026E+03
4  3.0000   0.0000   0.227927E+04  -0.781729E+03   0.388295E+03
5  3.7500   1.0000   0.335684E+03   0.174298E+03  -0.278740E+03
6  4.5000   2.0000  -0.447373E+04  -0.160265E+04   0.784478E+03
7  4.0000   0.0000   0.261557E+04   0.159628E+03   0.640715E+03
8  4.5000   1.0000   0.784093E+03   0.142944E+04   0.578198E+02
9  5.0000   2.0000  -0.380112E+04   0.280064E+03   0.128932E+04

Stresses at nodal points by bi-/tri- linear extrapolation
1  2.0000   0.0000   0.282041E+04  -0.259052E+03  -0.231569E+03
2  3.0000   1.0000  -0.285518E+03  -0.408174E+03  -0.343411E+03
3  4.0000   2.0000  -0.339144E+04  -0.557295E+03  -0.455252E+03
4  3.0000   0.0000   0.298817E+04   0.391914E+02  -0.788689E+01
5  3.7500   1.0000  -0.338755E+02   0.391914E+02  -0.788689E+01
6  4.5000   2.0000  -0.305592E+04   0.391914E+02  -0.788689E+01
7  4.0000   0.0000   0.315593E+04   0.337435E+03   0.215796E+03
8  4.5000   1.0000   0.217767E+03   0.486556E+03   0.327637E+03
9  5.0000   2.0000  -0.272040E+04   0.635678E+03   0.439478E+03

Element No. = 4
Stresses at Gaussian points by shape function interpolation
1  4.7560   0.4226   0.137280E+04  -0.777477E+02   0.136796E+03
2  5.0893   1.5774  -0.137280E+04   0.195844E+03   0.136796E+03
3  6.2440   0.4226   0.137280E+04  -0.751904E+03   0.136796E+03
4  5.9107   1.5774  -0.137280E+04  -0.102550E+04   0.136796E+03

Stresses at nodal points by shape function interpolation
1  4.0000   0.0000   0.195505E+04   0.509471E+03   0.820774E+02
2  4.5000   1.0000   0.607650E+03   0.118298E+04   0.215453E+03
3  5.0000   2.0000  -0.364590E+04   0.235807E+04  -0.273591E+02
4  5.5000   0.0000   0.176130E+04  -0.767087E+03   0.569982E+02
5  5.5000   1.0000   0.317035E+03  -0.731861E+03   0.177834E+03
6  5.5000   2.0000  -0.422713E+04  -0.147161E+04  -0.102597E+03
7  7.0000   0.0000   0.195505E+04  -0.493697E+03   0.820774E+02
8  6.5000   1.0000   0.607650E+03  -0.321776E+03   0.215453E+03
9  6.0000   2.0000  -0.364590E+04  -0.651439E+03  -0.273592E+02

Stresses at nodal points by bi-/tri- linear extrapolation
1  4.0000   0.0000   0.237776E+04  -0.443914E+01   0.136796E+03
2  4.5000   1.0000  -0.132473E-05   0.405948E+03   0.136796E+03
3  5.0000   2.0000  -0.237776E+04   0.816335E+03   0.136796E+03
4  5.5000   0.0000   0.237776E+04  -0.414826E+03   0.136796E+03
5  5.5000   1.0000   0.135372E-05  -0.414826E+03   0.136796E+03
6  5.5000   2.0000  -0.237776E+04  -0.414826E+03   0.136796E+03
7  7.0000   0.0000   0.237776E+04  -0.825213E+03   0.136796E+03
8  6.5000   1.0000   0.403216E-05  -0.123560E+04   0.136796E+03
9  6.0000   2.0000  -0.237776E+04  -0.164599E+04   0.136796E+03
```

```
Element No. = 5
Stresses at Gaussian points by shape function interpolation
1  7.4673     0.4226     0.164293E+04   0.565468E+03  -0.190540E+03
2  7.0120     1.5774    -0.152659E+04   0.509270E+03  -0.176490E+03
3  9.3214     0.4226     0.159684E+04  -0.172086E+03  -0.615111E+01
4  9.1994     1.5774    -0.156566E+04  -0.115889E+03  -0.202004E+02

Stresses at nodal points by shape function interpolation
1  7.0000     0.0000     0.288783E+04  -0.726886E+02  -0.200036E+03
2  6.5000     1.0000    -0.295230E+03  -0.169948E+03  -0.285524E+03
3  6.0000     2.0000    -0.262314E+04  -0.534377E+01  -0.174614E+03
4  8.5000     0.0000     0.324210E+04   0.831613E+03  -0.162829E+02
5  8.2500     1.0000     0.843481E+01   0.605168E+03  -0.128021E+03
6  8.0000     2.0000    -0.235743E+04   0.672882E+03  -0.367985E+02
7  10.0000    0.0000     0.280236E+04  -0.144016E+04   0.141831E+03
8  10.0000    1.0000    -0.368487E+03  -0.134206E+04   0.750525E+01
9  10.0000    2.0000    -0.268724E+04  -0.103095E+04   0.817866E+02

Stresses at nodal points by bi-/tri- linear extrapolation
1  7.0000     0.0000     0.282087E+04   0.871059E+03  -0.266937E+03
2  6.5000     1.0000     0.737590E+02   0.786763E+03  -0.245863E+03
3  6.0000     2.0000    -0.267335E+04   0.702467E+03  -0.224789E+03
4  8.5000     0.0000     0.277872E+04   0.196691E+03  -0.983453E+02
5  8.2500     1.0000     0.368795E+02   0.196691E+03  -0.983453E+02
6  8.0000     2.0000    -0.270496E+04   0.196691E+03  -0.983453E+02
7  10.0000    0.0000     0.273658E+04  -0.477677E+03   0.702467E+02
8  10.0000    1.0000    -0.139030E-06  -0.393381E+03   0.491727E+02
9  10.0000    2.0000    -0.273658E+04  -0.309085E+03   0.280987E+02

Stop at 15:44:20.17 03/20/2003
    0.17 seconds costed
```

The cantilever beam is studied again using 3-D elements. Figure 10.3 shows the five-element meshes with node arrangements. The input and output sampling files are given using the element OPH-18$\beta(\alpha)$. Stress results are omitted for simplicity.

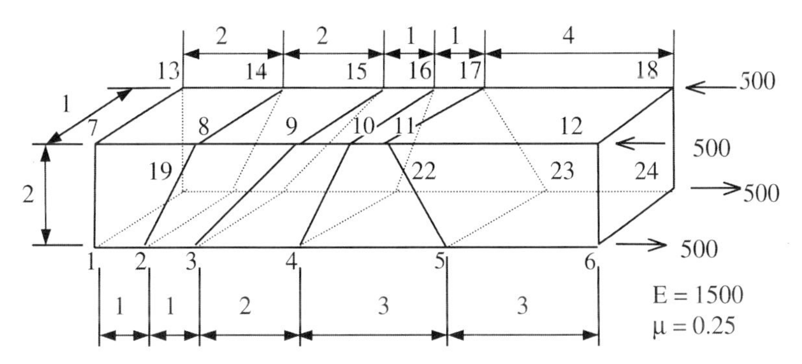

FIGURE 10.3
Cantilever beam with five irregular brick elements.

The input file:

```
Pure bending beam modeled by five 3-D elements
5    4    1    1    4    2    1
1    1    7    8    14   13   1    2    20   19
2    1    8    9    15   14   2    3    21   20
3    1    9    10   16   15   3    4    22   21
4    1    10   11   17   16   4    5    23   22
5    1    11   12   18   17   5    6    24   23
   1      0.00000       0.00000    0.00000
   2      1.00000       0.00000    0.00000
   3      2.00000       0.00000    0.00000
   4      4.00000       0.00000    0.00000
   5      7.00000       0.00000    0.00000
   6     10.00000       0.00000    0.00000
   7      0.00000       0.00000    2.00000
   8      2.00000       0.00000    2.00000
   9      4.00000       0.00000    2.00000
  10      5.00000       0.00000    2.00000
  11      6.00000       0.00000    2.00000
  12     10.00000       0.00000    2.00000
  13      0.00000       1.00000    2.00000
  14      2.00000       1.00000    2.00000
  15      4.00000       1.00000    2.00000
  16      5.00000       1.00000    2.00000
  17      6.00000       1.00000    2.00000
  18     10.00000       1.00000    2.00000
  19      0.00000       1.00000    0.00000
  20      1.00000       1.00000    0.00000
  21      2.00000       1.00000    0.00000
  22      4.00000       1.00000    0.00000
  23      7.00000       1.00000    0.00000
  24     10.00000       1.00000    0.00000
   1    111     0.000000     0.000000    0.000000
  19    111     0.000000     0.000000    0.000000
   7    100     0.000000     0.000000    0.000000
  13    100     0.000000     0.000000    0.00000
   1      1   101            2
   1   1500.    0.3
Pure bending loads
   4        0             0
   6        500.000       0.000      0.000
  12       -500.000       0.000      0.000
  18       -500.000       0.000      0.000
  24        500.000       0.000      0.000
```

The output file:

```
        Start at 16:14:49.96 03/20/2003

                 3-D SOLID

Pure bending beam modeled by five 3-D elements

nelem = 5  nvfix = 4  ntype = 4  nmats = 1  ngroup = 1

ndime = 3  nstre = 6  ndofn = 3  noutp = 2  nprop = 2   nload = 1

Element     Group No.  Node Numbers
   1            1       7   8  14  13  1  2  20  19
   2            1       8   9  15  14  2  3  21  20
   3            1       9  10  16  15  3  4  22  21
   4            1      10  11  17  16  4  5  23  22
   5            1      11  12  18  17  5  6  24  23


Node         X              Y               Z
   1     0.000000E+00  0.000000E+00   0.000000E+00
   2     1.000000E+00  0.000000E+00   0.000000E+00
   3     2.000000E+00  0.000000E+00   0.000000E+00
   4     4.000000E+00  0.000000E+00   0.000000E+00
   5     7.000000E+00  0.000000E+00   0.000000E+00
   6     1.000000E+01  0.000000E+00   0.000000E+00
   7     0.000000E+00  0.000000E+00   2.000000E+00
   8     2.000000E+00  0.000000E+00   2.000000E+00
   9     4.000000E+00  0.000000E+00   2.000000E+00
  10     5.000000E+00  0.000000E+00   2.000000E+00
  11     6.000000E+00  0.000000E+00   2.000000E+00
  12     1.000000E+01  0.000000E+00   2.000000E+00
  13     0.000000E+00  1.000000E+00   2.000000E+00
  14     2.000000E+00  1.000000E+00   2.000000E+00
  15     4.000000E+00  1.000000E+00   2.000000E+00
  16     5.000000E+00  1.000000E+00   2.000000E+00
  17     6.000000E+00  1.000000E+00   2.000000E+00
  18     1.000000E+01  1.000000E+00   2.000000E+00
  19     0.000000E+00  1.000000E+00   0.000000E+00
  20     1.000000E+00  1.000000E+00   0.000000E+00
  21     2.000000E+00  1.000000E+00   0.000000E+00
  22     4.000000E+00  1.000000E+00   0.000000E+00
  23     7.000000E+00  1.000000E+00   0.000000E+00
  24     1.000000E+01  1.000000E+00   0.000000E+00


Node     Code      Fixed Values
   1      111     0.000000 0.000000 0.000000
  19      111     0.000000 0.000000 0.000000
   7      100     0.000000 0.000000 0.000000
  13      100     0.000000 0.000000 0.000000
```

```
==================Element Properties===================
Element Group No. = 1
Material No. = 1 Element Model: OPH-18B(a) ngaus = 2


==================Material Properties==================
Material No. = 1
Young Modulus = 1.500000E+03 Poisson Ratio = 3.000000E-01


Load case 1
Pure bending loads
       4        0      0
 6   500.000 0.000  0.000
12  -500.000 0.000  0.000
18  -500.000 0.000  0.000
24   500.000 0.000  0.000

Load case 1
 Displacements
Node   X-coord       Y-coord       Z-coord       X-disp.       Y-disp.       Z-disp.
  1  0.000000E+00 10.000000E+00  0.000000E+00  0.000000E+00  0.000000E+00  0.000000E+00
  2  0.100000E+01  0.000000E+00  0.000000E+00  0.258468E+01  0.379369E+00  0.234798E+00
  3  0.200000E+01  0.000000E+00  0.000000E+00  0.527435E+01  0.282518E+00  0.278568E+01
  4  0.400000E+01  0.000000E+00  0.000000E+00  0.860076E+01  0.303885E+00  0.133609E+02
  5  0.700000E+01  0.000000E+00  0.000000E+00  0.132690E+02  0.298948E+00  0.476101E+02
  6  0.100000E+02  0.000000E+00  0.000000E+00  0.199347E+02  0.300502E+00  0.100906E+03
  7  0.000000E+00  0.000000E+00  0.200000E+01  0.000000E+00 -0.171234E+00 -0.106952E+01
  8  0.200000E+01  0.000000E+00  0.200000E+01 -0.329244E+01 -0.334391E+00  0.292679E+01
  9  0.400000E+01  0.000000E+00  0.200000E+01 -0.663694E+01 -0.291250E+00  0.132252E+02
 10  0.500000E+01  0.000000E+00  0.200000E+01 -0.930073E+01 -0.301979E+00  0.255078E+02
 11  0.600000E+01  0.000000E+00  0.200000E+01 -0.126342E+02 -0.299509E+00  0.350635E+02
 12  0.100000E+02  0.000000E+00  0.200000E+01 -0.199669E+02 -0.300216E+00  0.970919E+02
 13  0.000000E+00  0.100000E+01  0.200000E+01  0.000000E+00  0.171234E+00 -0.106952E+01
 14  0.200000E+01  0.100000E+01  0.200000E+01 -0.329244E+01  0.334391E+00  0.292679E+01
 15  0.400000E+01  0.100000E+01  0.200000E+01 -0.663694E+01  0.291250E+00  0.132252E+02
 16  0.500000E+01  0.100000E+01  0.200000E+01 -0.930073E+01  0.301979E+00  0.255078E+02
 17  0.600000E+01  0.100000E+01  0.200000E+01 -0.126342E+02  0.299509E+00  0.350635E+02
 18  0.100000E+02  0.100000E+01  0.200000E+01 -0.199669E+02  0.300216E+00  0.970919E+02
 19  0.000000E+00  0.100000E+01  0.000000E+00  0.000000E+00  0.000000E+00  0.000000E+00
 20  0.100000E+01  0.100000E+01  0.000000E+00  0.258468E+01 -0.379369E+00  0.234798E+00
 21  0.200000E+01  0.100000E+01  0.000000E+00  0.527435E+01 -0.282518E+00  0.278568E+01
 22  0.400000E+01  0.100000E+01  0.000000E+00  0.860076E+01 -0.303885E+00  0.133609E+02
 23  0.700000E+01  0.100000E+01  0.000000E+00  0.132690E+02 -0.298948E+00  0.476101E+02
 24  0.100000E+02  0.100000E+01  0.000000E+00  0.199347E+02 -0.300502E+00  0.100906E+03
```

```
Reactions
 Node  X-coord       Y-coord       Z-coord        X-reac.       Y-reac.        Z-reac.

 1   0.000000E+00 0.000000E+00  0.000000E+00 -0.500000E+03 -0.161144E+03 -0.368772E-11

 7   0.000000E+00 0.000000E+00  0.200000E+01  0.500000E+03  0.000000E+00  0.000000E+00

 13  0.000000E+00 0.100000E+01  0.200000E+01  0.500000E+03  0.000000E+00  0.000000E+00

 19  0.000000E+00 0.100000E+01  0.000000E+00 -0.500000E+03  0.161144E+03  0.359535E-11

 Stop at 16:14:50.07 03/20/2003
 0.11 seconds costed
```

Index

A

B

C

I

J

K

L

Q

R

S